Colloids - Types, Preparation and Applications

Edited by Mohamed Nageeb Rashed

Published in London, United Kingdom

IntechOpen

Supporting open minds since 2005

Colloids - Types, Preparation and Applications
http://dx.doi.org/10.5772/intechopen.92521
Edited by Mohamed Nageeb Rashed

Contributors
Ranganatha Sudhakar, Divya Bajpai Bajpai Tripathy, Anjali Gupta, Kajol Bhati, Kiran D. Pawar, Darshana V. Havaldar, Disha N. Moholkar, Rachana S. Potadar, Marina Stygar Lopes, Tatiana Yakhno, Vladimir Yakhno, Abhyarthana Pattanaik, Rayasam Venugopal, Roshan Nazir, Abhay Prasad, Ashish Parihar, Mohammed S. Alqahtani, Rabbani Syed, Jesus Garcia Ovejero, Alvaro Gallo-Cordova, Daniela Almeida, Puerto Morales, Jesica María José Santillán, David Muñeton Arboleda, Daniel Schinca, Lucía Scaffardi, Valeria Arce, Randhir Singh, Huda M. Mohammed, Abed Ahmed Mohammed, Stanley Onwubu, Chibuzor Stellamaris Okonkwo

First published in London, United Kingdom, 2021 by IntechOpen
IntechOpen is the global imprint of INTECHOPEN LIMITED, registered in England and Wales, registration number: 11086078, 5 Princes Gate Court, London, SW7 2QJ, United Kingdom
Printed in Croatia

British Library Cataloguing-in-Publication Data
A catalogue record for this book is available from the British Library

Additional hard and PDF copies can be obtained from orders@intechopen.com

Colloids - Types, Preparation and Applications
Edited by Mohamed Nageeb Rashed
p. cm.
Print ISBN 978-1-83962-969-3
Online ISBN 978-1-83962-979-2
eBook (PDF) ISBN 978-1-83962-980-8

For EU product safety concerns:
IN TECH d.o.o., Prolaz Marije Krucifikse Kozulić 3, 51000 Rijeka, Croatia, info@intechopen.com or visit our website at intechopen.com.

Meet the editor

Prof. Mohamed Nageeb Rashed is Professor of Analytical and Environmental Chemistry, Faculty of Science, Aswan University, Egypt, where he previously served as vice-dean for environmental affairs. In 1989, he received a Ph.D. in Environmental Analytical Chemistry from Assiut University, Egypt. His research interests are in analytical and environmental chemistry with special emphasis on monitoring and assessing biological trace elements and toxic metals in human blood, urine, water, crops, vegetables, and medicinal plants; relationships between environmental heavy metals and human diseases; uses of biological indicators for monitoring water pollution; environmental chemistry of lakes, rivers, and well water; water and wastewater treatment by adsorption and photocatalysis techniques; soil and water pollution monitoring, control, and treatment; and advanced oxidation treatments using photocatalysis, nanocatalysts, nanocomposites, and adsorption techniques for water and wastewater treatment. Dr. Rashed supervised several MSc and Ph.D. theses in the field of analytical and environmental chemistry. He was selected as an examiner for several Ph.D. theses in analytical chemistry from India, Kazakhstan, and Botswana. He has published about ninety scientific papers in peer-reviewed international journals, as well as several papers in national and international conferences. He participated as an invited speaker at thirty international conferences worldwide. Dr. Rashed is the editor-in-chief and an editorial board member for several international journals in the fields of chemistry and environment. His society membership includes several national and international societies. He received the Egyptian State Award for Environmental Researches in 2001 and the Aswan University Merit Award for Basic Science in 2020.

Contents

Preface

Colloids are types of mixtures in which a microscopic substance is dispersed throughout a dispersion medium. The particles of the dispersed phase have a diameter of 1–1000 nanometers.

The most common phenomena of colloids in the ecosystem appear in the blue color of the sky, which is due to the scattering of light by colloidal particles in the air, known as the Tyndall effect. In our body, blood is a colloidal solution. Other examples include milk, cream, gelatin, jelly, colored glass, river mud, and butter.

Colloids and colloidal systems are useful in human health as well as commercial and industrial situations. The important applications of colloids are in medicines, sewage disposal, water purification, formation of the delta, cleansing action of soap, industry, mining, smoke precipitation, photography, electroplating, agriculture, rubber industry, artificial rain, and more.

Colloids - Types, Preparation and Applications deals with several aspects of colloid morphology, synthesis, and applications. It gathers recent research by outstanding experts in the field of colloids, including research into colloid synthesis, modification, and applications.

The book includes twelve chapters divided into three sections. The first section includes four chapters that discuss different techniques for synthesizing colloids: Chapter 1, "Optimization of Biogenic Synthesis of Colloidal Metal Nanoparticles"; Chapter 2, "Recent Progress in the Electrochemical Exfoliation of Colloidal Graphene: A Review"; Chapter 3, "A Simple and "Green" Technique to Synthesize Metal Nanocolloids by Ultrashort Light Pulses"; and Chapter 4 "Gemini Imidazolinium Surfactants: A Versatile Class of Molecules". The second section includes three chapters that explain the structure, dynamic and stability of colloids: Chapter 5, "Aerogels Utilization in Electrochemical Capacitors"; Chapter 6, "Structure and Dynamics of Aqueous Dispersions"; and Chapter 7, "Colloidal Stability of Cellulose Suspensions" The third section includes five chapters that discuss applications of colloidal particles: Chapter 8, "Removal of Copper Ions from Aqueous Solution Using Liquid Surfactant Membrane Technique"; Chapter 9 "Hydrocolloids in Dentistry: A Review"; Chapter 10, "Application of Colloids and Its Relevance in Mineral Engineering"; Chapter 11 "Magnetic Iron Oxide Colloids for Environmental Applications"; and Chapter 12, "Colloidal Nanocrystal-Based Electrocatalysts for Combating Environmental Problems and Energy Crisis".

I would like to express our sincere thanks to Dr. Haiam Morsy Aboul-Ela, Marine Biotechnology and Natural Products Lab, the National Institute of Oceanography and Fisheries, Alexandria branch, Egypt; Dr. Radwa El-Salamony, Egyptian Petroleum Research Institute; and Dr. Khalid ElWakeel, Faculty of Science, Port Said University, Egypt for their contributions in the scientific revision of the book chapters.

I would also like to express my thanks to IntechOpen Author Service Manager Ms. Kristina Kardum Cvitan. I am also grateful to the book authors for their hard work and worthy contributions.

Mohamed Nageeb Rashed
Faculty of Science,
Aswan University,
Aswan, Egypt

Section 1

Different Technique for Synthesis of Colloids

Chapter 1

Optimization of Biogenic Synthesis of Colloidal Metal Nanoparticles

Disha N. Moholkar, Darshana V. Havaldar, Rachana S. Potadar and Kiran D. Pawar

Abstract

Nanotechnology which deals with the synthesis and characterization of dispersed or solid particles in nano-metric range has emerged out to be a novel approach due to its ample applications in biomedical fields. The advancements in the field of nanotechnology and substantial evidences in biomedical applications have led the researchers to explore safe, ecofriendly, rapid and sustainable approaches for the synthesis of colloidal metal nanoparticles. This chapter illustrates superiority of biogenic route of synthesis of nanoparticles over the different approaches such as chemical and physical methods. In biogenic route, plants and microorganisms like algae, fungi, yeast, actinomycetes etc. act as "bio-factories" which reduce the metal precursors and play a crucial role in the synthesis of nanoparticles with distinct morphologies. Thus, the need of hazardous chemicals is eliminated and a safer and greener approach of nanoparticles synthesis can be adopted. This chapter also outlines the effect of optimization of different parameters mainly pH, temperature, time and concentration of metal ions on the nanoparticle synthesis. It is evident that the optimization of various parameters can yield nanoparticles with desired properties suitable for respective biomedical applications.

Keywords: colloidal metal nanoparticles, biogenic synthesis, biomedical applications, optimization, nanobiotechnology

1. Introduction

Ever since the origin of human civilization as early as 500 BC, nanomaterials (NMs) have been used for a range of applications, biomedicinal formulations being a crucial one [1]. Due to small size ranging from 1 to 100 nm, high aspect ratio, distinguished magnetic, optical, electrical, mechanical properties as compared with bulk materials of their same kind, MNs are being widely explored for their possible range of biomedical applications. In addition, ease of synthesis, control over size and morphology have revolutionized the field of nanobiotechnology [2]. The convergence of nanotechnology and biotechnology has led to the emergence of innovative and powerful field that explores the possibility of utilizing various NMs for biomedical applications [2]. The manipulations of macro materials resulting in unique properties of NMs have attracted biomedical researchers to utilize these properties in pharmaceutical fields such that the NMs would play a momentous role and indeed add to the functionality of original compound [3]. The NMs and nano-biomaterials are being extensively used in biomedical field for diagnostics, imaging, drug delivery and as prostheses and implants due to their superior biocompatibility to artificial polymeric materials [4]. The metallic and non-metallic nanoparticles (NPs) used extensively in biomedicines are derived from sources such as bulk metals, non-metals, chemicals, plants and microbes. Owing to well-defined and tunable size, shape, molecular weight and uniform dispersity of lipids and proteins based NMs, they are used for the fabrication of nanocarriers such as liposomes, micelles and dendrimers for drug and gene delivery [5–7]. Depending upon the type of NMs, the pharmaceutical ingredient can be either encapsulated or attached onto the surface of such nanocarriers in such a way that, irrespective of the water solubility, the pharmaceutical ingredient can be delivered to the target site and protected against degradation [2]. Presently, almost 175 exclusive nanomedicinal products for the treatment of cancer and infectious diseases are at different stages of clinical trials soon to be launched into the market [8]. Concurrently, surgical blades, suture needles, contrast-enhancing agents for magnetic resonance imaging, bone replacement materials, wound dressing materials, anti- microbial textiles, *in vitro* molecular diagnostic chips, microcantilevers, and microneedles are already out in the market [9].

The capsules PillCamESo and PillCam Colon, sized as that of a normal pill act as a substitute for the traditional endoscopy technique. These contain a flashlight and a camera which is swallowed by the patient and the images of the gastrointestinal system are captured and sent wirelessly for further diagnostic purposes [10]. Similarly, 'microbots' structurally similar to flagella equal to half the human hair diameter are fabricated using computer chip technology. These comprise a magnetic head and can be controlled via an external magnetic field which delivers medicine to destroy tumors [11, 12]. Microbots can also relieve diabetes patients from the pain to test their blood multiple times every day and the inconvenience of self-testing to ensure stable blood-glucose levels. These could be used to retrieve data from varied locations of the body at the same time allowing continuous blood sugar level monitoring [2, 13]. The field of nanobiotechnology has also assisted insulin delivery systems to detect fluctuations in blood glucose levels and spontaneously modulate the adequate insulin release thereby maintaining normoglycemiea [14, 15]. A major drawback of non-specific drug delivery associated with conventional delivery system for cancer therapies can be overcome by using various NMs using metal NPs. To this end, metal NPs can be surface functionalized by attaching specific targeting moiety and imaging agents to target the cancerous cells [16]. This approach enables and enhances the efficiency in terms of not only timely detection of the cancerous cells but also treatment of tumors via targeted and specific release of drugs to yield maximum effectiveness with lower cytotoxicity to healthy cells [2]. In addition, nanobiotechnology has also contributed a solution for the treatment of a significant worldwide problem of hard tissue repair and regeneration by means of artificial bone scaffolds which mimic natural bone composition and structure [17, 18]. The use of biomimetics nano-assembly technology and additive manufacturing techniques make the scaffolds, cells and growth factors mimic the natural bone [19, 20]. Such scaffolds can also be used to deliver growth factors by acting as an alternative to extracellular matrix and other bioactive factors including small molecules, cytokines, peptides, proteins and genes [21, 22] to achieve controlled release and enhanced osteoblast proliferation and differentiation for stimulation of bone regeneration [23, 24].

2. Colloidal metal nanoparticles as important nanomaterials for various applications

In general terms, colloidal systems are heterogeneous systems in which very fine particles of one matter are scattered through another substance. Former is referred as "Dispersed Phase" while later as "Dispersion Medium" and both can be present in either of solid, liquid or gas states. Dispersed phase is completely insoluble in dispersion medium [25]. Colloidal NPs, as also called nanocolloids or solid colloidal particles, resemble a normal colloidal system where NPs act as dispersed phase. Being dispersed in the solvent medium, NPs are embroiled in some lively motions such as Brownian motions [26, 27]. As a consequence of their dominant characteristics over bulk correspondents the colloidal NPs, play vital role in number of applications [28]. The unique properties such as tunable size, configuration, structural arrangement, formulation, crystallinity and dimensions can deeply rectify the features of colloidal NPs according to the applications [29]. Colloidal NPs can be employed in prospective applications in the wide range of sectors including electronics, coatings, catalysis, packaging, biomedicine, biotechnology etc. In addition, the uses of colloidal NPs in biomedical field are increasing incredibly as they are being administrated with elegant attributes for healthier reactions with the biological circumstances and cope with on-demand requirements of *in vivo* diagnosis and therapies [30]. To boot, the fine size of NPs not only allows them to pass through the tissues or cells but also accesses them easily to target organs engrossing the novel biomedical applications at cellular level [31].

Magnetic NPs specifically iron oxide NPs are principally studied and utilized for their peculiar physicochemical, biological and magnetic features [32], remarkably stability, least perilous, significant magnetic vulnerability, severe saturation magnetism and biocompatibility [33]. Similarly, other magnetic NPs such as alloy, also known as bimetallic NPs of iron-cobalt (Fe-Co), iron-platinum (Fe-Pt) have high magnetic properties, super paramagnetism, high curie temperature [34, 35]. The exceedingly reported and mostly scrutinized uses of magnetic and bimetallic NPs are for target specific drug delivery [36, 37], in magnetic resonance imaging [38, 39] and to treat hyperthermia magnetically [40, 41].

Metallic NPs are the matter of curiosity that has been mesmerizing experts due to their extraordinary optical, electronic properties accompanied by its massive potential in nanotechnology. Nobel metal NPs of gold, silver, platinum, palladium, etc. have been used since ancient times for medicinal intents. Chemical inertness, ability to resist corrosion and oxidation even in moist air wholly justifies their uptake for biomedical applications [42]. Negative charge on the surface of gold NPs presents easy functionalization with organic compounds that offers further interactions with antibodies, drugs moieties or ligands for *in vitro* or *in vivo* drug delivery [43]. Likewise, silver NPs embrace distinct characteristics of being chemical inactivity, catalytic activity, high thermal and electrical conductive [44, 45]. The astonishing antimicrobial activity of silver NPs leads its utility in textile industries, wound healing dressings and as disinfectants [46, 47]. The employability of other metal NPs in bioimaging [48], biosensors [49], photothermal therapies [50] are growing day-by-day.

Metal oxide NPs such as titanium dioxide (TiO_2) and zinc oxide (ZnO) NPs are markedly used in paints, coatings, food coloring, beauty products, sunscreens etc. Equating with other metal oxide NPs, ZnO confers minimal toxicity to living cells so that there is increase in biomedical applications namely in diabetes treatment, wound healing, anti-inflammation treatment, anti-aging products, antibacterial activities, etc. [23, 51–53].

3. Methods for colloidal nanoparticles synthesis

Remarkable morphological, structural, magnetic, electronic and physicochemical characteristics of colloidal NPs render them extraordinary for their uses in various fields such as physical, electrochemical, optical, environmental, biomedical fields etc. These peculiar properties of colloidal NPs depend on their source and route of synthesis process. Unremitting research in the field of nanotechnology have invented a range of ways to fabricate NPs. On the whole, these fabrication methods are segregated into three major groups, notably physical methods, chemical methods, and bio-assisted (also called biological and biogenic) methods in which NPs' synthesis is performed either by top-down approach or bottom-up approach. The top-down approach induces gradual trimming of bulk counterparts which invariably leads to the mass production of NPs. On the contrary, bottom-up approach deals with the consolidation of atoms and molecules to yield NPs with series of dimensions [54].

3.1 Physical methods

Physical methods principally rely on top-down approach where high energy emissions, mechanical pressure, thermal or electrical powers are employed for melting, mitigation, abrasion of bulk materials to beget NPs. These techniques are devoid of solvent contamination, produce monodispersed and reproducible NPs making suitable for few specialized applications. However, generation of waste byproducts along synthesis is one of the flaws of physical methods [55]. Some of the most commonly used physical methods to generate NPs are high energy ball milling, laser ablation, electrospraying, inert gas condensation, physical vapor, deposition, flame spray pyrolysis etc. These methods are pictorially depicted in **Figure 1**.

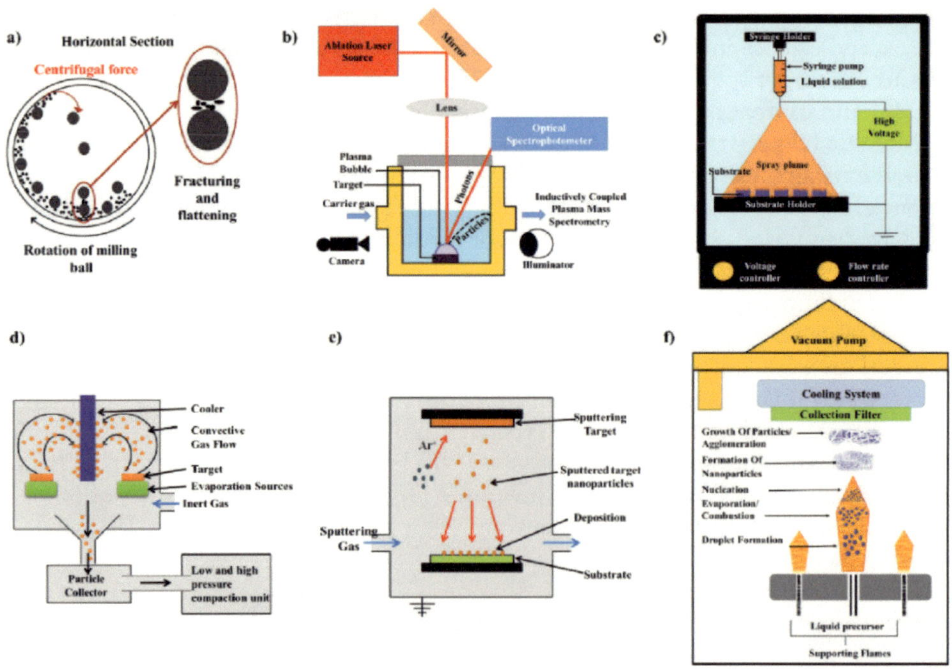

Figure 1.
Schematic representation of physical methods for synthesis of nanoparticles, (a) high energy ball milling, (b) laser ablation, (c) Electrospraying, (d) inert gas condensation, (e) physical vapor deposition, (f) flame spray pyrolysis.

High energy ball milling is a high pressure and thermal, sturdy and energy effective synthesizing manner in which immensely movable balls pass on their kinetic energy to the bulk materials. The crushing process disrupts the chemical bonds of the materials and rifts it into tiny particles to raise NPs with diverse conformation and dimensionalities [56]. High contamination prominently due to wear and tear crushing by balls, polydispersity in terms of irregular dimensions of synthesized NPs, aggregation and long milling time [57] are few of the disadvantages associated with high energy ball milling method.

Laser ablation is another physical method that either employs continuous laser or pulsed laser to strike on the material opted to break down into NPs. It is a flexible mode which involves series of melting, evaporation and ionization of material onto collector surface. The continuous bombardment of laser beam results into ablation of targeted material to micro and nanostructure materials [58]. Even though, NPs with high purity can be obtained through this method, its high cost, long operational time for production high input of power for extirpation of matter, difficulty in large scale production make this method not so popular [59].

The electrospraying mechanism is analogous to the electrospinning technique used to form fibers. In electrospraying, a blend of desired polymer solution and the solvent are filled in the syringe, subjected to high voltage electric field to split the solution into small charged nano-sized particles that are received by counter electrode. This technique provides flexibility over the size of NPs by varying the reaction conditions such as concentration of solution, electric field, conductivity, flow rate of liquid etc. [60]. Excess addition of cross linkages and low yield of NPs are some of the shortcomings of electrospraying technique [61].

Inert gas condensation is a very fundamental process that requires ultrahigh vacuum (UHV) conditions, inert gases like Helium (He) or Xenon (Xe) and a substrate cooled with liquid nitrogen. The target materials are first evaporated, then transferred along with inert gases and finally condensed on cooled substrate [62]. The agglomeration of condensed NPs, high cost associated with UHV conditions, difficulties related to maintaining clean vacuum situations, reproducibility and durability of working parameters etc. are some of the downsides of the techniques [63].

Physical vapor deposition is an ecologically compatible route that incorporates three successive vital steps such as pyrolysis of solid materials to convert into vapors, transmission of vaporized materials followed by nucleation and growth process. This integrated group of processes have been widely designed and used to fabricate NPs in addition to deposit thin films of nanometers to micrometers [64]. Despite the fact that the technique delivers marked advantages, the instability of precursor gas at ambient temperature as well as reaction temperature and high cost resulting from greatly controlled vacuum in chamber limits its use [65].

Flame spray pyrolysis is the recent and single step combustion process substantially operates to formulate compound and functional NPs. In this process, low volatile precursors are injected into highly sustainable flame with extreme temperature gradient where liquid precursor undergoes spray-to-particle or gas-to-particle pathway to form monodispersed NPs [55, 66]. The requirement of high stability and dispersibility of metal precursors and solvents, low volatility, relevant melting temperature limits the choice of materials and use of this technique [66].

3.2 Chemical methods

Chemical methods are certainly more favorable to synthesize colloidal NPs owing to their unaltered approach towards external stimuli. High yield and reproducibility make them highly recommended. There are number of chemical methods, most of them are based on bottom up approach [67]. The chemical methods of colloidal NPs synthesis are diagrammatically represented in **Figure 2**. Sol–gel, plasma enhanced, chemical vapor deposition, polyol synthesis and hydrothermal synthesis are some of the primarily used chemical methods for synthesizing monodispersed NPs.

Colloidal solution of solid particles in liquid i.e. sol and liquid containing polymer i.e. gel are the two constituents of sol–gel method. The basic steps of the process are explicitly hydrolysis whereby the chemical bonds of precursors are deteriorated by water to form gel continued by condensation for genesis of sols in the liquid. In the end, the leftover liquid is drained to finalize the morphology of NMs [55, 68]. Owing to few flaws such as low abrasion resistance, poor bonding, exalted permeability and difficult control over porosity of technique, it becomes difficult to realize its industrial scale up [69].

In plasma enhanced chemical vapor deposition, also titled as plasma assisted chemical vapor deposition, plasma triggers the chemical reactions for formation of thin films and formulation of NPs as well. It is a well-known process conducted at lower temperature. The system is assembled by vacuum process unit, power supply, heater and precursor. The wide range of NPs can be formed via this method, for instance gallium nitride and so forth [70]. The expensive instrumentation, instability in damp conditions, presence of poisonous gases in plasma stream and lengthy process are some of the shortfalls of the method [71].

Polyol synthesis method fabricates colloidal NPs by using poly ethylene glycol as a medium to conduct the reaction. It also performs as solvent, reducing agent and integrating agent simultaneously with addition of protecting or capping agents externally [72]. The process is used to synthesize range of NPs of metals (platinum, palladium, silver, cobalt, etc.), metal oxide NPs (Zinc oxide, Cobalt oxide etc.) and magnetic, hybrid NPs as well [55]. However, the confined propensity of polyol to reduce precursors and slender stabilization of nonpolar metal surfaces by polar polyol are two major inadequacies with which the process has to dealt with and which diminishes the efficacy of the process [73].

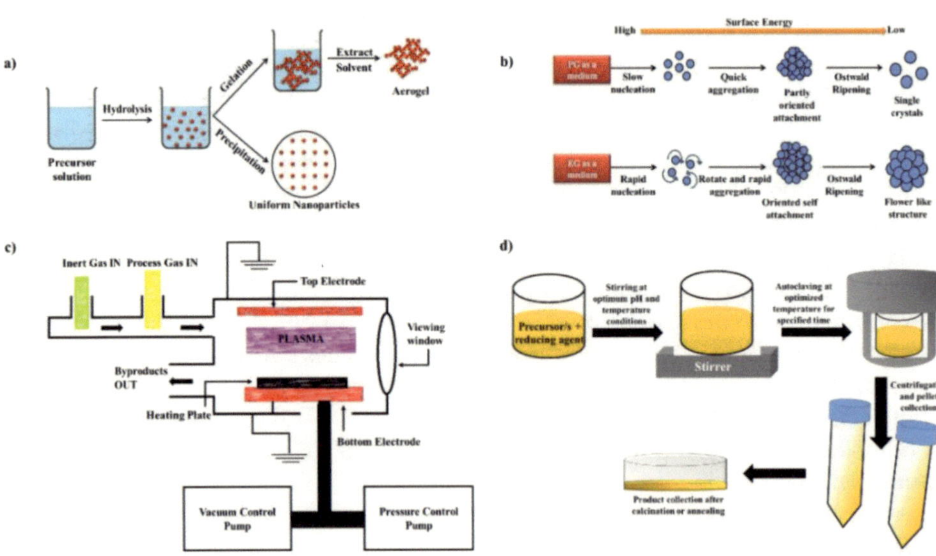

Figure 2.
Schematic representation of chemical methods for synthesis of nanoparticles, (a) sol–gel method, (b) polyol synthesis, (c) plasma enhanced chemical vapor deposition, (d) hydrothermal synthesis.

Hydrothermal synthesis method explores various temperatures and pressure environments to change the behavior of water in the vicinity. During synthesis, NPs are synthesized from colloidal system that comprises of two or more states of compound from solid, liquid or gas and added together with controlled conditions of pressure and temperature. This method is carried out either by batch hydrothermal process or continuous hydrothermal process to create NPs of metal oxide, lithium iron phosphate etc. The batch hydrothermal executes reaction optimal ratios of phases while other allows faster mode of reaction. One of the incredible advantages of the method is its capability to produce large quantities of NPs at a time with preferable properties [74, 75]. The reaction requires water in supercritical state, higher pressure and temperature which in turn limits the onsite examinations to get clarify with NPs synthesis [74].

3.3 Biological methods

Despite the fact that chemical and physical methods of colloidal NPs synthesis are awfully proficient, these methods anyway own copious shortcomings just like use of acutely life-threatening chemicals, non-polar organic solvents, diversified synthetic capping, reducing agents, etc. therefore, hamper their engagement in biomedical purpose. On top of this, synthesis via physicochemical routes fetches contamination on the exterior of NPs post synthesis that has brought up solemn disquietude regarding the unfavorable upshots of the chemically synthesized NPs on the environment and living cells [76, 77]. These limitations has forced researchers to look for novel, environment friendly alternatives to synthesize colloidal NPs [78, 79]. Green synthesis or biosynthesis is the most feasible substitute that makes the use of microorganisms and parts of plants instead of toxic and pernicious chemicals. Bacteria, fungi, algae and yeast are frequently used as bio-reactors that can hire a batch of anionic functional groups proteins, enzymes, reducing sugars, etc. to reduce metals salts to corresponding colloidal NPs [80, 81]. The different methods of biological synthesis of colloidal NPs are diagrammatically represented in **Figure 3**.

3.3.1 Advantage of nanoparticles synthesized via biological route

The routinely used NPs synthesis routes such as chemical and physical methods are not only energy and capital exhaustive but also employ the toxic chemicals and non-polar solvents for synthesis and synthetic additives or capping agents during the later process. These methods therefore rule out the application of such products in clinical and biomedical fields thereby creating a need for a safe, reliable, biocompatible and benign method for the production of NPs [82]. Worthy of the exceptional environment friendly nature, it has been reported that the NPs synthesis rate via biogenic methods are comparable to that of chemical methods [83–85].

For biomedical applications, it's obligatory that NPs must have depleted metal cytotoxicity and enhanced biocompatibility. Unlike physico-chemically synthesized NPs, green synthesized NPs are free from deleterious byproduct contamination that most often remain bound to the NPs surface and restraint their role in biomedical applications [86]. The decisive leverages that supports the biological routes for colloidal NPs synthesis are wide availability of key biological components, biocompatible reducing agents, capable of large scale synthesis with moderate temperature and pressure, dual working of enzymes or phytochemicals as reducing as well as stabilizing agents [87]. There are multiple superiorities concerned with bio-associated methods, uniquely expeditious and eco-friendly fabrication practices, less expensive and bio-tolerant nature of NPs. It does not demand for separate capping

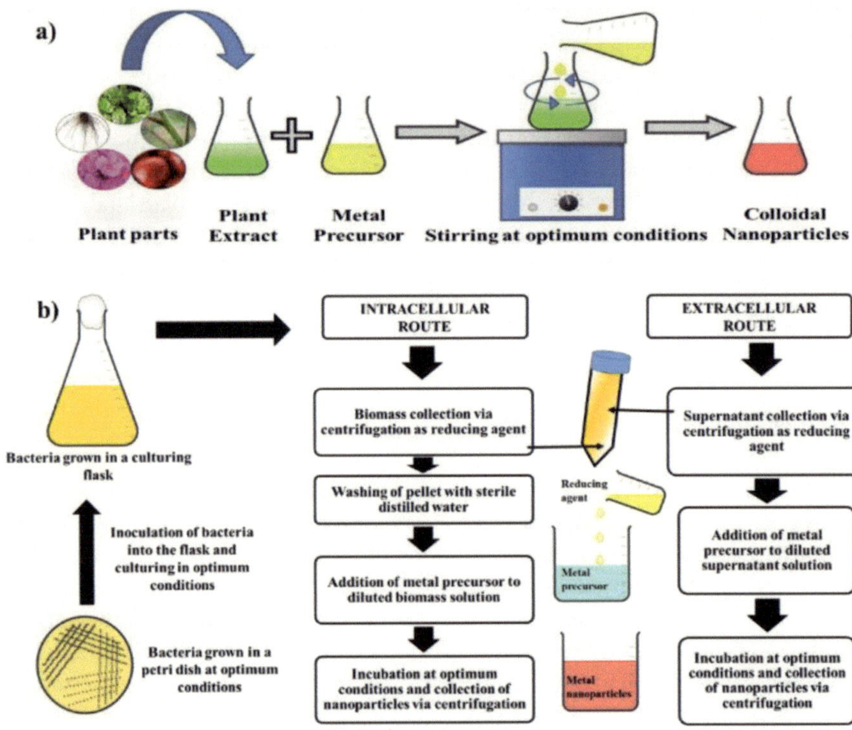

Figure 3.
Schematic representation of biological methods for synthesis of nanoparticles, (a) plant based synthesis, (b) microbial synthesis.

agents considering the potential of the plant's and the microorganism's components to act so [5]. On top of that, when NPs came in proximity of biological fluids while synthesis, they gradually and electively imbibe biomolecules establishing corona on the superficies that bestow additional potency and make them more efficient over uncovered NPs [88]. Precisely, medicinal plants are supposed to furnish the NPs with strengthened adequacy by entitling them with ample metabolites having pharmacological values [5, 89, 90]. As biosynthesized NPs are highly equipped with functional groups over the course of reaction, it eliminates additional steps required for physicochemical processes which automatically shorten the time period [91]. All of these supremacies make biosynthesis or green synthesis worth applicable.

3.3.2 General mechanisms of biological synthesis

3.3.2.1 Algal synthesis

Algae, either unicellular or multicellular, are autotrophic and aquatic photosynthetic organisms belonging to kingdom Protista. Depending upon their sizes that range from micrometer to macrometer, they are distinguished as microalgae or macroalgae and serve as extreme source of vitamins, minerals, and proteins. They have successfully drawn the utter attention by virtue of their competency to diminish the toxicity of metals accompanied by presence of bioactive components to stabilize the NPs; nonetheless the reports for algal synthesis of NPs are merely few, exclusively on iron oxide and zinc oxide [92]. Regardless the ongoing research on synthesis of NPs through different biological sources at greater extent, the detailed mechanism for the synthesis by algae is not revealed yet wholly. Studies so far disclose that cell walls of seaweed are comprised of polysaccharides that carry hydrophilic surface groups like carboxyl, hydroxyl and sulfate groups [93]. Further, it holds abundant biomolecules, intrinsically proteins and enzymes which play the role of biocatalyst to convert metal ions into NPs, meanwhile, the other larger amphiphilic biomolecules act as capping agents to stabilize the NPs [94–97]. Some of the reported examples of algal synthesized NPs include AuNPs synthesized by brown seaweeds *Fucus vesiculosus* [98], and *Turbinaria ornate* [99]. A report by Khanehzaei et al. [100] explains the algal synthesis of copper and copper oxide NPs by extract of red seaweed *Kappaphycus alvarezii*.

3.3.2.2 Fungal synthesis

Fungal synthesis is the quite pertinent among remaining bio-synthesis methods, even than bacteria, in the wake of their phenomenal properties adeptness of NPs' synthesis with various dimensions [101]. Fungi viz. yeasts or molds are eukaryotic organisms that bear mycelia which allocate them extended surface area for metal ions acquaintance. Fortuitously, cell surface of fungi possesses chain of biomolecules and reducing agents which offers them numerous additional privileges [102]. Moreover, the NPs' configuration due to fungi is rapid considering the fact that fungi biomass proliferate rapidly than bacteria, and contrary to bacteria, fungi have superior endurance and metal bioaccumulation. In supplement, it provokes monodispersed synthesis of NPs with quite defined structures. Typically, fungal manufacture of metal NPs is judiciously cheaper, eco-friendly, engage uncomplicated down-streaming operation and no need of external stabilizing agent as fungal biomass itself function as capping agent as well [103, 104]. Bhainsa and D′ Souza in 2006 reported the synthesis of AgNPs by the fungus *Aspergillus fumigates* [105]. Metal oxide NPs have also been synthesized through the fungus synthesis for example, silicon dioxide (SiO_2), TiO_2 and ferric oxide (Fe_2O_3) NPs by fungus *Fusarium Oxysporum* [106].

Amidst other microorganisms that fall into kingdom fungi, yeast is the most examined species given the fact that extracellular synthesis is simpler to regulate and to manipulate in laboratory scenario [104]. As an example, Bharde and his coworkers have reported the reduction of TiO2 to NPs by means of fugal extract of *Saccharomyces cerevisiae* [106].

3.3.2.3 Bacterial synthesis

There are copious number of bacteria that smoothly sustain with harsh environmental conditions. Moreover, they can multiply and grow at extreme speed, their maintenance is cost effective and are easy to manipulate for synthesis. For this sake, they are being employed for the biogenic synthesis of colloidal NPs. Furthermore, the bacterial growth parameters especially temperature, oxygen supply and incubation time can be monitored with ease as they might affect the sizes of NPs [107].

The synthesis of NPs from bacteria is either intracellular or extracellular, depending on the site of synthesis. The intracellular synthesis deals with carrying of metal ions inside the microbial cell while in case of extracellular synthesis, metal ions are entrapped by the surface of cell to reduce it into corresponding NPs in presence of enzymes and other biomolecules [81]. The mechanism for formation of NPs differs with respect to the bacteria. When metal ions that are almost poisonous to bacteria, come in proximity, bacteria secrete specific proteins, enzymes and other biochemicals as a safeguard provision. To rectify the detrimental effect, bacteria modify metal ions into NPs by assorting not only dissolution of metal ions but also their redox behavior and extracellular sorption. Bacteria with S-layer and

Magnetotactic bacteria are best suited to harvest metal NPs whose cell wall surfaces are shielded with protein rich components [108, 109].

For *in vitro* synthesis of NPs using bacteria, initially convenient bacterial species is cultured for 1-2 days in shaking incubator or orbital shaker at optimal parameters incorporating temperature, pH, media concentration, shaking speed etc. The culture is then centrifuged to separate biomass. For intracellular synthesis biomass is collected, washed thoroughly with deionized water and dissolved in sterile water which in turn acts as a bacterial extract to reduce metal ions. Conversely, supernatant after centrifugation can also be used for extracellular synthesis of NPs [110].

3.3.2.4 Plant-based synthesis

As phytomining practices, plants with ability to hyper accumulate metals are planted on metal contaminated soils for uptake of metal ions. The metal ions disseminate into plant and travel to the convinced plant parts where primary and secondary metabolites such as terpenoids, flavonoids, phenolic acids, proteins, polysaccharides, organic acids etc. remold ions into metal NPs [5, 111, 112]. This approach is merely time consuming, tedious and retrieval of synthesized NPs is strenuous [111, 113]. The biogenic synthesis of NPs using plant extract or biomass is one of the most effective, rapid, absolute non-hazardous and ecofriendly methods. Nanoparticlesof noble metals, metal oxides, bimetallic alloys, etc. have been mainly synthesized *in vitro* by harnessing this method which is well reviewed by Iravani in 2011 [114].

For phytomediated biogenic synthesis, plant extracts are prepared from different parts of plants largely leaves, flowers, fruits, stem, roots, peels etc. [111, 113] and used as a source of reducing and capping agents. To accomplish this, the metal ion solution is subsequently added to extract where co-precipitation of metal ions with accessible functional groups is favored. The reaction parameters such as reaction time, temperature, pH, ration and concentration of metal salt influence the synthesis [115], therefore can be fine tunes. For example, ZnO NPs cane be formulated through leaf extract of *Corymbia citriodora* [116], peel extract of *Nephelium lappaceum* [117], root extract of *Polygala tenuifolia* [118].

4. Selection of biological agents for the synthesis of nanoparticles

Two main criteria for the selection of suitable plants for synthesis are selection of plant part on the basis of enzyme activities and biochemical pathways (for example: plants with heavy metal accumulations and detoxification properties); and setting the optimal conditions for enhanced cell growth and high enzyme activity [114].

According to Das and Brar [119], the plants are majorly preferred due to their exceptional reduction ability, yet, only the ethanobotanical conclusions are not the only basis for the selection of plants for the synthesis of NPs. These authors pointed out the fact that, the bio-reduction of the metallic cations can be a part of the plant's defensive reaction towards ionic stress. The chemical evolution of phytochemicals must thus be reconsidered to probe the possibility of exploiting different plant groups for biogenic synthesis. It was therefore suggested that, some representatives for each plant group must be picked and the protocol be standardized keeping in mind the process parameters and laboratory scale to commercial scale scaling up. It is important that the plant encompassing the desired properties must not fail at large scale level [119]. Some of the NPs synthesized from plants have the extreme potential in biomedical fields and should be considered for scaling up purposes [120–122]. Das and Brar [119], also mentioned that instead of focusing on the advantages and disadvantages of biogenic synthesis routes such as efficiency, dual functionality, propensity, broad application etc.; it is important to have a broader perspective about the following parameters:

1. Clinical relevance must be checked by studying the previous well documented scientific reports.

2. Phylogenetic studies to set a reference plant.

3. In vitro study including cellular damage studies.

4. Precision in identifying the part of plant and mechanism

5. Geographical distribution studies to select a plant that does not have a very narrow distribution.

6. Genetic aspects, which is still an unexplored area.

7. This selection criteria is very feasible and applicable to all the plant groups and can bridge the gap exploitation of nature's ability and possibility to make biogenic synthesis more scalable.

Some bacterial species such as *Pseudomonas stutzeri* and *P. aeruginosa* have the ability to recourse specific defense mechanisms in order to deal with stress conditions like toxicity of heavy metal ions to survive and grow at high metal ion concentrations [123, 124]. Algae are economical contenders for the bioremediation and bioconversion of precious toxic metals into non-toxic nano forms due to their ability to accumulate and reduce metal ions into NPs [125]. Algae are preferred as they are convenient to handle, pose lower toxic effects to the environment and synthesize NPs at lower temperatures with great efficiency. Different algae widely used for the synthesis of NPs are: *Lyngbya majuscule, Spirulina platensis, Rhizoclonium hieroglyphicum, Phaeophyceae, Cyanophyceae, Rhodophyceae,* and *Chlorella vulgaris* [126, 127]. Fungi act as ideal biocatalysts for NPs synthesis and are preferred over bacteria due to greater potential of biologically active substances production [128]. Furthermore, fungal biomass are suitable for use in bioreactors as they can resist flow pressure, agitation and harsh conditions in chambers such as bioreactors and can exude extracellular reductive proteins suitable for employment in further steps of synthesis [102]. *Fusarium oxysporum* is one such fungi used for manufacturing NPs at industrial scale [129].

5. Preparation of extract and biomass for the synthesis of nanoparticles

The potential of phytosynthesis, a "green" synthesis approach is not yet completely utilized in full throttle for the colloidal NPs synthesis. As plants harbor a wide range of metabolites, it is possible to utilize plant tissue culture methods and optimizing the downstream processing techniques for the industrial production of NPs [130]. The part of the plant is chosen on the basis of desired application and the widely used plant parts of the part for extract preparation are leaf, seed, stem, fruit, root and flower. Initially, the test plant samples are collected, washed, dried and weighed. These are then chopped down into smaller pieces and soaked into sterile distilled water. This mixture is eventually incubated at optimized conditions such as temperature, stirring speed etc. After a defined time span, the mixture is centrifuged at high speed, filtered using muslin cloth or syringe filters and stored in chilled conditions until future use. The filtrate is then diluted according to optimized conditions and used as a source of reducing and capping agents for the synthesis of NPs [131]. The plant extract thus prepared is mixed with defined ratio of metal salts at optimum conditions for defined time period resulting in NPs [132]. Not only the reaction conditions, but also the nature of extract and its concentration has a significant effect on the NPs synthesis and its quality [133].

Microbial route of synthesis of NPs has garnered enormous interest of researchers in the field of nanobiotechnology. Microorganisms including bacteria, fungi, actinomycetes, yeasts, and viruses are considered as bio-factories, owing to their inherent potential to produce NPs via. Extracellular or intracellular route of synthesis [102]. In case of extracellular synthesis, the microorganisms after subsequent growth of 1-2 days in shaking condition and optimum growth conditions are centrifuged to remove the biomass. The filter-sterilized metal salt solution is then added to the supernatant and incubated. The mixture is then centrifuged to collect the NPs pellet. For intracellular synthesis, the biomass is collected by centrifuging the micro-organisms culture grown in optimum conditions. The biomass pellet is washed and mixed with filter-sterilized solution of metal salt. Color changes in the reaction mixture are observed as a preliminary confirmation of NPs synthesis and further confirmed by spectrophotometric observations and highly sophisticated techniques. Further, similar to that of extracellular synthesis, the mixture is centrifuged to collect the NPs pellet [134].

In case of algae-mediated synthesis of metal NPs, the algal extract is prepared in sterile distilled water or an appropriate organic solvent by boiling it for specified duration. Further, the algal extract and the metal precursors are stirred at optimum conditions. Finally depending upon the mode of synthesis of NPs via algae, i.e. extracellular or an intracellular, the supernatant and biomass are used for the further process [135]. The bioactive agents such as polysaccharides, polyphones, proteins, and/or other reducing factors reduce the metal ions in case of extracellular synthesis of NPs [96, 98, 135, 136] while in case of intracellular synthesis, the algal metabolism via photosynthesis and respiration causes reduction of metal ions [135, 137, 138]. Eventually, the chromatic changes determine the synthesis of NPs as preliminary confirmation. In mycosynthesis i.e. fungi based synthesis, the metal precursors are used to treat fungus mycelium resulting in production of fungi metabolites and enzymes. These bioactive substances reduce toxic metal ions into non-toxic metal NPs [129]. The fungi are usually cultured on an agar plate and further transferred into a liquid medium. Depending upon the route of synthesis, either the biomass or the supernatant is mixed along with metal precursor to yield NPs [139, 140].

6. Effect of different parameters on synthesis of metal nanoparticles

6.1 pH

The pH is one of the most important biogenic synthesis reaction parameters that influence the particle size and morphology of NPs [141, 142]. The NPs can be tailored to the desired size by altering the pH of the reaction mixture which causes changes in the charge over secondary metabolites which has significant effect on their ability to adsorb the metal ions [86]. In case of microbial synthesis of NPs, the culture conditions play a significant role. The small-sized and monodispersed metal oxide NPs are formed in alkaline conditions rather than acidic conditions. This is because more functional groups are available at higher pH that increase the binding ability and stability during nucleation and growth stages favoring the formation of less aggregated NPs [112]. Singh and Srivastava [143] observed a gradual blue shift (towards lower wavelength) in absorption maxima as the pH was increased from 3 to 7 indicating decrease in sizes of the NPs. Also, a red shift (towards higher wavelength) was observed when the pH was increased further from 7 to 11 The further increase in the pH was found to increase the NPs size. The reaction pH also has significant effect on particle morphology in terms of shape of the synthesized NPs [144]. Gericke and Pinches et al. [145] synthesized gold NPs from fungal cultures and observed that at pH 3, uniform sized spherical NPs of 10 nm were synthesized. When the pH was increased to 5, fewer smaller spherical particles were obtained, the morphology of most of the NPs changed to larger well defined triangular, hexagonal, spherical and rod like structures were also obtained. At higher reaction mixture pH 7 and 9, similar undefined structures were observed. Abeywardena et al. [146] employed sucrose solution based extraction of calcium to precipitate calcium carbonate nanostructures to study the effect of pH to yield nanostructures with different morphologies and sizes. The precipitation reaction was carried out at pH values of 7.5, 10.5 and 12.5 using CO_2 bubbling for carbonation as it promoted formation of smaller particles. Different morphologies such as catkin-like structure, spherical particles and rod-like formations; and tiny particles aggregated into large spheres were observed at pH 12.5, 10.5 and 7.5 respectively. Thus concluding that the alkaline pH is suitable for the formation of stable and less agglomerated nanoparticles. Aguilar et al. [147] studied the effect of different pH to yield stable silver nanoparticles using sugar cane bagasse extract. It was observed that acidic conditions (pH 3.5) were not favorable for the production of nanoparticles as the reaction yields mixture of submicron-sized silver (Ag) and silver chloride (AgCl) particles. At neutral pH, though the size of resultant nanoparticles dropped down in the range 8-30 nm, the mixture of Ag and AgCl particles still existed. In alkaline conditions i.e. pH 12, pure silver nanoparticles were obtained exhibiting excellent bactericidal and bacteriostatic properties against Gram positive and Gram negative bacteria. After the thorough inspection of the X-ray diffraction patterns and X-ray energy dispersive spectra (EDS) of the biosynthesized silver nanoparticles, it was evident that the Cl and S in the bagasse induces formation of side products such as AgCl and Ag@AgCl nanoparticles. While, at alkaline pH, the formation of such side products is avoided due to the interaction of Na ions of NaOH with Cl ions cane bagasse.

6.2 Temperature

Temperature is an equally important reaction parameters that significantly influence the biogenic synthesis of NPs [86]. For instance, the rate of synthesis increases at an elevated temperature as compared to that at room temperature eventually leading to higher product yield and crystalline NPs [148]. At elevated temperatures, the rate of reduction of metal ions increases and homogeneous nucleation of metal nuclei occurs facilitating the synthesis of NPs [149, 150]. Noteworthy, the required temperatures in physical methods of synthesis are greater than 350°C, while chemical reaction take place at temperatures as high as 350°C. On contrary, biogenic synthesis occur at considerably low temperatures in the range of 37°C-100°C [151]. In case of NPs synthesis using microorganisms, it is recommended that the microorganisms must be grown at the maximum possible temperature for optimal growth as the enzyme responsible for NPs synthesis shows enhanced catalytic activity at high temperatures and thus is more active [152]. Jameel et al. [153] highlighted the fact that reaction temperature affects the size, morphology, and synthesis rate of platinum NPs. Also, higher number of nucleation centers are produced at elevated temperatures which enhances the biosynthesis rates. Temperature controls the rate of formation of NPs i.e. at higher reaction temperature yields faster rate of particle growth. As majority of NPs were synthesized within an hour, lower reaction temperatures were reported suitable to tune the size of NPs [145]. Harshiny et al. [154] studied the effect of temperature over the range 40 to 70°C, on the antioxidant activity of iron nanoparticles using *Amaranthus dubius* leaf extract. Initially, the antioxidant activity (AA%) increases with increase in temperature up to 50°C due to higher DPPH radical scavenging activities while antioxidant activity decreases beyond 50°C due to the degradation of the active constituent amino acids.

6.3 Time

Longer incubation time yield larger NPs with well-defined shapes, while smaller incubation periods cause smaller sized NPs [145]. Moreover, time has two distinct effects on the quality and potential of NPs synthesized via biogenic route. For instance, if the reaction mixture is incubated for longer time than the optimum, the NPs tend to aggregate causing increased particle size. Moreover, some NPs may even shrink upon longer storage [155, 156]. Sangaonkar et al. [157] studied the effect of incubation time by incubating the reaction mixture at different time periods ranging from 2 to 120 h using UV spectroscopy studies to conclude that 24 h was the optimum time for the synthesis of silver NPs using fruit extract of *Garcinia indica*. Similarly, the reaction set up by Krishnaprabha et al. [158] required two hours for the complete reduction of Au precursors into AuNPs using *Garcinia indica* fruit rind extract as a reducing agent. Thus, the parameter 'incubation time' is codependent on other reaction factors such as concentration of precursor and the biological agent used for preparation of extract. Manzoor et al. [159] studied the effect of nucleation time to reveal that increase in nucleation time results in increase in particle size and wider particle size distribution. It is also evident that intermediate stirring offers tunable particle size and narrow size distribution. Though the synthesis time varies with the precursor and extract used, a keen observation of the color of reaction mixture and analysis of SPR peaks can reveal the optimum time for the reaction. Increase in reaction time and color intensity of the reaction mixture along with prominent SPR peaks can reveal that large amount of metal ions get converted into ($M+$) zerovalent metal NPs (M^0) [160].

6.4 Concentration of metal ions

Concentration of metal ions is one of the key factors influencing the size of synthesized NPs. Usually, the reactions mixtures require just the right quantity of reactants, if the concentrations are slightly increased, the reduction mechanisms are hindered and accumulation of NPs would result in noticeable large aggregates of NPs [149]. Tuning the concentration metal ions in the reaction can be performed either by changing the volume of solvent or the amount of precursor. While, changing the concentration by varying the volume via dilution method is a straightforward method, changing the precursor quantity is subjected to maintaining the ratio between surfactant and precursor [161]. Moreover, researchers have reported that increase in precursor concentration may lead to either increase [162–165] or decrease [166, 167] in particle size. Recently, it has been experimentally proven that nanoparticle growth can be controlled as growth rate is dependent upon the surface reactions occurring at NPs, while at low concentrations, as the diffusion constant increases and the mass transferred is reduced, the growth rate is also reduced [161].

Lakkappa et al. [168], demonstrated the effect of silver precursor concentration on the formation of silver NPs using *Capparis Moonii* as a reducing agent. Their study concluded that higher concentration of the solution resulted in smaller sized NPs yet in wide range of size distribution. At higher metal ion concentrations, bathochromic shift causing change in intensity leads to broad SPR band, lower size dispersion and high aggregation; while, at lower concentrations yield high intensity, better absorbance and narrow SPR bands [160]. Thus, lower concentrations are preferred for the synthesis of metal NPs [169]. Sibiya and Moloto [170] carried out an experiment wherein two precursor salts were equipped for the formation of NPs. They found out that when the ratio of precursors was increased from 1:1 to 1:10, two distinct nanoparticle shapes: spherical and rod-like respectively were obtained. This change in morphology was attributed to the fact that, at higher precursor concentrations, the time required for NPs growth is longer, which therefore subsequently leads to different morphologies.

6.5 Other factors

The phytoconstituents (phenol, polyphenols, polysaccharides, tannins and anthocyanins), their quantity and volume of extract, influence the reduction of metal ions, average particle size, processing, synthesis time and stability of NPs. As the plant extracts act as reducing agent, their volume up to a certain extent works efficiently for the formation of stable metal NPs [149]. In plant based synthesis, as the composition of metabolites varies vastly in different plant parts of same species, the size of synthesized NPs varies with respect to part of plant used for extract preparation [171]. Singh and Srivastava [143] reported that as the concentration of black cardamom extract as a reducing agent was decreased, the size of resultant NPs increased.

In case of microbial route of synthesis, the enzymes and proteins existing in either the cell walls or cytoplasm reduce the precursor ions thereby aggregating atoms leading to formation of NPs [172]. Thus, such activity specific enzymes and proteins can be identified and isolated to facilitate reactions to be carried out in a cell-free environment producing NPs with tunable size and shape. Such experiments often yield triangular and hexagonal thin plate-like structures irrespective of source, being plant part or microorganisms [145, 173]. The pressure applied to the reaction mixture is also known to influence the shape and size of the resultant NPs [174]. Ambient pressure conditions accelerate the rate of reduction of metal ions using biological agents [175]. Plants are rich in various secondary metabolites which act as reducing and stabilizing agents and thus affect the NPs synthesis. The composition of such metabolites differ with different types of plant, plant part, and the protocol followed for the preparation of extract [176].

7. Conclusion

This chapter summarizes the fundamental introduction of NMs and NPs, the significance of colloidal metal NPs for range of applications, diverse physicochemical and biological pathways for synthesis of colloidal NPs and the parameters affecting the synthesis of NPs. Colloidal metal NPs, notably noble metal NPs such as gold, are being utilized since ancient times for multiple applications due to eminent and unique properties which make them superior compared with molecules or fellow bulk materials. Further, these NPs are preferred for biomedical applications due to their effectiveness to attenuate the shortcomings of traditional provisions, as it can be manipulated to deal healthily with the *in vivo* biological environment. Alongside, the size of NPs compliments them with access through cells and tissues which helps not only to diagnose the disease proficiently but also aids for site specific treatment. Such prospective usages encourage the production of NPs in massive quantity. The colloidal metal NPs have been synthesized dominantly by using chemical and physical methods for many years, but the shortfalls of the methods confine their practice in biological fields. Hence researchers have opted a biological friendly way to synthesize NPs from natural environmental sources. Even though the mechanism of green production of NPs from microorganisms and plants is not yet entirely revealed, the synthesizing parameters, namely pH, temperature, time, concentration of metal salt affect the synthesis of NPs to a great extent and their optimization is very much necessary to yield the NPs in bulk quantity with desired properties. Besides the ongoing research on colloidal metal NPs from last few decades, its actual implementation in clinical field is at primary stage and many aspects such as distribution of NPs inside body, their accumulation and clearance from the body after treatment etc. need to be addressed for better involvement of NPs in biomedical applications.

Conflict of interest

All the authors declare that there is no conflict of interest.

Author details

Disha N. Moholkar†, Darshana V. Havaldar†, Rachana S. Potadar and Kiran D. Pawar*
School of Nanoscience and Biotechnology, Shivaji University, Kolhapur, India

*Address all correspondence to: kdp.snst@unishivaji.ac.in; pawarkiran1912@gmail.com

† These authors contributed equally.

References

[1] Krukemeyer M, Krenn V, Huebner F, Wagner W, Resch R. History and possible uses of nanomedicine based on nanoparticles and nanotechnological Progress. J. Nanomed. Nanotechnol. 2015;**06** https://doi.org/10.4172/2157-7439.1000336

[2] Kalangutkar PK. Nanotechnology – Advancing the field of biomedical engineering, Int. J. Curr. Res. Rev. 2015;7:66-70

[3] Rivera Gil P, Hühn D, del Mercato LL, Sasse D, Parak WJ. Nanopharmacy: Inorganic nanoscale devices as vectors and active compounds. Pharmacological Research. 2010;**62**:115-125 https://doi.org/10.1016/j.phrs.2010.01.009

[4] Ramos AP, Cruz MAE, Tovani CB, Ciancaglini P. Biomedical applications of nanotechnology. Biophysical Reviews. 2017;**9**:79-89 https://doi.org/10.1007/s12551-016-0246-2

[5] Makarov VV, Love AJ, Sinitsyna OV, Makarova SS, Yaminsky IV, Taliansky ME, et al. "Green" nanotechnologies: Synthesis of metal nanoparticles using plants. Acta Naturae. 2014;**6**:35-44 https://doi.org/10.32607/20758251-2014-6-1-35-44

[6] Malik A, Chaudhary S, Garg G, Tomar A. Dendrimers : A Tool for Drug Delivery. Adv. Biol. Res. (Rennes). 2012;**6**:165-169 https://doi.org/10.5829/idosi.abr.2012.6.4.6254

[7] Allen TM, Cullis PR. Liposomal drug delivery systems: From concept to clinical applications. Advanced Drug Delivery Reviews. 2013;**65**:36-48 https://doi.org/10.1016/j.addr.2012.09.037

[8] Weissig V, Pettinger TK, Murdock N. Nanopharmaceuticals (part 1): Products on the market. International Journal of Nanomedicine. 2014;**9**:4357-4373 https://doi.org/10.2147/IJN.S46900

[9] Roszek B, De Jong WH, Geertsma RE. Nanotechnology in medical applications : State-of-the-art in materials and devices, RIVM rep. In: 265001001/2005. 2005

[10] Basar MR, Malek F, Juni KM, Idris MS, Saleh MIM. Ingestible wireless capsule technology: A review of development and future indication. Int. J. Antennas Propag. 2012;**2012** https://doi.org/10.1155/2012/807165

[11] Yang J, Zhang C, Wang XD, Wang WX, Xi N, Liu LQ. Development of micro- and nanorobotics: A review. Science China Technological Sciences. 2019;**62**:1-20 https://doi.org/10.1007/s11431-018-9339-8

[12] Ceylan H, Giltinan J, Kozielski K, Sitti M. Mobile microrobots for bioengineering applications. Lab on a Chip. 2017;**17**:1705-1724 https://doi.org/10.1039/c7lc00064b

[13] Ornes S. Medical microrobots have potential in surgery, therapy, imaging. and diagnostics, Proc. Natl. Acad. Sci. U. S. A. 2017;**114**:12356-12358 https://doi.org/10.1073/pnas.1716034114

[14] Bratlie KM, York RL, Invernale MA, Langer RL, Anderson DG. Materials for diabetes therapeutics. Advanced Healthcare Materials. 2012;**1**:267-284 https://doi.org/10.1002/adhm.201200037

[15] Ravaine V, Ancla C, Catargi B. Chemically controlled closed-loop insulin delivery. Journal of Controlled Release. 2008;**132**:2-11 https://doi.org/10.1016/j.jconrel.2008.08.009

[16] Thomas R, Park IK, Jeong YY. Magnetic iron oxide nanoparticles for multimodal imaging and therapy

of cancer. International Journal of Molecular Sciences. 2013;**14**:15910-15930 https://doi.org/10.3390/ijms140815910

[17] Atala A, Kurtis Kasper F, Mikos AG. Engineering complex tissues. Science Translational Medicine. 2012;**4** https://doi.org/10.1126/scitranslmed.3004890

[18] Atala A, Jeremy M, Antonios M, Gordana V-N. Translational Approaches in Tissue Engineering and Regenerative Medicine, in: Transl. Artech: Approaches Tissue Eng. Regen. Med; 2007 https://ieeexplore.ieee.org/document/9100170 (accessed September 1, 2020

[19] Wu T, Yu S, Chen D, Wang Y. Bionic design, materials and performance of bone tissue scaffolds. Materials (Basel). 2017;**10** https://doi.org/10.3390/ma10101187

[20] Cosgriff-Hernandez E, Mikos AG. New biomaterials as scaffolds for tissue engineering. Pharmaceutical Research. 2008;**25**:2345-2347 https://doi.org/10.1007/s11095-008-9666-4

[21] Sun G, Mao JJ. Engineering dextran-based scaffolds for drug delivery and tissue repair. Nanomedicine. 2012;**7**:1771-1784 https://doi.org/10.2217/nnm.12.149

[22] Oest ME, Dupont KM, Kong H-J, Mooney DJ, Guldberg RE. The effect of a subject-specific amount of lateral wedge on knee. J. Orthop. Res. Sept. 2007;**25**:1121-1127 https://doi.org/10.1002/jor

[23] Lee EJ, Huh BK, Kim SN, Lee JY, Park CG, Mikos AG, et al. Application of materials as medical devices with localized drug delivery capabilities for enhanced wound repair. Progress in Materials Science. 2017;**89**:392-410 https://doi.org/10.1016/j.pmatsci.2017.06.003

[24] Samorezov JE, Alsberg E. Spatial regulation of controlled bioactive factor delivery for bone tissue engineering. Advanced Drug Delivery Reviews. 2015;**84**:45-67 https://doi.org/10.1016/j.addr.2014.11.018

[25] Dukhin AS, Goetz PJ. Fundamentals of Interface and colloid. Science. 2017 https://doi.org/10.1016/b978-0-444-63908-0.00002-8

[26] Tresa Sunny A, Velayudhan P, Thomas S. Physics and Chemistry of Colloidal Metal Oxide Nanocrystals and their Applications to Nanotechnologies and Microsystems: An Introduction. Elsevier Inc.; 2020 https://doi.org/10.1016/b978-0-12-813357-6.00001-2

[27] Kreuter J. Nanoparticles-a historical perspective. International Journal of Pharmaceutics. 2007;**331**:1-10 https://doi.org/10.1016/j.ijpharm.2006.10.021

[28] Issa B, Obaidat IM, Albiss BA, Haik Y. Magnetic nanoparticles: Surface effects and properties related to biomedicine applications. International Journal of Molecular Sciences. 2013;**14**:21266-21305 https://doi.org/10.3390/ijms141121266

[29] Ashok A, Kumar A, Tarlochan F. Colloidal Metal Oxide Nanocrystals in Catalysis. Elsevier Inc.; 2020 https://doi.org/10.1016/b978-0-12-813357-6.00012-7

[30] Zhang K, Wang S, Zhou C, Cheng L, Gao X, Xie X, et al. Advanced smart biomaterials and constructs for hard tissue engineering and regeneration. Bone Res. 2018;**6** https://doi.org/10.1038/s41413-018-0032-9

[31] Dhillon GS, Brar SK, Kaur S, Verma M. Green approach for nanoparticle biosynthesis by fungi: Current trends and applications. Critical Reviews in Biotechnology. 2012;**32**:49-73

https://doi.org/10.3109/07388551.2010.550568

[32] Karimi Z, Karimi L, Shokrollahi H. Nano-magnetic particles used in biomedicine: Core and coating materials. Materials Science and Engineering: C. 2013;**33**:2465-2475 https://doi.org/10.1016/j.msec.2013.01.045

[33] Tural B, Özkan N, Volkan M. Preparation and characterization of polymer coated superparamagnetic magnetite nanoparticle agglomerates. Journal of Physics and Chemistry of Solids. 2009;**70**:860-866 https://doi.org/10.1016/j.jpcs.2009.04.007

[34] Sun S. Recent advances in chemical synthesis, self-assembly, and applications of FePt nanoparticles. Advanced Materials. 2006;**18**:393-403 https://doi.org/10.1002/adma.200501464

[35] Shi Y, Lin M, Jiang X, Liang S. Recent advances in FePt nanoparticles for biomedicine. Journal of Nanomaterials. 2015;**2015** https://doi.org/10.1155/2015/467873

[36] Ding W, Guo L. Immobilized transferrin Fe3O4@SiO2 nanoparticle with high doxorubicin loading for dual-targeted tumor drug delivery. International Journal of Nanomedicine. 2013;**8**:4631-4639 https://doi.org/10.2147/IJN.S51745

[37] Chen CL, Kuo LR, Lee SY, Hwu YK, Chou SW, Chen CC, et al. Photothermal cancer therapy via femtosecond-laser-excited FePt nanoparticles. Biomaterials. 2013;**34**:1128-1134 https://doi.org/10.1016/j.biomaterials.2012.10.044

[38] Hong RY, Feng B, Chen LL, Liu GH, Li HZ, Zheng Y, et al. Synthesis, characterization and MRI application of dextran-coated Fe3O4 magnetic nanoparticles. Biochemical Engineering Journal. 2008;**42**:290-300 https://doi.org/10.1016/j.bej.2008.07.009

[39] Liang SY, Zhou Q, Wang M, Zhu YH, Wu QZ, Yang XL. Water-soluble l-cysteine-coated FePt nanoparticles as dual MRI/CT imaging contrast agent for glioma. International Journal of Nanomedicine. 2015;**10**:2325-2333 https://doi.org/10.2147/IJN.S75174

[40] Shah RR, Davis TP, Glover AL, Nikles DE, Brazell CS. Impact of magnetic field parameters and iron oxide nanoparticle properties on heat generation for use in magnetic hyperthermia. Journal of Magnetism and Magnetic Materials. 2015:96-106 https://doi.org/10.1016/j.physbeh.2017.03.040

[41] Seehra MS, Singh V, Dutta P, Neeleshwar S, Chen YY, Chen CL, et al. Size-dependent magnetic parameters of fcc FePt nanoparticles: Applications to magnetic hyperthermia. Journal of Physics D: Applied Physics. 2010;**43** https://doi.org/10.1088/0022-3727/43/14/145002

[42] Patra CR, Bhattacharya R, Mukhopadhyay D, Mukherjee P. Fabrication of gold nanoparticles for targeted therapy in pancreatic cancer. Advanced Drug Delivery Reviews. 2010;**62**:346-361 https://doi.org/10.1016/j.addr.2009.11.007

[43] Lee J, Chatterjee DK, Lee MH, Krishnan S. Gold nanoparticles in breast cancer treatment: Promise and potential pitfalls. Cancer Letters. 2014;**347**:46-53 https://doi.org/10.1016/j.canlet.2014.02.006

[44] Ge L, Li Q, Wang M, Ouyang J, Li X, Xing MMQ. Nanosilver particles in medical applications: Synthesis, performance. and toxicity, Int. J. Nanomedicine. 2014;**9**:2399-2407 https://doi.org/10.2147/IJN.S55015

[45] Wei L, Lu J, Xu H, Patel A, Chen Z-S, Chen G. Silver nanoparticles: Synthesis, properties, and therapeutic applications. Drug Discovery Today. 2015:595-601 https://doi.org/10.1016/j.physbeh.2017.03.040

[46] Paladini F, Pollini M. Antimicrobial silver nanoparticles for wound healing application: Progress and future trends. Materials (Basel). 2019;**12** https://doi.org/10.3390/ma12162540

[47] Deshmukh SP, Patil SM, Mullani SB, Delekar SD. Silver nanoparticles as an effective disinfectant: A review. Materials Science and Engineering: C. 2019;**97**:954-965 https://doi.org/10.1016/j.msec.2018.12.102

[48] Meir R, Shamalov K, Betzer O, Motiei M, Horovitz-Fried M, Yehuda R, et al. Nanomedicine for Cancer immunotherapy: Tracking Cancer-specific T-cells in vivo with gold nanoparticles and CT imaging. ACS Nano. 2015;**9**:6363-6372 https://doi.org/10.1021/acsnano.5b01939

[49] Han X, Fang X, Shi A, Wang J, Zhang Y. An electrochemical DNA biosensor based on gold nanorods decorated graphene oxide sheets for sensing platform. Analytical Biochemistry. 2013;**443**:117-123 https://doi.org/10.1016/j.ab.2013.08.027

[50] Curry T, Kopelman R, Shilo M, Popovtzer R. Multifunctional theranostic gold nanoparticles for targeted CT imaging and photothermal therapy. Contrast Media & Molecular Imaging. 2014;**9**:53-61 https://doi.org/10.1002/cmmi.1563

[51] Mishra PK, Mishra H, Ekielski A, Talegaonkar S, Vaidya B. Zinc oxide nanoparticles: A promising nanomaterial for biomedical applications. Drug Discovery Today. 2017;**22**:1825-1834 https://doi.org/10.1016/j.drudis.2017.08.006

[52] Zhang ZY, Xiong HM. Photoluminescent ZnO nanoparticles and their biological applications. Materials (Basel). 2015;**8**:3101-3127 https://doi.org/10.3390/ma8063101

[53] Xiong HM. ZnO nanoparticles applied to bioimaging and drug delivery. Advanced Materials. 2013;**25**:5329-5335 https://doi.org/10.1002/adma.201301732

[54] Daraio C, Jin S. Synthesis and patterning methods for nanostructures useful for biological applications. Nanotechnol. Biol. Med. 2014:27-45 https://doi.org/10.1007/978-0-387-31296-5

[55] Dhand C, Dwivedi N, Loh XJ, Jie Ying AN, Verma NK, Beuerman RW, et al. Methods and strategies for the synthesis of diverse nanoparticles and their applications: A comprehensive overview. RSC Advances. 2015;**5**:105003-105037 https://doi.org/10.1039/c5ra19388e

[56] Xing T, Sunarso J, Yang W, Yin Y, Glushenkov AM, Li LH, et al. Ball milling: A green mechanochemical approach for synthesis of nitrogen doped carbon nanoparticles. Nanoscale. 2013;**5**:7970-7976 https://doi.org/10.1039/c3nr02328a

[57] Piras CC, Fernández-Prieto S, De Borggraeve WM. Ball milling: A green technology for the preparation and functionalisation of nanocellulose derivatives. Nanoscale Adv. 2019;**1**:937-947 https://doi.org/10.1039/c8na00238j

[58] Chen YH, Yeh CS. Laser ablation method: Use of surfactants to form the dispersed Ag nanoparticles. Colloids Surfaces A Physicochem. Eng. Asp. 2002;**197**:133-139 https://doi.org/10.1016/S0927-7757(01)00854-8

[59] Barcikowski S, Meńndez-Manjón A, Chichkov B, Brikas M, Račiukaitis G. Generation of nanoparticle colloids by picosecond and femtosecond

laser ablations in liquid flow. Applied Physics Letters. 2007;**91** https://doi.org/10.1063/1.2773937

[60] Sridhar R, Lakshminarayanan R, Madhaiyan K, Barathi VA, Limh KHC, Ramakrishna S. Electrosprayed nanoparticles and electrospun nanofibers based on natural materials: Applications in tissue regeneration, drug delivery and pharmaceuticals. Chemical Society Reviews. 2015;**44**:790-814 https://doi.org/10.1039/c4cs00226a

[61] Tapia-Hernández JA, Torres-Chávez PI, Ramírez-Wong B, Rascón-Chu A, Plascencia-Jatomea M, Barreras-Urbina CG, et al. Micro- and nanoparticles by electrospray: Advances and applications in foods. Journal of Agricultural and Food Chemistry. 2015;**63**:4699-4707 https://doi.org/10.1021/acs.jafc.5b01403

[62] Ward MB, Brydson R, Cochrane RF. Mn nanoparticles produced by inert gas condensation. Journal of Physics Conference Series. 2006;**26**:296-299 https://doi.org/10.1088/1742-6596/26/1/071

[63] Suryanarayana C, Prabhu B. Synthesis of nanostructured materials by inert-gas condensation methods, nanostructured mater. Process. Prop. Appl. Second Ed. 2006:47-90 https://doi.org/10.1016/B978-081551534-0.50004-X

[64] Okuyama K, Lenggoro WW. Preparation of nanoparticles via spray route. Chemical Engineering Science. 2003;**58**:537-547 https://doi.org/10.1016/S0009-2509(02)00578-X

[65] Moore B, Asadi E, Lewis G. Deposition methods for microstructured and nanostructured coatings on metallic bone implants: A review. Advances in Materials Science and Engineering. 2017;**2017** https://doi.org/10.1155/2017/5812907

[66] Teoh WY, Amal R, Mädler L. Flame spray pyrolysis: An enabling technology for nanoparticles design and fabrication. Nanoscale. 2010;**2**:1324-1347 https://doi.org/10.1039/c0nr00017e

[67] Hyeon T. Chemical synthesis of magnetic nanoparticles Taeghwan. Chemical Communications. 2003:927-934 http://faculty.washington.edu/markeh/MSE599/Huyen_ChemComm_2003.pdf

[68] Brinker CJ, Scherer GW. Sol-gel science: The physics and chemistry of sol-gel processing. Sol-Gel Sci. Phys. Chem. Sol-Gel Process. 2013:1-908 https://doi.org/10.1016/C2009-0-22386-5

[69] Kumar A, Yadav N, Bhatt M, Mishra NK, Chaudhary P, Singh R. Sol-gel method, Res. Journal of Chemical Sciences. 2015;**5**:98-105 https://doi.org/10.11470/oubutsu1932.62.1248

[70] Shimada M, Wang WN, Okuyama K. Synthesis of gallium nitride nanoparticles by microwave plasma-enhanced CVD. Chemical Vapor Deposition. 2010;**16**:151-156 https://doi.org/10.1002/cvde.200906811

[71] Hamedani Y, Macha P, Bunning TJ, Naik RR, Vasudev MC. Plasma-Enhanced Chemical Vapor Deposition: Where we Are and the Outlook for the Future. Cells Solid State Devices: Chem. Vap. Depos. - Recent Adv. Appl. Opt. Sol; 2016 https://doi.org/10.5772/64654

[72] Rahman P, Green M. The synthesis of rare earth fluoride based nanoparticles. Nanoscale. 2009;**1**:214-224 https://doi.org/10.1039/b9nr00089e

[73] Dong H, Chen YC, Feldmann C. Polyol synthesis of nanoparticles: Status and options regarding metals, oxides, chalcogenides, and non-metal elements. Green Chemistry. 2015;**17**:4107-4132 https://doi.org/10.1039/c5gc00943j

[74] Hayashi H, Hakuta Y. Hydrothermal synthesis of metal oxide nanoparticles in supercritical water. Materials (Basel). 2010;**3**:3794-3817 https://doi.org/10.3390/ma3073794

[75] Abedini A, Daud AR, Hamid MAA, Othman NK, Saion E. A review on radiation-induced nucleation and growth of colloidal metallic nanoparticles. Nanoscale Research Letters. 2013;**8**:1-10 https://doi.org/10.1186/1556-276X-8-474

[76] Gangula A, Podila R, M R, Karanam L, Janardhana C, Rao AM. Catalytic reduction of 4-nitrophenol using biogenic gold and silver nanoparticles derived from breynia rhamnoides. Langmuir. 2011;**27**:15268-15274 https://doi.org/10.1021/la2034559

[77] Devi TB, Ahmaruzzaman M, Begum S. A rapid, facile and green synthesis of Ag@AgCl nanoparticles for the effective reduction of 2,4-dinitrophenyl hydrazine. New Journal of Chemistry. 2016;**40**:1497-1506 https://doi.org/10.1039/c5nj02367j

[78] Duan H, Wang D, Li Y. Green chemistry for nanoparticle synthesis. Chemical Society Reviews. 2015;**44**:5778-5792 https://doi.org/10.1039/c4cs00363b

[79] Arabi M, Ghaedi M, Ostovan A. Development of a lower toxic approach based on Green synthesis of water-compatible molecularly imprinted nanoparticles for the extraction of hydrochlorothiazide from human urine. ACS Sustainable Chemistry & Engineering. 2017;**5**:3775-3785 https://doi.org/10.1021/acssuschemeng.6b02615

[80] Nanda A, Saravanan M. Biosynthesis of silver nanoparticles from Staphylococcus aureus and its antimicrobial activity against MRSA and MRSE. Nanomedicine Nanotechnology, Biol. Med. 2009;**5**:452-456 https://doi.org/10.1016/j.nano.2009.01.012

[81] Zhang X, Yan S, Tyagi RD, Surampalli RY. Synthesis of nanoparticles by microorganisms and their application in enhancing microbiological reaction rates. Chemosphere. 2011;**82**:489-494 https://doi.org/10.1016/j.chemosphere.2010.10.023

[82] Jain N, Bhargava A, Majumdar S, Tarafdar JC, Panwar J. Extracellular biosynthesis and characterization of silver nanoparticles using Aspergillus flavus NJP08: A mechanism perspective. Nanoscale. 2011;**3**:635-641 https://doi.org/10.1039/c0nr00656d

[83] Mandal D, Bolander ME, Mukhopadhyay D, Sarkar G, Mukherjee P. The use of microorganisms for the formation of metal nanoparticles and their application. Applied Microbiology and Biotechnology. 2006;**69**:485-492 https://doi.org/10.1007/s00253-005-0179-3

[84] Bhattacharya D, Gupta RK. Nanotechnology and potential of microorganisms. Critical Reviews in Biotechnology. 2005;**25**:199-204 https://doi.org/10.1080/07388550500361994

[85] Shankar SS, Ahmad A, Pasricha R, Sastry M. Bioreduction of chloroaurate ions by geranium leaves and its endophytic fungus yields gold nanoparticles of different shapes. Journal of Materials Chemistry. 2003;**13**:1822-1826 https://doi.org/10.1039/b303808b

[86] Baker S, Rakshith D, Kavitha KS, Santosh P, Kavitha HU, Rao Y, et al. Plants: Emerging as nanofactories towards facile route in synthesis of nanoparticles. BioImpacts: BI. 2013;**3**:111-117 https://doi.org/10.5681/bi.2013.012

[87] Dauthal P, Mukhopadhyay M. Noble metal nanoparticles: Plant-mediated synthesis, mechanistic aspects of synthesis, and applications.

Industrial and Engineering Chemistry Research. 2016;**55**:9557-9577 https://doi.org/10.1021/acs.iecr.6b00861

[88] Monopoli MP, Åberg C, Salvati A, Dawson KA. Biomolecular coronas provide the biological identity of nanosized materials. Nature Nanotechnology. 2012;7:779-786 https://doi.org/10.1038/nnano.2012.207

[89] Sintubin L, Verstraete W, Boon N. Biologically produced nanosilver: Current state and future perspectives. Biotechnology and Bioengineering. 2012;**109**:2422-2436 https://doi.org/10.1002/bit.24570

[90] Mukherjee S, Sushma V, Patra S, Barui AK, Bhadra MP, Sreedhar B, et al. Green chemistry approach for the synthesis and stabilization of biocompatible gold nanoparticles and their potential applications in cancer therapy. Nanotechnology. 2012;**23** https://doi.org/10.1088/0957-4484/23/45/455103

[91] Singh P, Kim Y, Zhang D, Yang D. Biological synthesis of nanoparticles from plants and microorganisms. Trends in Biotechnology. 2016;**34**:588-599 https://doi.org/10.1016/j.tibtech.2016.02.006

[92] Fawcett D, Verduin JJ, Shah M, Sharma SB, Poinern GEJ. A review of Current research into the biogenic synthesis of metal and metal oxide nanoparticles via marine algae and seagrasses. J. Nanosci. 2017;**2017**:1-15 https://doi.org/10.1155/2017/8013850

[93] Venkatesan J, Anil S, Kim SK, Shim MS. Seaweed polysaccharide-based nanoparticles: Preparation and applications for drug delivery. Polymers (Basel). 2016;**8**:1-25 https://doi.org/10.3390/polym8020030

[94] Kumar P, Senthamil Selvi S, Govindaraju M. Seaweed-mediated biosynthesis of silver nanoparticles using Gracilaria corticata for its antifungal activity against Candida spp. Applied Nanoscience. 2013;**3**:495-500 https://doi.org/10.1007/s13204-012-0151-3

[95] Ermakova S, Kusaykin M, Trincone A, Tatiana Z. Are multifunctional marine polysaccharides a myth or reality? Frontiers in Chemistry. 2015;**3**:1-4 https://doi.org/10.3389/fchem.2015.00039

[96] Mahdavi M, Namvar F, Bin Ahmad M, Mohamad R. Green biosynthesis and characterization of magnetic iron oxide (Fe 3O4) nanoparticles using seaweed (Sargassum muticum) aqueous extract. Molecules. 2013;**18**:5954-5964 https://doi.org/10.3390/molecules18055954

[97] Stalin Dhas T, Ganesh Kumar V, Stanley Abraham L, Karthick V, Govindaraju K. Sargassum myriocystum mediated biosynthesis of gold nanoparticles. Spectrochim. Acta - Part A Mol. Biomol. Spectrosc. 2012;**99**:97-101 https://doi.org/10.1016/j.saa.2012.09.024

[98] Mata YN, Torres E, Blázquez ML, Ballester A, González F, Muñoz JA. Gold(III) biosorption and bioreduction with the brown alga Fucus vesiculosus. Journal of Hazardous Materials. 2009;**166**:612-618 https://doi.org/10.1016/j.jhazmat.2008.11.064

[99] Ashokkumar T, Vijayaraghavan K. Brown seaweed-mediated biosynthesis of gold nanoparticles. J. Environ. Biotechnol. Res. 2016;**2**:45-50 http://www.vinanie.com/jebr/articles/v2n1p45.html

[100] Khanehzaei H, Ahmad MB, Shameli K, Ajdari Z. Synthesis and characterization of Cu@Cu2O core shell nanoparticles prepared in seaweed Kappaphycus alvarezii media. International Journal of Electrochemical Science. 2015;**10**:404-413

[101] Jeevanandam J, Chan YS, Danquah MK. Biosynthesis of metal and metal oxide nanoparticles. ChemBioEng Rev. 2016;**3**:55-67 https://doi.org/10.1002/cben.201500018

[102] Narayanan KB, Sakthivel N. Biological synthesis of metal nanoparticles by microbes. Advances in Colloid and Interface Science. 2010;**156**:1-13 https://doi.org/10.1016/j.cis.2010.02.001

[103] Taherzadeh MJ, Fox M, Hjorth H, Edebo L. Production of mycelium biomass and ethanol from paper pulp sulfite liquor by Rhizopus oryzae. Bioresource Technology. 2003;**88**:167-177 https://doi.org/10.1016/S0960-8524(03)00010-5

[104] Mohanpuria P, Rana NK, Yadav SK. Biosynthesis of nanoparticles: Technological concepts and future applications. Journal of Nanoparticle Research. 2008;**10**:507-517 https://doi.org/10.1007/s11051-007-9275-x

[105] Bhainsa KC, D'Souza SF. Extracellular biosynthesis of silver nanoparticles using the fungus Aspergillus fumigatus. Colloids Surfaces B Biointerfaces. 2006;**47**:160-164 https://doi.org/10.1016/j.colsurfb.2005.11.026

[106] Bharde A, Rautaray D, Bansal V, Ahmad A, Sarkar I, Yusuf SM, et al. Extracellular biosynthesis of magnetite using fungi. Small. 2006;**2**:135-141 https://doi.org/10.1002/smll.200500180

[107] Pantidos N. Biological synthesis of metallic nanoparticles by Bacteria, Fungi and plants, J. Nanomed. Nanotechnol. 2014;**05** https://doi.org/10.4172/2157-7439.1000233

[108] Iravani S. Bacteria in nanoparticle synthesis: Current status and future prospects. Int. Sch. Res. Not. 2014;**2014**:1-18 https://doi.org/10.1155/2014/359316

[109] Agarwal A, Rao S. Creative pathology teaching with word puzzles until students learn: A study in a medical university. Asian J. Res. Med. Pharm. Sci. 2017;**2**:1-7 https://doi.org/10.9734/ajrimps/2017/38416

[110] Singh P, Kim YJ, Zhang D, Yang DC. Biological synthesis of nanoparticles from plants and microorganisms. Trends in Biotechnology. 2016;**34**:588-599 https://doi.org/10.1016/j.tibtech.2016.02.006

[111] Imran Din M, Rani A. Recent advances in the synthesis and stabilization of nickel and nickel oxide nanoparticles: A green adeptness. Int. J. Anal. Chem. 2016;**2016** https://doi.org/10.1155/2016/3512145

[112] Khan SA, Shahid S, Sajid MR, Noreen F, Kanwal S, BIOGENIC SYNTHESIS OF CuO. NANOPARTICLES AND THEIR BIOMEDICAL APPLICATIONS: A CURRENT REVIEW., Int. Journal of Advanced Research. 2017;**5**:925-946 https://doi.org/10.21474/ijar01/4495

[113] Gardea-Torresdey JL, Parsons JG, Gomez E, Peralta-Videa J, Troiani HE, Santiago P, et al. Formation and growth of Au nanoparticles inside live alfalfa plants. Nano Letters. 2002;**2**:397-401 https://doi.org/10.1021/nl015673

[114] Iravani S. Green synthesis of metal nanoparticles using plants. Green Chemistry. 2011;**13**:2638-2650 https://doi.org/10.1039/c1gc15386b

[115] Singh J, Kumar V, Kim KH, Rawat M. Biogenic synthesis of copper oxide nanoparticles using plant extract and its prodigious potential for photocatalytic degradation of dyes. Environmental Research. 2019;**177**:108569 https://doi.org/10.1016/j.envres.2019.108569

[116] Zheng Y, Fu L, Han F, Wang A, Cai W, Yu J, et al. Green biosynthesis

and characterization of zinc oxide nanoparticles using Corymbia citriodora leaf extract and their photocatalytic activity. Green Chemistry Letters and Reviews. 2015;**8**:59-63 https://doi.org/10.1080/17518253.2015.1075069

[117] Yuvakkumar R, Suresh J, Nathanael AJ, Sundrarajan M, Hong SI, Novel green synthetic strategy to prepare ZnO nanocrystals using rambutan. (Nephelium lappaceum L.) peel extract and its antibacterial applications, mater. Sci. Eng. C. 2014;**41**:17-27 https://doi.org/10.1016/j.msec.2014.04.025

[118] Nagajyothi PC, Minh An TN, Sreekanth TVM, Il Lee J, Joo DL, Lee KD. Green route biosynthesis: Characterization and catalytic activity of ZnO nanoparticles. Materials Letters. 2013;**108**:160-163 https://doi.org/10.1016/j.matlet.2013.06.095

[119] Das RK, Brar SK. Plant mediated green synthesis: Modified approaches. Nanoscale. 2013;**5**:10155-10162 https://doi.org/10.1039/c3nr02548a

[120] Shameli K, Bin Ahmad M, Zamanian A, Sangpour P, Shabanzadeh P, Abdollahi Y, et al. Green biosynthesis of silver nanoparticles using Curcuma longa tuber powder. International Journal of Nanomedicine. 2012;**7**:5603-5610 https://doi.org/10.2147/IJN.S36786

[121] Ho R, Teai T, Bianchini J-P, Lafont R, Raharivelomanana P. Chapter 23 ferns: From traditional uses to pharmaceutical development, chemical identification of active principles. In: Work. With Ferns Issues Appl. 2010. pp. 1-386 https://doi.org/10.1007/978-1-4419-7162-3

[122] Nune SK, Chanda N, Shukla R, Katti K, Kulkarni RR, Thilakavathy S, et al. Green nanotechnology from tea: Phytochemicals in tea as building blocks for production of biocompatible gold nanoparticles. Journal of Materials Chemistry. 2009;**19**:2912-2920 https://doi.org/10.1039/b822015h

[123] Abd El-Aziz M, Badr Y, Mahmoud MA. Biosynthesis of gold nanoparticles using pseudomonas aeruginosa. AIP Conf. Proc. 2007;**888**:177-181 https://doi.org/10.1063/1.2711108

[124] Desai MP, Patil RV, Pawar KD. Selective and sensitive colorimetric detection of platinum using pseudomonas stutzeri mediated optimally synthesized antibacterial silver nanoparticles. Biotechnol. Reports. 2020;**25**:e00404 https://doi.org/10.1016/j.btre.2019.e00404

[125] Mehta SK, Gaur JP. Use of algae for removing heavy metal ions from wastewater: Progress and prospects. Critical Reviews in Biotechnology. 2005;**25**:113-152 https://doi.org/10.1080/07388550500248571

[126] Chakraborty N, Banerjee A, Lahiri S, Panda A, Ghosh AN, Pal R. Biorecovery of gold using cyanobacteria and an eukaryotic alga with special reference to nanogold formation - a novel phenomenon. Journal of Applied Phycology. 2009;**21**:145-152 https://doi.org/10.1007/s10811-008-9343-3

[127] Sharma A, Sharma S, Sharma K, Chetri SPK, Vashishtha A, Singh P, et al. Algae as crucial organisms in advancing nanotechnology: A systematic review. Journal of Applied Phycology. 2016;**28**:1759-1774 https://doi.org/10.1007/s10811-015-0715-1

[128] Li G, He D, Qian Y, Guan B, Gao S, Cui Y, et al. Fungus-mediated green synthesis of silver nanoparticles using aspergillus terreus. International Journal of Molecular Sciences. 2012;**13**:466-476 https://doi.org/10.3390/ijms13010466

[129] Zielonka A, Klimek-Ochab M. Fungal synthesis of size-defined

nanoparticles, Adv. Nat. Sci. Nanosci. Nanotechnol. 2017;**8** https://doi.org/10.1088/2043-6254/aa84d4

[130] Jha AK, Prasad K, Prasad K, Kulkarni AR. Plant system: Nature's nanofactory. Colloids Surfaces B Biointerfaces. 2009;**73**:219-223 https://doi.org/10.1016/j.colsurfb.2009.05.018

[131] Lade BD, Shanware AS. Phytonanofabrication. IntechOpen: Methodology and Factors Affecting Biosynthesis of Nanoparticles; 2020. p. 38 https://doi.org/10.1016/j.colsurfa.2011.12.014

[132] Li M, Noriega-Trevino ME, Nino-Martinez N, Marambio-Jones C, Wang J, Damoiseaux R, et al. Synergistic bactericidal activity of Ag-TiO 2 nanoparticles in both light and dark conditions. Environmental Science & Technology. 2011;**45**:8989-8995 https://doi.org/10.1021/es201675m

[133] Dwivedi AD, Gopal K. Biosynthesis of silver and gold nanoparticles using Chenopodium album leaf extract, colloids surfaces a Physicochem. Eng. Asp. 2010;**369**:27-33 https://doi.org/10.1016/j.colsurfa.2010.07.020

[134] Singh P, Kim YJ, Zhang D, Yang DC. Biological synthesis of nanoparticles from plants and microorganisms. Trends in Biotechnology. 2016;**34**:588-599 https://doi.org/10.1016/j.tibtech.2016.02.006

[135] Dahoumane SA, Yéprémian C, Djédiat C, Couté A, Fiévet F, Coradin T, et al. A global approach of the mechanism involved in the biosynthesis of gold colloids using micro-algae. Journal of Nanoparticle Research. 2014;**16**:1-12 https://doi.org/10.1007/s11051-014-2607-8

[136] Greene B, Hosea M, McPherson R, Henzl M, Alexander MD, Darnall DW. Interaction of gold(I) and gold(III) complexes with algal biomass. Environmental Science & Technology. 1986;**20**:627-632 https://doi.org/10.1021/es00148a014

[137] Barwal I, Ranjan P, Kateriya S, Yadav SC. Cellular oxido-reductive proteins of Chlamydomonas reinhardtii control the biosynthesis of silver nanoparticles. J. Nanobiotechnology. 2011;**9**:1-12 https://doi.org/10.1186/1477-3155-9-56

[138] Jeffryes C, Agathos SN, Rorrer G. Biogenic nanomaterials from photosynthetic microorganisms. Current Opinion in Biotechnology. 2015;**33**:23-31 https://doi.org/10.1016/j.copbio.2014.10.005

[139] Costa Silva LP, Pinto Oliveira J, Keijok WJ, da Silva AR, Aguiar AR, Guimarães MCC, et al. Extracellular biosynthesis of silver nanoparticles using the cell-free filtrate of nematophagous fungus Duddingtonia flagrans. International Journal of Nanomedicine. 2017;**12**:6373-6381 https://doi.org/10.2147/IJN.S137703

[140] Ottoni CA, Simões MF, Fernandes S. J.G. dos Santos, E.S. da Silva, R.F.B. de Souza, a.E. Maiorano, screening of filamentous fungi for antimicrobial silver nanoparticles synthesis. AMB Express. 2017;**7** https://doi.org/10.1186/s13568-017-0332-2

[141] Armendariz V, Herrera I, Peralta-Videa JR, Jose-Yacaman M, Troiani H, Santiago P, et al. Size controlled gold nanoparticle formation by Avena sativa biomass: Use of plants in nanobiotechnology. Journal of Nanoparticle Research. 2004;**6**:377-382 https://doi.org/10.1007/s11051-004-0741-4

[142] Gardea-Torresdey JL, Tiemann KJ, Gamez G, Dokken K, Cano-Aguilera I, Furenlid LR, et al. Reduction and accumulation of gold(III) by Medicago sativa alfalfa biomass: X-ray absorption spectroscopy, pH. and temperature

dependence, Environ. Sci. Technol. 2000;**34**:4392-4396 https://doi.org/10.1021/es991325m

[143] Singh AK, Srivastava ON. One-step Green synthesis of gold nanoparticles using black cardamom and effect of pH on its synthesis. Nanoscale Research Letters. 2015;**10** https://doi.org/10.1186/s11671-015-1055-4

[144] Nair B, Pradeep T. Coalescence of nanoclusters and formation of submicron crystallites assisted by lactobacillus strains. Crystal Growth & Design. 2002;**2**:293-298 https://doi.org/10.1021/cg0255164

[145] Gericke M, Pinches A. Microbial production of gold nanoparticles. Gold Bulletin. 2006;**39**:22-28 https://doi.org/10.1007/BF03215529

[146] Abeywardena MR, Elkaduwe RKWHMK, Karunarathne DGGP, Pitawala HMTGA. Surfactant assisted synthesis of precipitated calcium carbonate nanoparticles using dolomite : Effect of pH on morphology and particle size. Advanced Powder Technology. 2019 https://doi.org/10.1016/j.apt.2019.10.018

[147] Aguilar NM, Arteaga-Cardona F, Estévez JO, Silva-González NR, Benítez-Serrano JC, Salazar-Kuri U. Controlled biosynthesis of silver nanoparticles using sugar industry waste, and its antimicrobial activity, J. Environ. Chem. Eng. 2018;**6**:6275-6281 https://doi.org/10.1016/j.jece.2018.09.056

[148] Pereira L, Mehboob F, Stams AJM, Mota MM, Rijnaarts HHM, Alves MM. Metallic nanoparticles: Microbial synthesis and unique properties for biotechnological applications. bioavailability and biotransformation, Crit. Rev. Biotechnol. 2013;**35**:114-128 https://doi.org/10.3109/07388551.2013.819484

[149] Lade BD, Shanware AS. Phytonanofabrication. IntechOpen: Methodology and Factors Affecting Biosynthesis of Nanoparticles; 2020. p. 38 https://doi.org/10.1016/j.colsurfa.2011.12.014

[150] Lade BD. Biochemical and Molecular Approaches for Characterization of Wound Stress Induced Antimicrobial Secondary Metabolites in *Passiflora foetida* Linn (PhD Thesis), Sant Gadge Baba Amravati University. Amravati, MS: India; 2017

[151] Rai A, Singh A, Ahmad A, Sastry M. Role of halide ions and temperature on the morphology of biologically synthesized gold nanotriangles. Langmuir. 2006;**22**:736-741 https://doi.org/10.1021/la052055q

[152] Gurunathan S, Kalishwaralal K, Vaidyanathan R, Venkataraman D, Pandian SRK, Muniyandi J, et al. Biosynthesis, purification and characterization of silver nanoparticles using Escherichia coli. Colloids Surfaces B Biointerfaces. 2009;**74**:328-335 https://doi.org/10.1016/j.colsurfb.2009.07.048

[153] Jameel MS, Aziz AA, Dheyab MA. Green synthesis: Proposed mechanism and factors influencing the synthesis of platinum nanoparticles. Green Process. Synth. 2020;**9**:386-398 https://doi.org/10.1515/gps-2020-0041

[154] Harshiny M, Iswarya CN, Matheswaran M. Biogenic synthesis of iron nanoparticles using Amaranthus dubius leaf extract as a reducing agent. Powder Technology. 2015;**286**:744-749 https://doi.org/10.1016/j.powtec.2015.09.021

[155] Darroudi M, Bin Ahmad M, Zamiri R, Zak AK, Abdullah AH, Ibrahim NA. Time-dependent effect in green synthesis of silver nanoparticles. International Journal of Nanomedicine.

2011;**6**:677-681 https://doi.org/10.2147/IJN.S17669

[156] Baer DR. Surface characterization of nanoparticles, J. Surf. Anal. 2011;**17**:163-169 https://doi.org/10.1384/jsa.17.163

[157] Sangaonkar GM, Pawar KD. Garcinia indica mediated biogenic synthesis of silver nanoparticles with antibacterial and antioxidant activities. Colloids Surfaces B Biointerfaces. 2018;**164**:210-217 https://doi.org/10.1016/j.colsurfb.2018.01.044

[158] Krishnaprabha M, Pattabi M. Synthesis of gold nanoparticles using garcinia indica fruit rind extract. International Journal of Nanoscience. 2016;**15**:1-6 https://doi.org/10.1142/S0219581X16600152

[159] Manzoor U, Tuz Zahra F, Rafique S, Moin MT, Mujahid M. Effect of synthesis temperature, nucleation time, and postsynthesis heat treatment of ZnO nanoparticles and its sensing properties. Journal of Nanomaterials. 2015;**2015** https://doi.org/10.1155/2015/189058

[160] Dada AO, Inyinbor AA, Idu EI, BelloOM,OluyoriAP,Adelani-AkandeTA, et al. Effect of operational parameters, characterization and antibacterial studies of green synthesis of silver nanoparticles using Tithonia diversifolia. PeerJ. 2018;**6**:1-17 https://doi.org/10.7717/peerj.5865

[161] Sharifi Dehsari H, Halda Ribeiro A, Ersöz B, Tremel W, Jakob G, Asadi K. Effect of precursor concentration on size evolution of iron oxide nanoparticles. CrystEngComm. 2017;**19**:6694-6702 https://doi.org/10.1039/c7ce01406f

[162] Cabrera LI, Somoza Á, Marco JF, Serna CJ, Puerto Morales M. Synthesis and surface modification of uniform MFe 2O 4 (M = Fe, Mn, and Co) nanoparticles with tunable sizes and functionalities. Journal of Nanoparticle Research. 2012;**14** https://doi.org/10.1007/s11051-012-0873-x

[163] Miguel-Sancho N, Bomati-Miguel O, Roca AG, Martinez G, Arruebo M, Santamaria J. Synthesis of magnetic nanocrystals by thermal decomposition in glycol media: Effect of process variables and mechanistic study. Industrial and Engineering Chemistry Research. 2012;**51**:8348-8357 https://doi.org/10.1021/ie3002974

[164] Huang JH, Parab HJ, Liu RS, Lai TC, Hsiao M, Chen CH, et al. Investigation of the growth mechanism of iron oxide nanoparticles via a seed-mediated method and its cytotoxicity studies. Journal of Physical Chemistry C. 2008;**112**:15684-15690 https://doi.org/10.1021/jp803452j

[165] Zheng H, Rice P, Wang S, Sun S. Superparamagnetic properties of MnFe 2 O 4 nanoparticles. J. Am. Chem. Soc. Commun. 2004;**126**:11458-11459 https://doi.org/10.1021/ja045911d

[166] Hufschmid R, Arami H, Ferguson RM, Gonzales M, Teeman E, Brush LN, et al. Synthesis of phase-pure and monodisperse iron oxide nanoparticles by thermal decomposition. Nanoscale. 2015;**7**:11142-11154 https://doi.org/10.1039/c5nr01651g

[167] Li Y, Liu J, Wang Y. Zhong Lin Wang, preparation of monodispersed Fe-Mo nanoparticles as the catalyst for CVD synthesis of carbon nanotubes. Chemistry of Materials. 2001;**13**:1008-1014 https://doi.org/10.1021/cm000787s

[168] Anigol LB, Charantimath JS, Gurubasavaraj PM. Effect of concentration and pH on the size of silver nanoparticles synthesized by Green chemistry. Org. Med. Chem. Int. J. Biosynth. 2017;**3**:1-5 https://doi.org/10.19080/OMCIJ.2017.03.555622

[169] Ahmad N, Ang BC, Amalina MA, Bong CW. Influence of precursor concentration and temperature on the formation of nanosilver in chemical reduction method. Sains Malaysiana. 2018;**47**:157-168 https://doi.org/10.17576/jsm-2018-4701-19

[170] Sibiya PN, Moloto MJ. Effect of precursor concentration and pH on the shape and size of starch capped silver selenide (Ag2Se) nanoparticles. Chalcogenide Letters. 2014;**11**:577-588

[171] Mittal J, Batra A, Singh A, Sharma MM. Phytofabrication of nanoparticles through plant as nanofactories, Adv. Nat. Sci. Nanosci. Nanotechnol. 2014;**5** https://doi.org/10.1088/2043-6262/5/4/043002

[172] P. Mukherjee, A. Ahmad, D. Mandal, S. Senapati, S.R. Sainkar, M.I. Khan, R. Ramani, R. Parischa, P. V. Ajayakumar, M. Alam, M. Sastry, R. Kumar, Bioreduction of AuCl4– Ions by the Fungus, Verticillium sp. and Surface Trapping of the Gold Nanoparticles Formed D.M. and S.S. thank the Council of Scientific and Industrial Research (CSIR), Government of India, for financial assistance., Angew. Chemie Int. Ed. 40 (2001) 3585. https://doi.org/10.1002/1521-3773(20011001)40:19<3585::aid-anie3585>3.0.co;2-k.

[173] Shankar SS, Rai A, Ahmad A, Sastry M. Rapid synthesis of Au, Ag, and bimetallic Au core-Ag shell nanoparticles using neem (Azadirachta indica) leaf broth. Journal of Colloid and Interface Science. 2004;**275**:496-502 https://doi.org/10.1016/j.jcis.2004.03.003

[174] Pandey DBA. Synthesis of zinc-based nanomaterials: A biological perspective. IET Nanobiotechnology. 2012;**6**:144-148 https://doi.org/10.1049/iet-nbt.2011.0051

[175] Tran QH, Nguyen VQ, Le AT. Silver nanoparticles: Synthesis, properties, toxicology, applications and perspectives, Adv. Nat. Sci. Nanoscience and Nanotechnology. 2013;**4** https://doi.org/10.1088/2043-6262/4/3/033001

[176] Park Y, Hong YN, Weyers A, Kim YS, Linhardt RJ. Polysaccharides and phytochemicals: A natural reservoir for the green synthesis of gold and silver nanoparticles. IET Nanobiotechnology. 2011;**5**:69-78 https://doi.org/10.1049/iet-nbt.2010.0033

Chapter 2

Recent Progress in the Electrochemical Exfoliation of Colloidal Graphene: A Review

Randhir Singh

Abstract

Graphene is a wonder nanomaterial which is used in a wide variety of electronics applications because of its excellent electrical, optical, chemical and mechanical properties. For the efficient use of graphene in the preparation of modern electronics devices it is imperative to first prepare a colloidal solution of graphene. Although various techniques are being used for the synthesis of colloidal form of graphene, the synthesis of colloidal graphene via electrochemical exfoliation is time saving and easy, facile method which can be easily performed in the laboratory without any expensive and sophisticated equipment as required in other techniques. Through electrochemical exfoliation of colloidal graphene, high quality graphene can be obtained within short time. Further, after the electrochemical exfoliation of colloidal graphene, the colloidal solution is stable in the organic solvent for few weeks. The conducting electrodes prepared by this colloidal solution of graphene have wide application in the areas of flexible energy storage devices and sensors fabrication.

Keywords: colloids: graphene synthesis, electrochemical exfoliation, energy storage device

1. Introduction

Graphene is the wonder nanomaterial discovered in 2004, most widely investigated because of its excellent electrical, mechanical, optical, chemical properties [1, 2]. The main properties are it is 97.7% transparent and is used for making transparent conducting electrodes. Its high carrier mobility (200,000 $cm^2 v^{-1} s^{-1}$), Young's modulus of 1.0 TPa is another important properties of graphene. It is considered 200 times more conductive than copper and 100 times stronger than steel [3–7]. In addition to this it is very flexible in nature as it can be stretched to 20% of its original length. These exceptional properties of graphene are highly suitable for the fabrication of various modern electronics device applications such as energy storage devices and sensors etc. However, for the efficient use of graphene in these modern devices, the colloidal dispersion of graphene has to be prepared for using in solution phase. The colloidal solution of graphene has some advantages in comparison to the other forms of graphene to be utilized for the formation of various devices.

Nowadays the modern electronics devices are being fabricated using printable electronics process. For the fabrication of the electronics devices such as sensors, energy storage devices using printable electronics the graphene has to be available in the colloidal solution form so that it will be easy to fabricate these graphene based flexible electronics devices using spray coating, brush coating, screen printing techniques [8]. Therefore, it is absolutely necessary to produce the colloidal solution of graphene.

2. Limitations of graphene synthesis methods

Various methods have been used for the preparation of graphene each having its own limitations as compared to others. Some of the prominent methods are mechanical exfoliation, Hummer's method, liquid phase exfoliation, epitaxial growth, chemical vapor deposition etc. First, the mechanical exfoliation method by which the graphene was first exfoliated from solid graphite source is a very time consuming technique of producing graphene from graphite [1]. It is a hit and trial method in which the researcher is not sure whether the graphene exfoliated on the scotch tape is single layered or multi layered graphene sheet. So, it is not a controllable and leads to a lot of wastage of time. Similarly the hummer's method uses very harmful acids and is also very time consuming method. The quality of graphene obtained from this method is not of high quality as indicated from the TEM results of graphene sheets.

Chemical vapor deposition requires very expensive and sophisticated equipment, hence the synthesis cannot be performed in all the laboratories [9, 10]. Liquid phase exfoliation is a technique where long hours of sonication process is required which is very time consuming process [11]. Further the long hours of sonication deteriorates the quality and size of graphene nanosheets in the dispersion. Therefore, among other methods of graphene synthesis it is found that the colloidal dispersion of graphene prepared by electrochemical exfoliation is time saving method, economical and easy to use in any laboratory without any sophisticated and expensive equipment. The relative advantages of electrochemical exfoliation method in comparison to others methods is shown in **Table 1**.

S.No.	Graphene synthesis method	Relative advantage/disadvantage
1	Liquid phase exfoliation	Requires Long hours of sonication, Time consuming
2	Epitaxial growth	Low yield, Difficult in graphene layer transfer
3	Chemical vapor deposition	Requires expensive and sophisticated equipment
4	Mechanical exfoliation	Not suitable for large scale production of graphene
5	Hummer's method	Harmful chemicals, acids used, graphene not pure
6	Electrochemical exfoliation	FASTER, HIGH YIELD, ENVIRONMENT FRIENDLY

Table 1.
Relative advantages of electrochemical exfoliation method in comparison to others.

3. Colloidal graphene

A colloidal solution of graphene (**Figure 1**) has graphene nanoparticles evenly distributed throughout the solution. In the colloidal solution of graphene the graphene nanoparticles remain dispersed in the solution without settling to the bottom, for quite a long time. Further, The colloidal solution of graphene has very large surface area and exhibits high electrochemical behavior. Further, the advantage of using solution phase to form various graphene devices such as sensors, electrodes, energy storage devices makes it more prominent method as compared

Figure 1.
Colloidal solution of graphene in aqueous electrolyte.

to other synthesis methods of graphene such as chemical vapor deposition, liquid phase exfoliation, mechanical exfoliation method, epitaxy and hummer's method. Due to large surface area of colloidal dispersion of graphene, it has been suitably utilized for the preparation of flexible energy storage devices [12–15].

4. Electrochemical exfoliation

The colloidal graphene has high specific surface area and does not exhibit aggregation. The colloidal graphene was produced by the intercalation of the sulfate ions in between the individual graphene nanosheets present in the graphite rod. This intercalation process separates the individual graphene nanosheets which accumulate in the electrolyte solution at the end of electrochemical exfoliation process to form the colloidal solution of graphene [14]. Various steps involved in the formation of colloidal solution of graphene through electrochemical exfoliation process are shown in **Figure 2**.

The quality of colloidal graphene produced by the electrochemical exfoliation depends upon the type of the aqueous electrolyte used. Therefore, to improve the quality of colloidal graphene various electrolytes have been studied by researchers [8]. Some of the electrolytes used to prepare colloidal solution of graphene are ammonium sulfate, phosphoric acid, potassium sulfate, sodium sulfate, sulfuric acid electrolytes. In addition of these aqueous electrolytes lithium sulfate has also been observed to produce colloidal solution of graphene via electrochemical route [16].

The electrochemical exfoliation of graphene is performed by using one graphite rod and one platinum wire immersed in any aqueous solution containing sulfate ions. A DC voltage source is used for the exfoliation process [8]. Usually, a DC voltage of 10 V is applied for 1 to 1.5 hour for the exfoliation of graphene to complete. After 1.5 hour, the colloidal solution of graphene is obtained in the aqueous electrolyte solution (**Figure 3**). When the exfoliation process is completed, the graphite rod has been completely converted into graphene colloids in the solution. Later, the colloidal solution of graphene is used for the preparation of the graphene based electrodes.

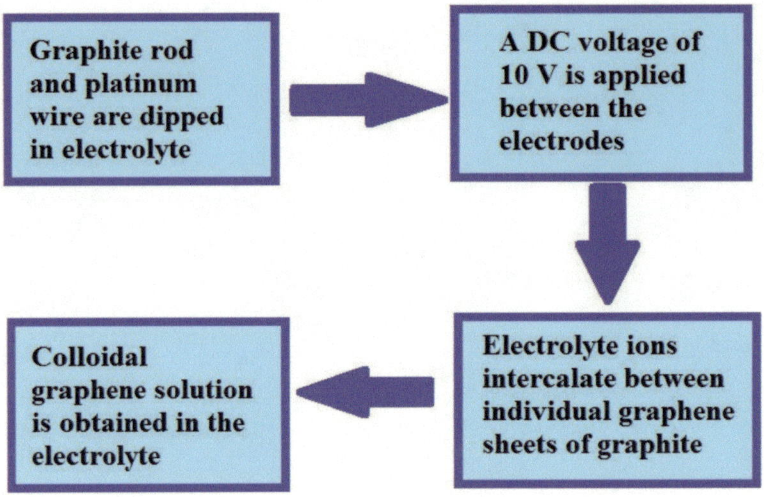

Figure 2.
Various steps involved in the formation of colloidal solution of graphene through electrochemical exfoliation process.

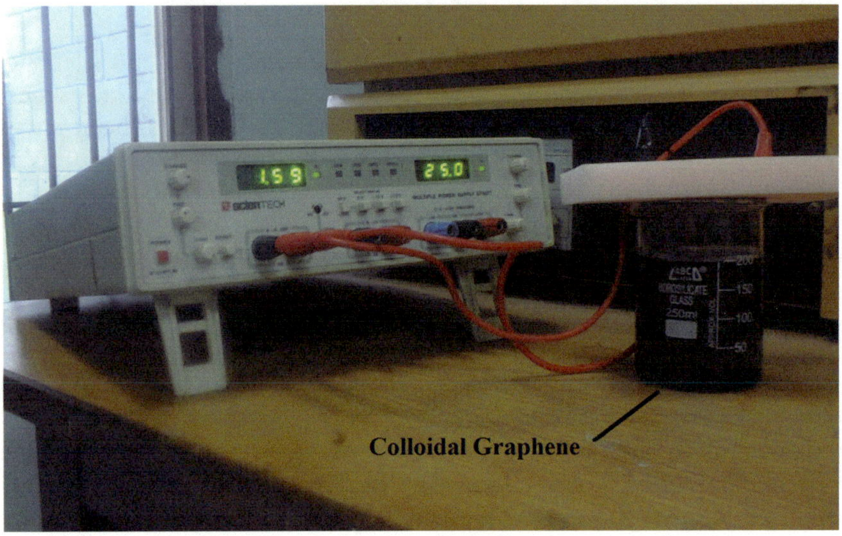

Figure 3.
Electrochemical exfoliation of graphite for the formation of colloidal graphene.

5. Recent works on electrochemical exfoliation of graphene

Due to the various advantages of the electrochemical exfoliation technique in comparison to others, it has been extensively utilized recently for the preparation of graphene nanosheets for various applications. Recently industrial scale synthesis of few-layer graphene nanosheets have been synthesized where the electrochemical experiment shows that rate of graphene exfoliation increases with the higher concentration of intercalates and accordingly the colloidal conductivity changes [17]. In comparison to graphene, graphene oxide is more environment friendly because of the presence of oxygen-containing functional groups. In Comparison to standard Hummers' method for synthesis of GO, electrochemical exfoliation of graphite is considered facile and green with better crystallinity and higher oxidation degree [18].

The role of electrolytes is very important in the quality of the graphene produced. The graphene synthesis mechanism in acidic (0.5 M H_2SO_4), neutral (0.5 M Li_2SO_4) is compared to non-destructive intercalation in organic electrolyte (1 M $NaClO_4$ in acetonitrile) [19]. Another important advantage of electrochemical exfoliation is that it can be utilized for the synthesis of graphene from electronic waste because it is a great threat to the environment due to difficulty in recycling, difficulty in conversion of waste to useful materials. So, researchers have reported a facile and fast production method of electrochemical exfoliation of graphene from graphite of used Zn–C batteries [20]. The quality of graphene produced by the electrochemical exfoliation can be enhanced by taking mixtures of various electrolytes with pencil rods as electrodes. For the synthesis of graphene sheets, different mixtures of H_2SO_4 and HNO_3 solution were investigated with different volume ratios of H_2SO_4: HNO_3, with maximum oxidation was achieved at 1:1 ratio of both mixtures [21]. In another study, electrochemical exfoliation method was used for the preparation of high-quality water-dispersible graphene using molybdate aqueous solutions as the electrolyte [22]. Further, graphene is very effective against the cancer cells for treatment of cancer patients. Some researchers have demonstrated the cytotoxic effect of graphene with high content of nitrogen on colon cancer cells and antioxidant and protective properties on human endothelial cells [23]. Further, the researchers have investigated the role of Transition Metal Salts During Electrochemical Exfoliation of Graphite for Energy Storage Applications [24]. Recently, colloidal graphene produced by the electrochemical exfoliation of graphite in potassium sulphate and sulfuric acid electrolytes has been successfully utilized for the supercapacitor applications [25, 26]. **Table 2** shows various electrolytes used recently for the electrochemical exfoliation of graphene.

S.No.	Aqueous electrolyte	DC Voltage	Concentration	Graphene /GO Exfoliation	Reference
1	Sulfuric Acid	10 V, 3 V	0.5 M to 2 M	Yes	[17]
2	Sodium sulfate, oxalate acid	10 V, 15 V	0.05 M, 0.1 M	Yes	[18]
3	Sulfuric acid, Lithium sulphate	3 V, 5V	0.5 M, 0.5 M	Yes	[19]
4	poly(sodium 4-styrenesulfonate) (PSS)	5 V	0.5 M	Yes	[20]
5	Mixture of Nitric acid and Sulfuric Acid	10 V	0.6 M	Yes	[21]
6	molybdate aqueous solution	9 V, 10 V	0.1 M	Yes	[22]
7	Ammonium sulphate	9 V	0.1 M	Yes	[23]
8	Cobalt sulphate and Sodium sulphate	2 V	0.5 M, 0.05 M	Yes	[24]
9	Potassium Sulphate	10 V	0.1 M	Yes	[25]
10	Sulfuric acid	10 V	0.1 M	Yes	[26]
11	Ammonium sulphate	10 V	0.1 M	Yes	[8]
12	Sodium sulphate	10 V	0.1 M	Yes	[8]
13	Potassium sulphate	10 V	0.1 M	Yes	[8]
14	Ammonium chloride	10 V	0.1 M	NO	[8]

S.No.	Aqueous electrolyte	DC Voltage	Concentration	Graphene /GO Exfoliation	Reference
15	Sodium nitrate	10 V	0.1 M	POOR	[8]
16	Sodium perchlorate	10 V	0.1 M	NO	[8]
17	sodium dodecyl sulphate (SDS)	5 V	0.001 M to 0.1 M	Yes	[27]
18	Sodium Dodecyl Sulphate (SDS)	5 V to 9V	0.1 M to 0.01 M	Yes	[28]
19	TBA·H_2SO_4, NaOH	10 V	0.1 M	Yes	[29]
20	$(NH_4)_2SO_4$,CH_4N_2S	10 V	0.1 M	Yes	[30]
21	sodium saccharin	2 V to 10 V	0.1 M	Yes	[31]
22	NaCl (NaBr,NaI)	10 V	0.05 M	Yes	[32]
23	ionic liquids in acetonitrile	10 V-20 V	0.1 M	Yes	[33]

Table 2.
Recent electrolytes used for the electrochemical exfoliation of graphene.

6. Application of colloidal graphene

The colloidal dispersion of graphene produced by the electrochemical exfoliation method is ideal for the preparation of the graphene based flexible paper electrodes. For this, colloidal graphene is coated using brush coating method on A4 paper and dried. These graphene based flexible paper electrodes are used for the preparation of the flexible energy storage devices [8]. First a gel electrolyte is prepared by using PVA (polyvinyl alcohol) and water [34], then two graphene based paper electrodes are joined using this gel electrolyte in between electrodes. Later, two aluminum foils are joined to form the contacts with these electrodes for taking various measurements of these flexible energy storage devices (**Figure 4**). The actual fabricated flexible energy storage device using colloidal graphene is shown in **Figure 5**.

The performance of the flexible energy storage devices is measured by the CV Curves [8]. Further, the type of the gel electrolyte changes the shape of the CV curve measured and hence the performance of the energy storage device.

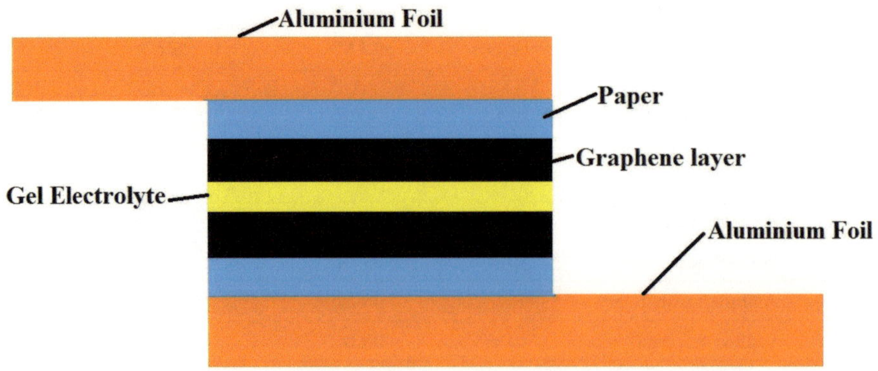

Figure 4.
Schematic diagram of the graphene based flexible energy storage device structure.

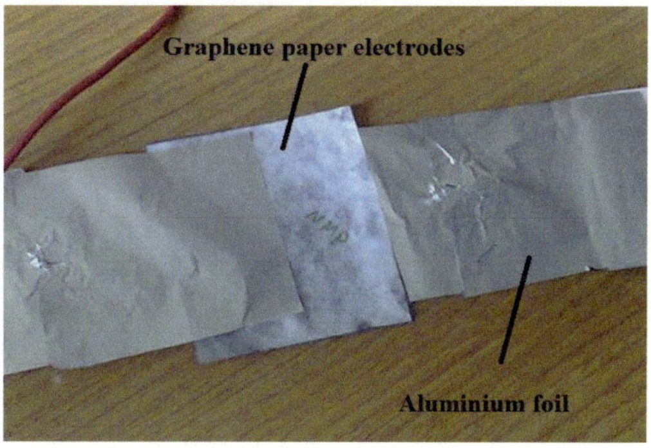

Figure 5.
The flexible energy storage device fabricated using colloidal graphene.

Different flexible energy storage devices can be prepared by using various gel electrolytes and their performance can be compared to obtain the most suitable gel electrolyte for the preparation of the flexible energy storage devices.

7. Conclusion

This review article discusses electrochemical synthesis of colloidal graphene. Colloidal form of graphene is better than other forms of graphene available because of its solution route is easy to be used for the formation of graphene based electrodes. It is observed that electrochemical exfoliation of graphite is the simple, time saving and economical method for the production of colloidal form of graphene. Various aqueous electrolytes can be used for the electrochemical exfoliation of graphene from graphite. After the colloidal solution is obtained it is used for the preparation of the flexible paper electrodes which are suitable for the formation of flexible energy storage devices.

Conflict of interests

The authors declare no conflict of interests.

Author details

Randhir Singh
Department of Electronics and Communication Engineering, University Institute of Engineering and Technology, Kurukshetra University, Haryana, India

Address all correspondence to: mr_randhir_singh@yahoo.co.in

References

[1] Novoselov K S, Geim A K, Morozov S V, Jiang D, Zhang Y, Dubonos S V, Grigorieva I V & Firosov A A. Electric field effect in atomically thin carbon films. Science. 2004; 306: 666-669.

[2] Novoselov K S, Jiang D, Schedin F, Booth T J, Khotkevich V V, Morozov S V & Geim A K. Two-dimensional atomic crystals. Proc Natl Acad Sci USA. 2005; 30:10451-10453.

[3] Bolotin K I, Sikes K J, Jiang Z, Klima M, Fudenberg G, Hone J, Kim P & Stormer H L. Ultrahigh electron mobility in suspended graphene. Solid State Commun. 2008;146:351-355.

[4] Lee C, Wei X, Kysar J W & Hone J. Measurement of the elastic properties and intrinsic strength of monolayer graphene. Science. 2008; 321: 385-388.

[5] Zhu Y, Murali S, Cai W, Li X, Suk J W, Potts J R & Ruoff R S. Graphene and graphene oxide: synthesis, properties, and applications. Adv Mater. 2010; 35: 3906-3924.

[6] Moser J, Barreiro A & Bachtold A, Current-induced cleaning of graphene. Appl Phys Lett. 2007; 91: 163513.

[7] Geim, A K & Novoselov, K S, The rise of graphene. Nature Mater. 2007; 6: 183-191.

[8] Khaled Parvez, Zhong-Shuai Wu, Rongjin Li, Xianjie Liu, Robert Graf, Xinliang Feng,Klaus Müllen. Exfoliation of Graphite into Graphene in Aqueous Solutions of Inorganic Salts. Journal of the American Chemical Society; 2014;136:6083-6091

[9] Zhang Y,Zhang L, Zhou C, Review of Chemical Vapor Deposition of Graphene and Related Applications. Acc. Chem. Res.; 2013; 46: 2329-2339.

[10] Mishra N, Boeckl J, Motta N, Iacopi F. Graphene growth on silicon carbide: A review. Phys. Status Solidi ;2016; 2013: 2277-2289.

[11] Hernandez Y, Nicolosi V, Lotya M, Blighe F.M, Sun Z, De S, McGovern I.T, Holland B, Byrne M, Gun'Ko Y.K, High- yield production of graphene by liquid-phase exfoliation of graphite. Nat. Nanotechnol; 2008; 3: 563-568.

[12] K. Chen, D. Xue, Ionic Supercapacitor Electrode Materials: A System-Level Design of Electrode and Electrolyte for Transforming Ions into Colloids. Colloids Interface Sci. Commun. 2014: 1: 39-42.

[13] K. Chen, D. Xue, $YbCl_3$ electrode in alkaline aqueous electrolyte with high pseudocapacitance. J. Colloid Interface Sci. 2014; 424: 84-89.

[14] K. Chen, D. Xue, Preparation of colloidal graphene in quantity by electrochemical exfoliation. J. Colloid Interface Sci. 2014; 436: 41-46.

[15] K. Chen, D. Xue. Formation of electroactive colloids via in situ coprecipitation under electric field: erbium chloride alkaline aqueous pseudocapacitor. J. Colloid Interface Sci. 2014;430:265-271.

[16] R. Singh. Synthesis of colloidal graphene by electrochemical exfoliation of graphite in Lithium Sulphate. Materials Today Proceedings. 2018 ;5: 973-979.

[17] S. K. Sahoo, A.K. Behera, R. Chandran, A. Mallik. Industrial scale synthesis of few-layer graphene nanosheets (FLGNSs): an exploration of electrochemical exfoliation approach. Journal of Applied Electrochemistry. 2020;6: 673-688.

[18] Duhong Chen, Zhen Lin, Matthew M Sartin, Teng-Xiang Huang, Jia Liu, Qiugen Zhang, Lianhuan Han, Jian-Feng

Li, Zhong-Qun Tian, Dongping Zhan. Photosynergetic Electrochemical Synthesis of Graphene Oxide. J Am Chem Soc. 2020 ;142:6516-6520.

[19] Zhenyuan Xia, Vittorio Bellani, Jinhua Sun, Vicenzo Palermo. Electrochemical exfoliation of graphite in H_2SO_4, Li_2SO_4 and $NaClO_4$ solutions monitored in-situ by Raman microscopy and spectroscopy. Faraday Discussions. 2020.

[20] Bagas Prakoso, Yuanyuan Ma, Ruth Stephanie, Naufal Hanif Hawari, Veinardi Suendo, Hermawan Judawisastra, Yun Zong, Zhaolin Liu, Afriyanti Sumboja. Facile synthesis of battery waste-derived graphene for transparent and conductive film application by an electrochemical exfoliation method. RSC Adv. 2020; 10: 10322-10328.

[21] Hamed Aghamohammadi, Reza Eslami-Farsani. An experimental investigation on the sulfur and nitrogen co-doping and oxidation of prepared graphene by electrochemical exfoliation of pencil graphite rods. Ceramics International. 2020;46: 28860-28869.

[22] Juan Wu; Hongfei Wang; Jun Qiu; Jingwen Shao; Kefu Zhang; Lifeng Yan. Electrochemical exfoliation for few-layer graphene in molybdate aqueous solution and its application for fast electrothermal film. Progress in Natural Science: Materials International. 2020; 30:312-320.

[23] Ioana Baldea, Diana Olteanu, Gabriela Adriana Filip, Florina Pogacean, Maria Coros, Maria Suciu, Septimiu Cassian Tripon, Mihai Cenariu, Lidia Magerusan, Raluca-Ioana Stefan-van Staden, Stela Pruneanu. Cytotoxicity mechanisms of nitrogen-doped graphene obtained by electrochemical exfoliation of graphite rods, on human endothelial and colon cancer cells. Carbon; 2019;158:267-281.

[24] Andinet Ejigu, Kazunori Fujisawa, Ben F. Spencer, Bin Wang, Mauricio Terrones, Ian A. Kinloch, Robert A. W. Dryfe. Electrochemical Exfoliation: On the Role of Transition Metal Salts During Electrochemical Exfoliation of Graphite: Antioxidants or Metal Oxide Decorators for Energy Storage Applications. Advanced Functional Materials: 2018; 28:1804357.

[25] Singh R. Electrochemical Exfoliation of Graphite into Graphene for Flexible Supercapacitor Application. Materials Today Proceedings. 2018;5:1125-1130

[26] Singh R. Study of Graphene based Flexible Supercapacitors with Different Gel Electrolytes Materials Today Proceedings. 2018;5:943-949

[27] Nurhafizah Md Disa, Suriani Abu Bakar, Suhufa Alfarisa, Azmi Mohamed, Illyas Md, AzlaN Kamari, Norhayati Hashim, Azira Abd. Aziz,Mohamad Rusop Mahmood. The Synthesis of Graphene Oxide via Electrochemical Exfoliation Method. Advanced materials research;2015;1109:55-59

[28] Wan Hazman Danial, Arunabhiram Chutia, Zaiton Abdul Majid, Riadh Sahnoun, Madzlan Aziz. Electrochemical Synthesis and Characterization of Stable colloidal Suspension of Graphene using Two-electrode Cell System. AIP Conference Proceedings;2015; 1669: 020020.

[29] Yang S, Ricciardulli AG, Liu S, et al. Ultrafast delamination of graphite into high-quality graphene using alternating currents. Angew Chem. 2017;56:6669-6675.

[30] Li L, Wang M, Guo J, et al. Regulation of radicals from electrochemical exfoliation of a double-graphite electrode to fabricate high-quality graphene. J Mater Chem C. 2018;6:6257-6263.

[31] Kumar MK P, Shanthini S, Srivastava C. Electrochemical exfoliation

of graphite for producing graphene using saccharin. RSC Adv. 2015;5:53865-53869.

[32] Munuera JM, Paredes JI, Enterría M. Electrochemical exfoliation of graphite in aqueous sodium halide electrolytes toward low oxygen content graphene for energy and environmental applications. ACS Appl Mater Interfaces; 2017; 9:24085-24099.

[33] Taheri Najafabadi A, Gyenge E. Synergistic production of graphene microsheets by simultaneous anodic and cathodic electro-exfoliation of graphitic electrodes in aprotic ionic liquids. Carbon; 2015;84:449-459.

[34] Farshad Barzegar, Julien K. Dangbegnon, Abdulhakeem Bello, Damilola Y. Momodu. Effect of conductive additives to gel electrolytes on activated carbon-based supercapacitors. AIP Advances.2015;5:97171-97179

Chapter 3

A Simple and "Green" Technique to Synthesize Metal Nanocolloids by Ultrashort Light Pulses

Jesica María José Santillán, David Muñetón Arboleda, Valeria Beatriz Arce, Lucía Beatriz Scaffardi and Daniel Carlos Schinca

Abstract

In this chapter Ag, Ni and Fe nanocolloids synthesized by "green" ultrashort pulse laser ablation of solid metal targets using different pulse energies and liquid media are characterized by different techniques. Optical extinction spectroscopy (OES), micro-Raman spectroscopy (MRS), transmission electron microscopy (TEM) and electron diffraction (ED) were independently used to analyze optical, morphological and compositional properties of the generated nanocolloids. In a deeper way, the stability characteristics of Ag nanocolloids in aqueous solutions with different stabilizers were studied owing to their potential use in biocompatible compounds. Besides, due to their interesting applications, few atoms Ag nanoclusters (NCs) were synthesized using the same ablation technique, analyzing their fluorescent and photocatalytic properties. On the other hand, to expand the characterization of the nanocolloids, their magnetic behavior was inspected for the Ni and Fe by vibrating sample magnetometry (VSM).

Keywords: nanocolloids, green synthesis, metal nanoparticles, nanoclusters, femtosecond laser ablation

1. Introduction

Interest in metal nanomaterials synthesis has grown rapidly in the last years due to their particular physical and chemical properties arising from atom interaction and quantum confinement at the nanoscale. Their applicability spans different fields of science and technology [1–3]. It is known that chemical synthesis methods tend to yield highly monodisperse colloidal suspensions, but mixed with unwanted chemical precursors, which often leads to purification steps to remove the chemical by-products and may derive in expensive and complicated procedures. For this reason, femtosecond laser ablation synthesis in solution (FLASiS) has emerged as a competitive and alternative method for synthesizing metallic nanomaterials without the intervention of unwanted chemical compounds. Besides, it has the ability of producing small spherical nanoparticles (NPs) [4, 5] as well as few atoms metal NCs [6].

In this chapter, different independent techniques are used to characterize the ablated nanomaterials. Altogether, they retrieve complementary and interrelated information about different NPs characteristics. TEM provides knowledge about morphology, internal structure, sphericity and size distribution in a small piece of sample. ED is an established technique that can identify different phase compositions and crystallinity type. Optical spectroscopy techniques (absorption, extinction, scattering and fluorescence) have the ability to interact with a very large number of NPs (on the order of $10^{12}\,cm^{-3}$), enhancing statistics. OES together with Mie theory yields information related to size distribution, sphericity, configuration and composition of the NPs in the colloidal sample. MRS retrieves information about possible interactions between stabilizer solution molecules adsorbed to the NPs walls and the NP itself. In particular, for Ag nanocolloids generated in aqueous solutions with small concentrations of stabilizers, long term stability characteristics were studied, aiming to possible applications in biocompatible antibacterial compounds. For the case of magnetic metals NPs, magnetic nanocolloid properties were studied using VSM. Finally, fluorescent and photocatalytic properties of few atoms Ag NCs were analyzed.

2. FLASiS as a green route for NPs and few atoms NCs synthesis

Traditional techniques for metal NPs synthesis have relied on chemical reaction associated to metallic salt dissociation [7], which suitably reduce to form metallic atoms.

FLASiS is based on the incidence of a focused laser pulse on a bulk target immersed in a liquid [8, 9] (**Figure 1**). The produced plasma plume that contains the ablated material expands into the surrounding liquid and generates a cavitation bubble, which acts as a reactor for NPs formation through condensation of atoms [8]. This process produces ions and atoms that reach different nucleation stages, and generates large NPs (radii >20 nm), medium NPs (2 nm < radii <20 nm), small NPs (1 nm < radii <2 nm) together with very small NCs (radii <1 nm) [6–10].

Since FLASiS is capable of synthesizing NPs directly in a selected liquid without producing unwanted compounds in the solution, it is considered a "green" technique. As the schematic in **Figure 1** depicts, the NPs generated during FLASiS

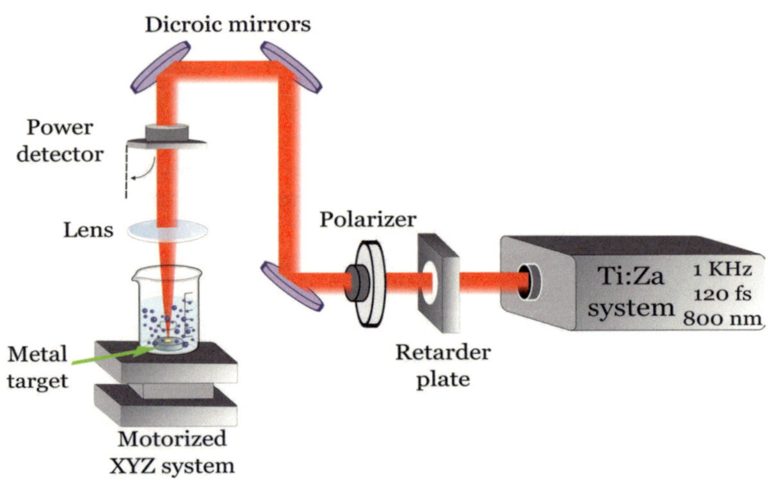

Figure 1.
General schematic of the experimental setup for FLASiS. The cell containing the target is placed over a XYZ micro translational platform.

remain in the liquid, forming a suspension with a NPs concentration dependent on laser pulse energy and ablation time.

2.1 Characterization of Ag nanocolloids synthesized by FLASiS and chemical route in aqueous solutions of trisodium citrate (TSC) and starch (st)

Ag NPs have attractive scientific interest due of their broad perspectives in biosensors [11], food production [12], water purification [13], antimicrobial and antiviral agents [14, 15], among others. In these areas of research and development, the ability to control size, shape, functionalization and stability of Ag NPs is essential for expanding their possible applicability. For this goal, typical methods are based on chemical reduction of salts in solution, commonly used for providing good size control and resulting in final spherical shape. However, this approach leaves chemical residuals in the final colloidal suspension, which may be toxic for certain applications, thus adding an extra difficulty in sample purification. In this sense, FLASiS has become an alternative method for overcoming the mentioned drawback.

Figure 2 shows experimental and theoretical extinction spectra of Ag nanocolloids prepared by salt reduction (short dashed line) as well as those generated by FLASiS with 100 μJ (dashed line) and 500 μJ (dashed dotted line) pulse energies, with two different stabilizers 1 mM TSC (a) and 1% soluble st (b) solutions. These spectra are normalized to plasmon maximum. The theoretical fits (dotted lines) are determined using OES, which is based in the calculation of the Mie theory for metal spherical NPs [17, 18] with log-normal size distributions (insets).

Nanocolloids stabilized with TSC show a small redshift in peak position respect to the typical plasmonic resonance band at 395 nm, indicating the existence of $Ag@Ag_2O$ NPs [19, 20] with a log-normal shell thickness distribution in the three samples obtained. Nanocolloids obtained by FLASiS have a larger contribution of $Ag@Ag_2O$ NPs than the sample generated by chemical synthesis (**Figure 2(a)**). This finding agrees with the known fact that, during laser ablation of a metal target in liquid media, an oxidation-reduction process occurs, producing an oxide coating growth around the NPs [20, 21].

On the other hand, the st stabilized nanocolloids obtained by salt reduction have a large redshift at the peak position compared to FLASiS (**Figure 2(b)**), due to a larger oxide shell thickness around the NPs.

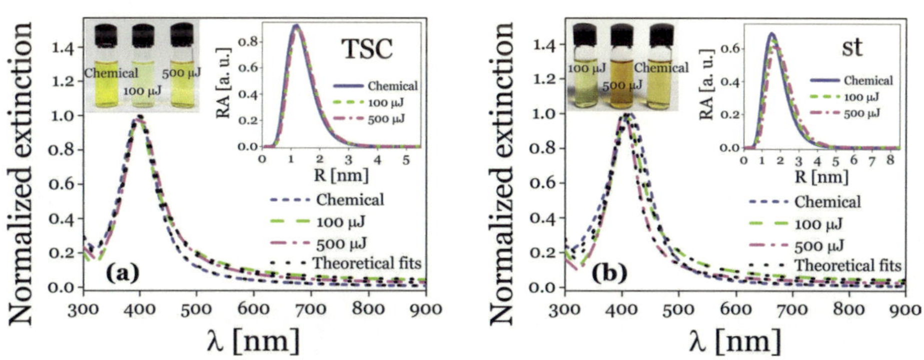

Figure 2.
Normalized experimental optical extinction spectra together with theoretical fits (dotted lines) of freshly prepared Ag nanocolloids obtained by silver salt reduction and FLASiS with 100 μJ and 500 μJ pulse energies, using 1 mM TSC (a) and 1% soluble st (b) solutions. Insets exhibit the relative abundances (RA) of the NPs present in the nanocolloids used to fitting the experimental spectra (reprinted with permission from [16] copyright 2018 Elsevier).

For the fitting of the spectra using Mie theory, three types of species were considered: Ag, Ag@Ag_2O, and in a smaller amount hollow Ag NPs. It is important to recall that these types of nanostructures are prone to occur for pulse laser ablation conditions [9, 22].

The morphological characterization and sizing of NPs synthesized by both methods was performed using TEM.

Figure 3 presents TEM images of Ag nanocolloids obtained by FLASiS with 500 μJ pulse energy in 1 mM TSC (a) and 1% soluble st (c) solutions, and prepared by salt reduction in 1 mM TSC (b) and 1% soluble st (d) solutions. All images are typical panoramic views with predominant spherical shape NPs. Right insets in panels (a) and (b) are lattice-resolved images of a single NP with Bragg planes identified as (200) of Ag FCC crystal, whereas the right inset in panel (c) exhibits the presence of hollow NPs. Left insets in the panels (a) - (d) are size histograms taken from several images. The results given by TEM are in good agreement with those obtained by OES from the fitting of the spectra.

Micro-Raman spectroscopy was conducted on the FLASiS samples to assess the existence of silver oxide species. **Figure 4** shows Raman spectra of a dried drop of Ag nanocolloids prepared by FLASiS with 500 μJ pulse energy in 1 mM TSC (a) and

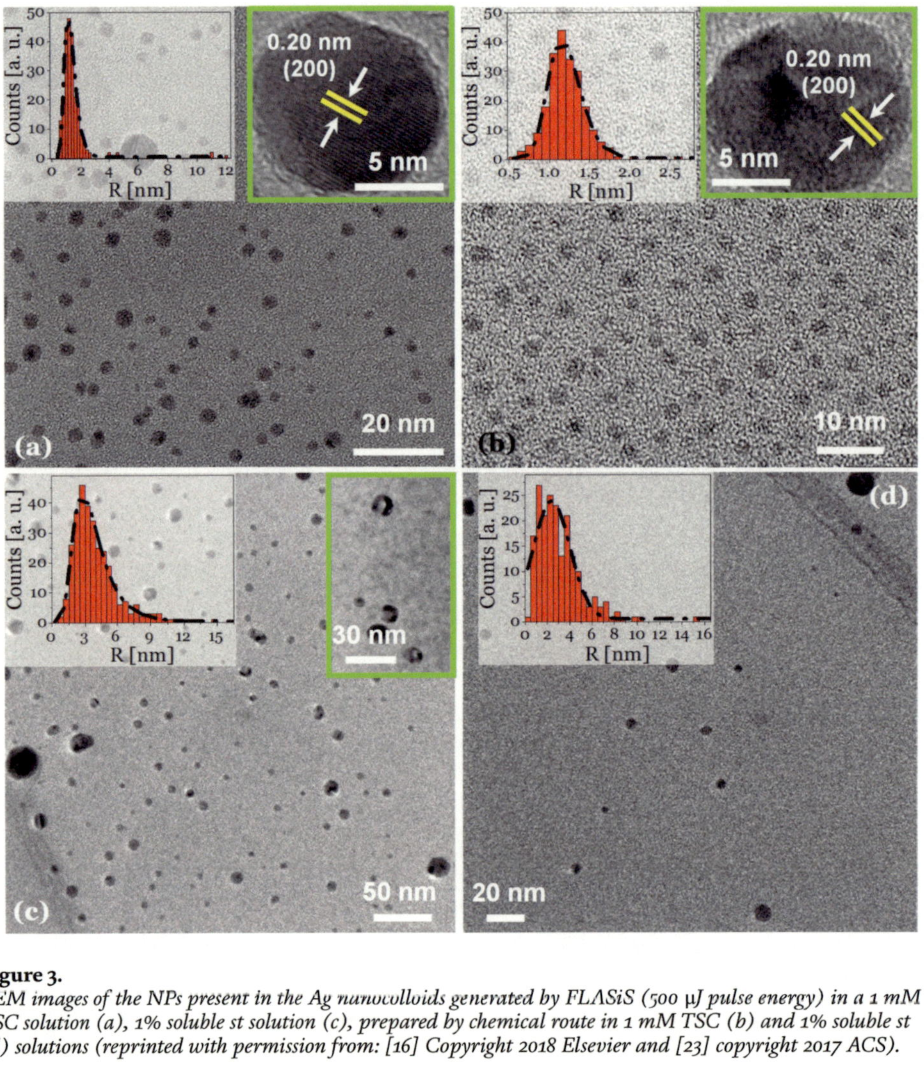

Figure 3.
TEM images of the NPs present in the Ag nanocolloids generated by FLASiS (500 μJ pulse energy) in a 1 mM TSC solution (a), 1% soluble st solution (c), prepared by chemical route in 1 mM TSC (b) and 1% soluble st (d) solutions (reprinted with permission from: [16] Copyright 2018 Elsevier and [23] copyright 2017 ACS).

1% soluble st (b) solutions, acquired in different regions of the sample. Several peaks corresponding to Ag_2O (asterisk) may be recognized in both samples. Raman signals at 240 cm^{-1} and 490 cm^{-1} are typical Ag-O stretching/bending modes in Ag_2O [24]. Besides, characteristic peaks of the metallic NP interaction with each stabilizer (TSC (diamond) and st (full circle)) are also observed [25–27].

Stability analysis was conducted on Ag nanocolloids produced by FLASiS (100 μJ and 500 μJ pulse energies) and chemical route with 1 mM TSC (**Figure 5(a)**) and 1% soluble st (**Figure 5(b)**) solutions. Plasmon resonance position and full width at half maximum (FWHM) for experimental spectra, were measured for freshly prepared samples and followed during several weeks (even up to one-year old samples with TSC stabilizer). The shift of plasmon peak (upper panel) and the behavior of FWHM (lower panel) of the experimental spectra in **Figure 5**, are represented by geometric symbols, while the lines are drawn to visually follow the evolution of both parameters.

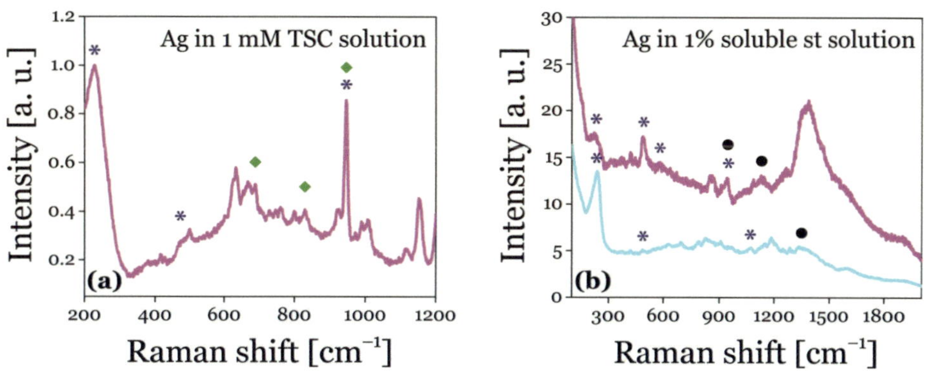

Figure 4.
Raman spectra of Ag nanocolloids obtained by FLASiS with 500 μJ pulse energy in 1 mM TSC (a) and 1% soluble st (b) solutions acquired at different sites in the samples (reprinted with permission from: [16] Copyright 2018 Elsevier and [23] copyright 2017 ACS).

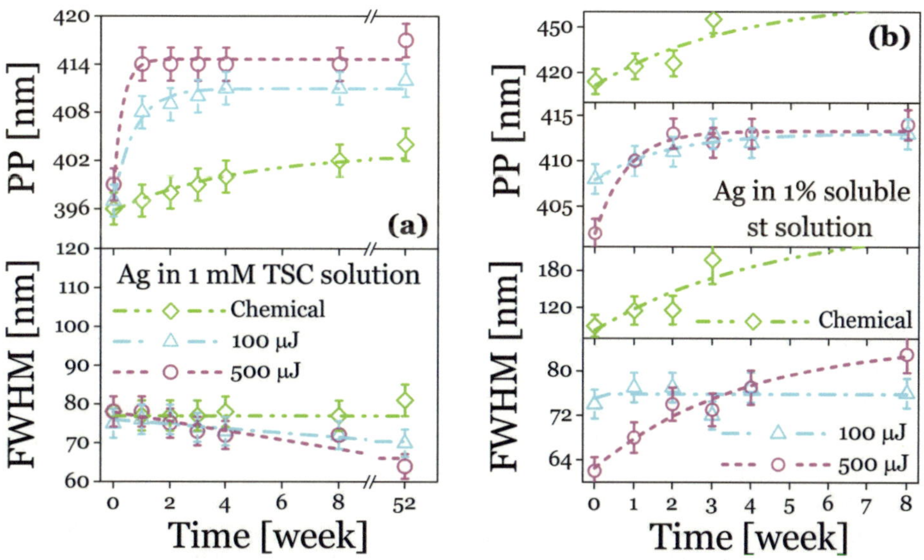

Figure 5.
Peak position (PP) and FWHM of the experimental extinction spectra of Ag nanocolloids synthesized by FLASiS (100 μJ and 500 μJ pulse energies) and salt reduction in 1 mM TSC (a) and 1% soluble st (b) solutions. These plasmonic characteristics were monitored during several weeks (reprinted with permission from: [16] Copyright 2018 Elsevier and [23] Copyright 2017 ACS).

From the plots in the upper panel (a), it can be seen that samples obtained by FLASiS in 1 mM TSC solution, reach their saturation regime at the second week (fast rate). However, nanocolloid produced by salt reduction method seems to reach saturation at times beyond one year (slow rate).

Nanocolloids synthesized by FLASiS still after one year show a clear plasmonic band without signs of agglomeration, indicating the excellent properties of TSC as stabilizer. Long-term stability results are similar to those for salt reduction chemical synthesis using TSC as stabilizer.

On the other hand, FWHM (lower panel (a)) reveals a very stable regime for the salt reduction synthesis during one year, whereas for FLASiS samples it presents a decrease, indicating a slight narrowing of the plasmon resonance.

For the case of st used as stabilizer (b), the monitoring of plasmon resonance peak position and FWHM show that FLASiS samples stabilize much faster than those produced by salt reduction. It is also observed that low energy pulses seem to produce more stable nanocolloids than higher energy pulses. This larger stability could be due to some kind of laser-induced NPs surface modification with amylose that avoids coalescence and sets a limit to their size. In contrast to all these cases, no stable suspensions are obtained for salt reduction, because plasmon peak is continuously red-shifted and FWHM increases without showing stationary behavior.

2.2 Synthesis and characterization of metal nanocolloids with magnetic properties (Ni and Fe)

In recent years, one of the most active topics in nanotechnology is the synthesis, characterization and functionalization of magnetic NPs. The interest in this type of NPs is due to their wide applications in areas of diagnosis and therapy in biomedicine [28, 29], as contrast agents in magnetic resonance imaging [30], for drug administration [31], as catalysts [32, 33], among others.

In this Subsection the characterization of Ni and Fe nanocolloids synthesized by FLASiS with different pulse energies and in different liquid media is addressed. Independent characterization techniques such as OES, TEM, ED, MRS, and VSM are used, which provide complementary and interrelated information.

2.2.1 Ni nanocolloids synthesized by FLASiS in n-heptane and water

In the synthesis of nanomaterials, Ni nanocolloids have attracted scientific interest because of their extensive prospects in catalysts [32, 33], information storage [32], magnetic behavior [34], biomedicine [35], among others.

Although there are different studies of the production of Ni colloidal suspensions by laser ablation, few of these have been in the femtosecond regime. Experimental and theoretical extinction spectra of Ni nanocolloids synthesized by FLASiS with 100 μJ pulse energy in *n*-heptane and water are observed in **Figure 6**. These spectra are recorded immediately after synthesis and normalized at $\lambda = 340\,\text{nm}$. For the case of nanocolloid in *n*-heptane, the log-normal distribution (inset in **Figure 6(a)**) is formed by Ni NPs with modal radius of 2.5 nm and 10 nm, together with hollow Ni NPs with external modal radius of 6.6 nm (10% shell), 12 nm (20% shell) and 15.6 nm (4% shell). However, for the case of the nanocolloid obtained in water, the multimodal size distribution (inset in **Figure 6(b)**) shows the presence of Ni@NiO NPs (short dashed line), NiO@Ni (dashed dotted line) and hollow Ni (short dashed dotted line), with sizes similar to that determined for *n*-heptane but shifted to higher values of external radii due to the presence of oxide shells around the NPs. Each one of the mentioned species influences the extinction spectrum in specific and distinct regions, in such a way that the combination of

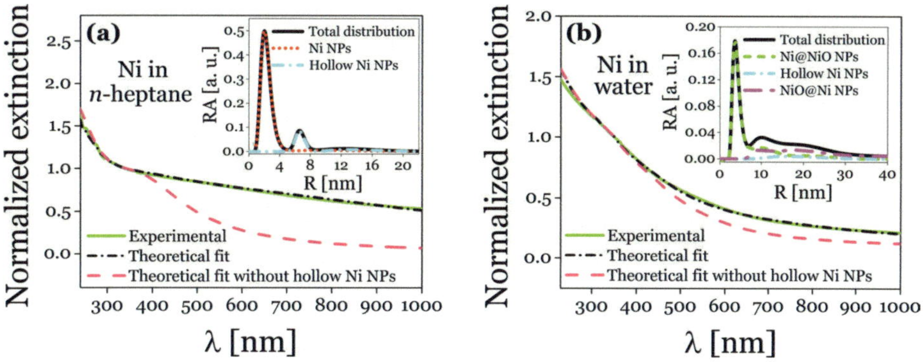

Figure 6.
Experimental (solid line) and theoretical (dashed dotted line) extinction spectra of the Ni nanocolloids in (a) n*-heptane and (b) water. The dashed line represents the theoretical fit without considering the presence of hollow Ni NPs. The insets show the RA of the different species of NPs present in the nanocolloids (reprinted with permission from [18] Copyright 2015 ASC).*

structures, sizes, and relative abundances derived from the optimum fit constitute a unique set of fitting parameters, showing the high sensitive of OES technique [18].

Morphology analysis of NPs in the Ni nanocolloids was performed using TEM. **Figure 7** presents TEM images of the nanocolloids synthesized in *n*-heptane ((a) and (b)) and water ((c) and (d)). Panel (a) is a panoramic view of Ni and hollow Ni NPs together with an enlargement of a NP with the latter structure indicated by the dotted line. Panel (b) images a Ni NP from a different region. Panel (c) is a panoramic view where NPs with different structures are observed, and panel (d) contains enlargements where the Bragg planes of NiO can be seen. ED is performed to phase identification of the NPs. Panel (e) shows a representative ED pattern indexed with the reflection lines of NiO (cubic, $Fm\,\bar{3}\,m$, JCPDS #75-0197) and Ni (cubic, $Fm\bar{3}m$, JCPDS # 04-0850) for the Ni nanocolloid in water. ED rings are marked according to the panel table (f), where the Miller indices (h, k, l) and interplane distances (d) are indicated.

Insets in panels (b) and (c) present histograms of radii corresponding to a statistic performed on several TEM images, where the results are fitted by two log-normal size distributions that describe the most prominent characteristics of the size distribution histogram. The results obtained through TEM analysis for both nanocolloids corroborate the morphological determinations achieved through OES.

Magnetic response of Ni nanocolloids synthesized by FLASiS in *n*-heptane and water was determined by the VSM technique. **Figure 8** exhibits this magnetic response as a function of the applied field. It is observed that the nanocolloid in *n*-heptane exhibits a greater magnetic response than in water. The above is concluded comparing the saturation magnetization and coercivity of 7.5 emu g^{-1} and 90 Oe in *n*-heptane, and 4.4 emu g^{-1} and 61 Oe in water, respectively.

The fitting curve in **Figure 8** agrees with the superparamagnetic behavior of the nanocolloids, corresponding to a Langevin function weighted with a log-normal distribution of magnetic moments ($g(\mu)$) and a linear contribution proportional to the susceptibility (χ_p) of the field. This function is given by Eq. 1:

$$M = N_d \int_0^\infty \mu \left[\coth\left(\frac{\mu\mu_o H}{k_B T} \right) - \frac{k_B T}{\mu\mu_0 H} \right] g(\mu)\, d\mu + \chi_p H \qquad (1)$$

where, the mean magnetic moment ($\mu = \mu_0 e^{\sigma^2/2}$) is obtained from the fitted parameters of the log-normal (median μ_0 and dispersion σ) and the saturation magnetization is given by $M_S = N_d \mu$, where N_d is the number density of NPs.

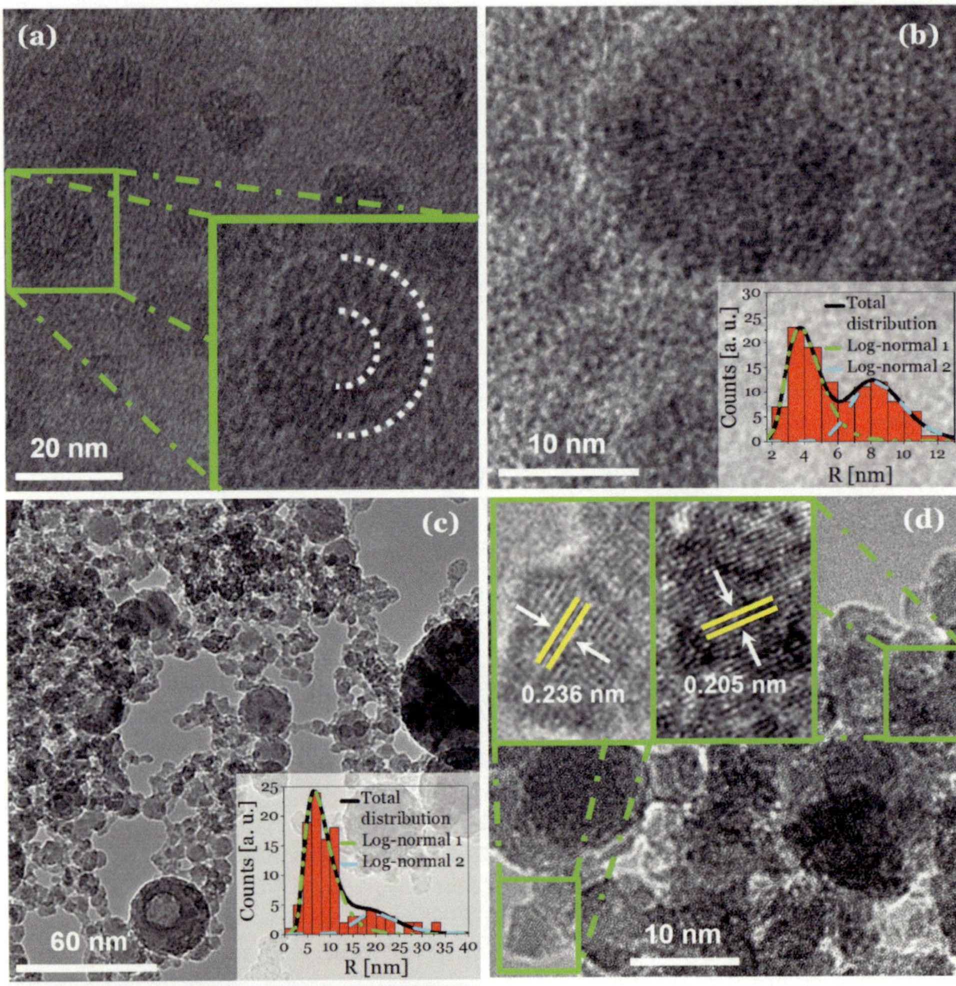

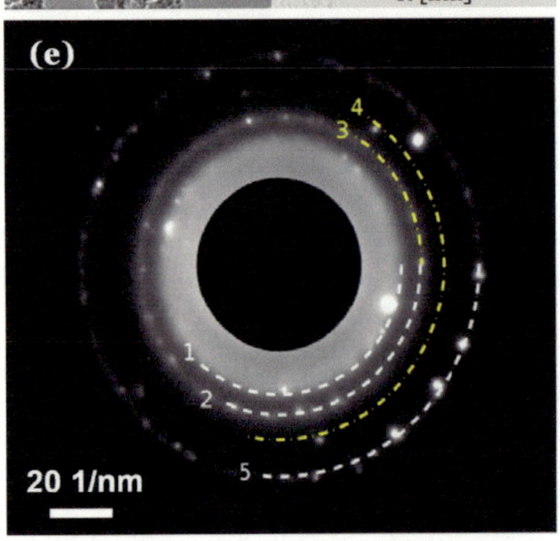

(f)

Line	d_{Ni} [nm]	d_{NiO} [nm]	(*h*, *k*, *l*)
1		0,2369	(1, 1, 1)
2		0,2052	(2, 0, 0)
3	0,2034		(1, 1, 1)
4	0,1762		(2, 0, 0)
5		0,1451	(2, 2, 0)

Figure 7.
TEM images of the NPs present in the Ni nanocolloids synthesized in n-*heptane (a) and (b), and water (c) and (d). (e) ED pattern indexed with the reflection lines belonging to NiO (lines 1, 2 and 5) and Ni (lines 3 and 4) according to table (f) for the Ni nanocolloid in water (reprinted with permission from [18] Copyright 2015 ASC).*

From the fit of the experimental data, the log-normal size distributions are determined (inset) considering that each NP of volume V is magnetized as $M_S = \mu / V$. Furthermore, r_M and r_T are calculated taking into account the

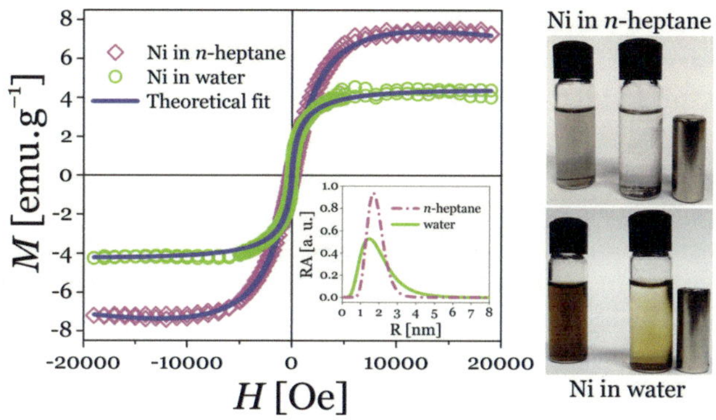

Figure 8.
Experimental magnetization curves and Langevin fitting of Ni nanocolloids in n-heptane and in water. The inset shows the size distribution for each nanocolloid. At the right: photographs of the colloidal suspensions showing the magnetic effect on the NPs exerted by a NdFeB magnet.

experimental and theoretical magnetization, respectively. The difference between these two values corresponds to the size of the magnetically NiO oxide shell frustrated layer, which does not present any type of magnetization. The size distributions are determined without considering hollow NPs or structures that present higher oxidation. Therefore, this determination is an estimative measurement that complements the characterization by OES and TEM.

2.2.2 Fe nanocolloids synthesized by FLASiS in water and ethanol

Synthesis of metallic Fe NPs and their dispersion in various liquid media is of great interest in the field of nano-magnetic materials, owing to great potential in biomedical applications [28–31]. The main routes for the synthesis of Fe NPs have been through wet chemistry [36–38]. Depending on the technique, NPs with different morphological and physicochemical characteristics can be obtained. In the case of FLASiS, the different processes that eventually occur lead to the formation of self-organized spherical nanostructures with different morphological, structural, compositional, and size characteristics compared with those generated with chemical techniques. **Figure 9** presents the experimental extinction spectra of the different Fe nanocolloids obtained by FLASiS using pulse energies of 70 μJ, 300 μJ and 700 μJ in water and ethanol. All spectra show an overall decrease in optical extinction as laser energy decreases, indicating, as expected, a lesser amount of ablated material. This is also qualitatively supported by the decreasing coloration of the colloidal suspensions (insets in **Figure 9**).

The extinction spectra of the Fe nanocolloids lack the characteristic plasmonic resonance exhibited by some metals, as seen above for the case of Ag (subsection 2.1.) and Ni (subsection 2.2.1.). This fact makes it difficult to fully characterize these suspensions by OES.

However, it can be observed that the spectra decrease monotonically in both media, except for the region from 300 nm up to 400 nm, in which the spectral behavior remains shoulder-shaped. This may be due to the presence of NPs with sizes greater than 20 nm. Although the number density of these NPs may be low, they have large enough cross-section, so their contribution is observable in the extinction spectra.

Furthermore, for ethanol, the formation of NPs with Fe_3C is highly probable, due to the binding of free carbons to Fe NPs during the ablation process at the plasma-liquid interface [8]. This is concluded from the fact that Fe_3C has an absorption band in the range of 300 nm - 400 nm, as can be seen in panel (b). This is due to the reactivity that such solvent has with Fe atoms at the high temperatures present in the plasma during FLASiS. Similar results are reported by other authors [40] who assign this band to the presence of Fe_3C in colloids.

From the spectroscopic results it can be concluded that the colloids have very similar composition. Therefore, from this point only the samples obtained with 700 μJ pulse energy are analyzed.

Figure 10 exhibits Raman spectra of the Fe nanocolloids in water (a) and ethanol (b), recorded in different regions of the sample. In both cases, it can be seen that depending on the local measurement area, the spectra show Raman signals of magnetite (Fe_3O_4 , full circle), hematite ($\alpha-Fe_2O_3$, asterisk) or mixtures of them. For the case of water, maghemite ($\lambda-Fe_2O_3$, diamond) signals are also detected.

Figure 11 shows TEM images of the Fe nanocolloids in water (a) and ethanol (b). Panel (a) is a panoramic view that includes core@shell NPs indicated with dashed line circles. An enlargement of a hollow Fe NP is observed in the inset. Panel (b) presents a group of NPs of typical size in ethanol. ED measurements are

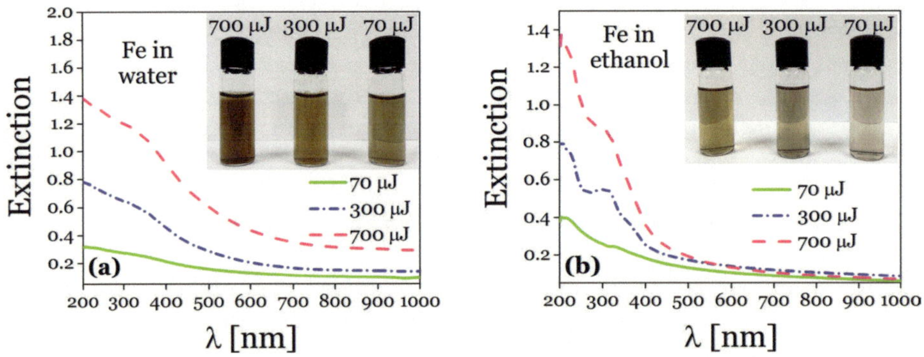

Figure 9.
Experimental extinction spectra of Fe nanocolloids synthesized by FLASiS in (a) water and (b) ethanol, with three different pulse energies. The insets exhibit photographs of the nanocolloids in each solvent (reprinted with permission from [39] copyright 2017 Wiley-VCH).

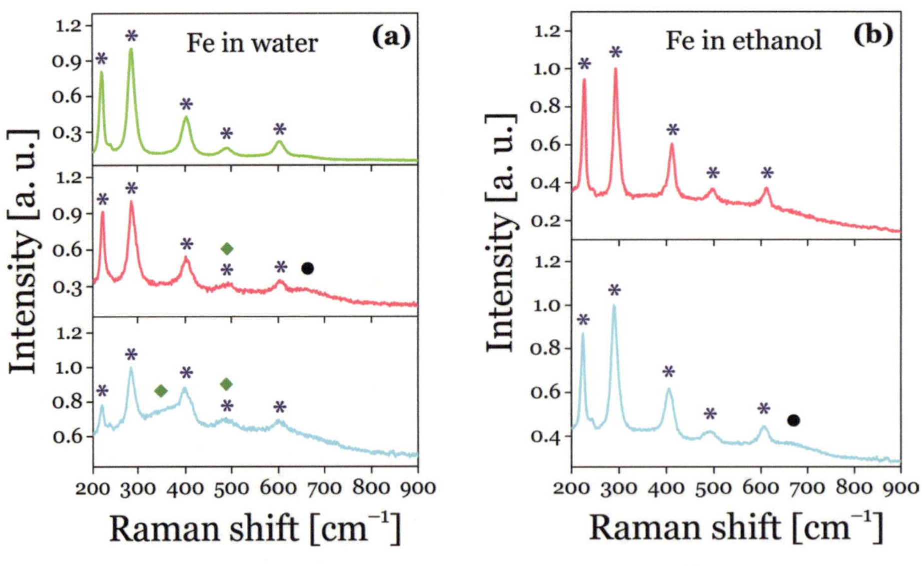

Figure 10.
Raman spectra of Fe nanocolloids in water (a) and ethanol (b) recorded at different points in the samples (reprinted with permission from [39] copyright 2017 Wiley-VCH).

performed to phase identification. Panel (c) exhibits a typical pattern for the Fe nanocolloid in water, where there are faint halos suggesting a high crystallinity of the NPs. Table in panel (d) presents the labeled ED rings, according to the inter-plane distances $d_{\alpha-Fe}$, $d_{Fe_2O_3}$ and $d_{Fe_3O_4}$, and to the corresponding Miller indices (h, k, l), indexed with reflection lines (JCPDS #06-0696 to $\alpha-Fe$, #75-0033 to Fe_3O_4 and #39-1346 to $\gamma-Fe_2O_3$). The crystallographic parameters of Fe_3O_4, $\gamma-Fe_2O_3$ and $\alpha-Fe$ are obtained from TEM simulation software (JEMS).

The field-dependent magnetization curves at room temperature for the Fe nanocolloids in water and in ethanol are presented in **Figure 12**. The cycles show the common behavior for NPs in a superparamagnetic state.

After a theoretical fitting using the Langevin function (Eq. 1), the log-normal distributions of radii centered at 1.7 nm for water and 3.7 nm for ethanol, are determined (inset in **Figure 12**). In this Figure it is observed that for the nanocolloid in water, the regime of magnetic saturation is not reached even at 18 kOe. This is probably due to surface effects and non-magnetic shells on NPs.

The magnetic saturation of the Fe nanocolloid in water (49.3 e mu g^{-1}) is higher than for the case of ethanol (26.5 emu g^{-1}). In both cases, the saturation magnetization is less than for Fe in bulk size (217 emu g^{-1}) [41]. However, the Fe NPs obtained

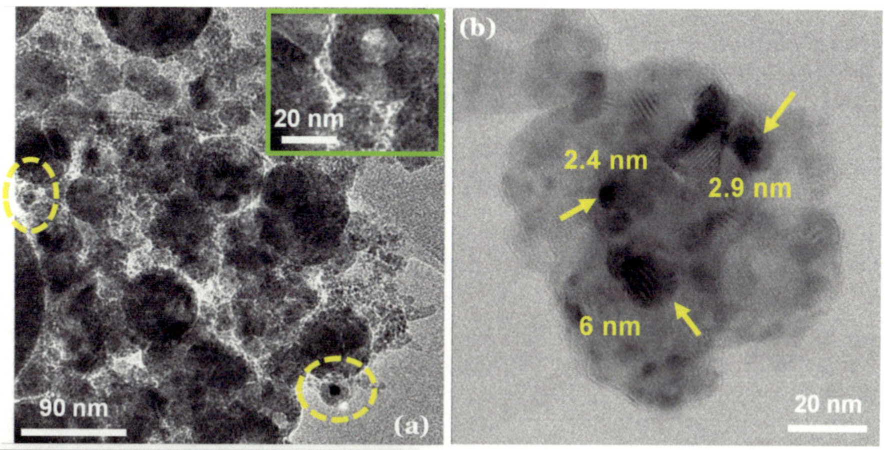

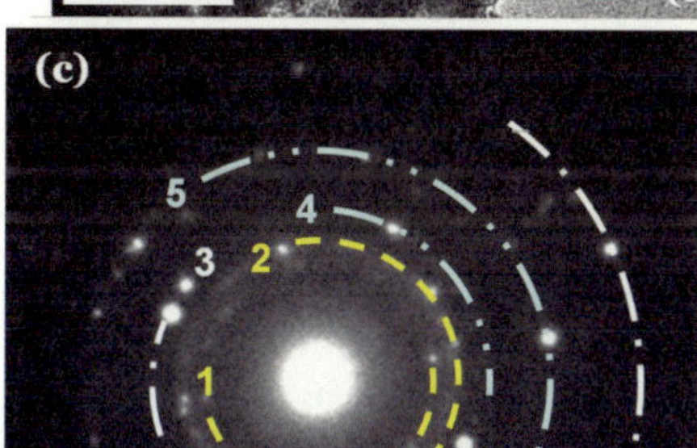

(d)

Line	$d_{\alpha\text{-}Fe}$ [nm]	$d_{Fe_2O_3}$ [nm]	$d_{Fe_3O_4}$ [nm]	(h, k, l)
1			0,299	(0, 2, 2)
2			0,255	(1, 3, 3)
3	0,201			(0, 1, 1)
4		0,184		(0, 2, 4)
5		0,147		(3, 0, 0)

Figure 11.
TEM images of Fe nanocolloids in water (a) and ethanol (b). The images show NPs with spherical structure where the numbers indicate their radius. (c) ED pattern indexed with the reflection lines belonging to Fe_3O_4 *(lines 1 and 2),* $\alpha-Fe$ *(lines 3 and 6) and* Fe_2O_3 *(lines 4 and 5) according to table (d) for the Fe nanocolloid in water (reprinted with permission from [39] copyright 2017 Wiley-VCH).*

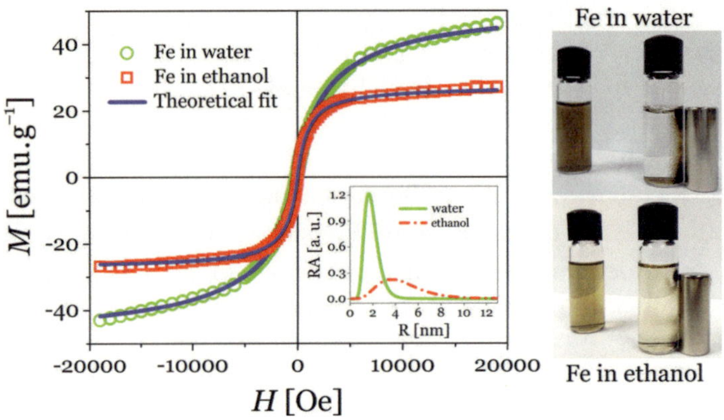

Figure 12.
Experimental magnetization curves and Langevin fitting of Fe nanocolloids in water and in ethanol. The inset shows the size distribution for each nanocolloid. At right: photographs of the colloidal suspensions where the magnetic effect on the NPs exerted by a NdFeB magnet may be observed.

have saturation magnetization greater than that determined by Maneeratanasarn *et al.* [42], who published on the synthesis of magnetic NPs by laser ablation of a target of α - Fe_2O_3 in ethanol, deionized water and acetone.

On the other hand, the smooth change in the magnetization slope in the region close to zero of the applied field, evidences the presence of different oxide phases as previously revealed in the MRS and ED studies.

2.3 Synthesis and characterization of few atoms Ag nanoclusters

Metallic clusters are known as few nanometer sized particles made up of subunits, which can be atoms of a single element (mono metal), or of several elements (alloys). Their novel chemical and physical properties are dependent only on the number of atoms they contain. These size-dependent properties, which make them suitable for applications in catalysis [43], photoluminescence [44], biomedical [45], magnetism [46], among others, show significant deviations from their bulk and large NPs counterparts. There are different procedures for clusters synthesis, which rely on the use of microemulsions [47], thiol cappings [48], vesicles [49] and electrochemistry techniques [46].

Particularly, Ag NCs have received much attention as novel fluorophores due to their good photostability, high quantum yield emitters and low toxicity. These properties make them suitable for microscopy settings, with potential biocompatibility, applications to sensing and bio-labelling when DNA is used as template [50]. With the experimental setup shown in **Figure 1**, Ag colloidal suspensions containing different sized NPs were obtained. To separate small clusters from the large Ag NPs, the nanocolloids were centrifuged varying centrifugation speed and time [6].

Fluorescence spectra of the as-prepared nanocolloids, normalized to their absorbance at 220 nm, yield band structures in the range 250 nm to 625 nm (**Figure 13(a)**). These bands correspond to transitions arising from a discretization of the energy bands when bulk metal downscale to few atoms structures. According to the jellium model, the HOMO-LUMO bandgap energy ($\mathbf{E}_g$) of the metal cluster, the Fermi energy ($\mathbf{E}_F$) of the bulk metal and the number of atoms (N) in the cluster are related by the expression $\mathbf{E}_g = \mathbf{E}_F \times \mathbf{N}^{-1/3}$ [51, 52]. Considering $\mathbf{E}_F$ = 5.49 eV for Ag, full line in **Figure 13(b)** shows the $\mathbf{E}_g$ relation with N according to this expression, showing a

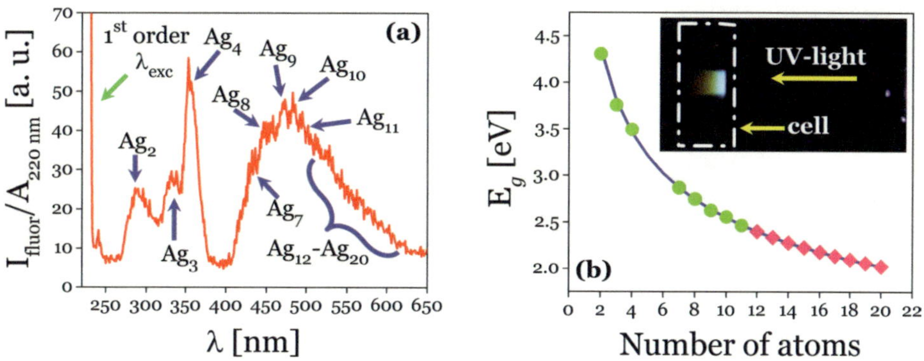

Figure 13.
(a) Fluorescence spectra in the UV-visible region. (b) E_g *vs. N according to the jellium model (full line). Symbols denote observed experimental fluorescence band peaks. Inset shows a snapshot of the induced visible fluorescence on the sample cell [6].*

monotonically decreasing dependence on cluster atoms number. Besides, it is clear that, for increasing N, the $\mathbf{E}_g$ values get closer to each other.

Experimental $\mathbf{E}_g$ values may be estimated from the peak wavelength of the fluorescence UV-visible bands in **Figure 13(a)**. With these values, clusters atom number may be determined from the curve in **Figure 13(b)**, which enables to assign the different band peaks to specific cluster atom numbers (circles and diamonds). In this way, isolated peaks corresponding to $\mathbf{Ag_2}$, $\mathbf{Ag_3}$ and $\mathbf{Ag_4}$ are clearly identified. It is also observed the overlapping of bands for increasing N, giving rise to the observed wide band. Arrows indicate the wavelengths corresponding to $\mathbf{Ag_7}$ to $\mathbf{Ag_{11}}$ NCs. $\mathbf{Ag_{12}}$ to $\mathbf{Ag_{20}}$ are indicated by the curly bracket. Inset in **Figure 13(b)** is a photograph of the visible fluorescence observed in the sample cell when it is illuminated by UV light (220 nm), corresponding to the white band between 400 nm and 625 nm.

When the as-prepared nanocolloids are centrifuged at 15000 rpm with increasing centrifugation times, the bands corresponding to larger NCs disappear, remaining a dominant band at 284 nm, which increases in intensity as centrifugation time increases (**Figure 14**). Inset presents spectra of the as-prepared and 20 min centrifugation samples as well as that for pure water for comparison.

HRTEM analysis using HAADF-STEM mode for image quality improvement on selected parts of the sample is shown in **Figure 15**. Panel (a) shows a panoramic of clusters in different aggregation stages. Coexistence of 1 nm radius NPs together with few atoms NCs is readily observed. In the NPs pointed by yellow arrows, Bragg planes can be observed. Agglomeration of atomic NCs is indicated by dashed green line, while few atoms clusters of 0.1 nm in size are pointed by white arrows. Panel (b) exhibits another site of the sample, with similar formations, including a cluster in a proto-particle stage with crystalline structure but without a defined morphology, enclosed by full blue circle.

Reactivity of NCs is high compared to their bulk counterparts, due to their high surface to volume ratio. Photocatalytic activity of Ag nanocolloid containing mainly small NCs was assessed by degradation of freshly prepared MB. **Figure 16(a)** presents the absorption spectrum of pure MB solution at different time intervals while illuminated by a white light lamp. The main MB absorption band at 660 nm decreases gradually as exposure time elapses showing native dye photocatalytic degradation. **Figure 16(b)** shows the same experiment, but when MB is mixed with the few atoms NCs colloid.

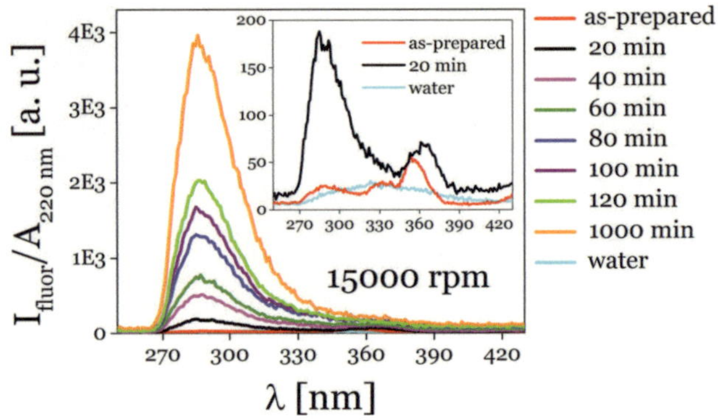

Figure 14.
Fluorescence spectra of nanocolloid obtained by FLASiS (600 μJ pulse energy) of Ag solid target in water for different centrifugation times. Excitation wavelength is $\lambda_{exc} = 220\ \text{nm}$ *. A dominant band at 284 nm is readily seen [6].*

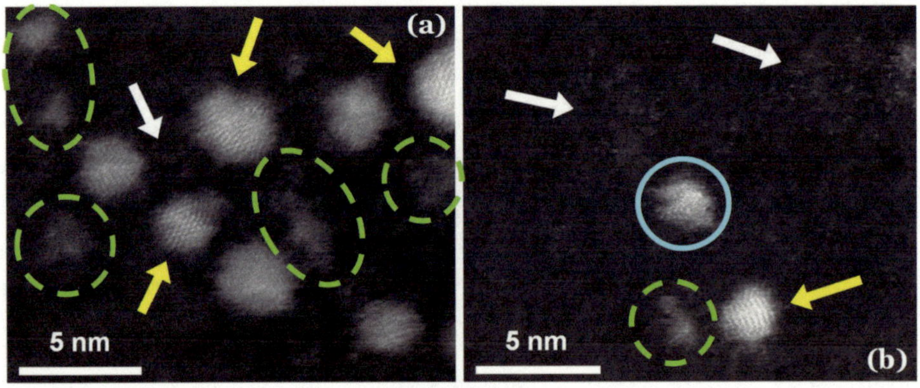

Figure 15.
High resolution and double-corrected electron microscopy analysis in HAADF-STEM mode for NCs observation [6].

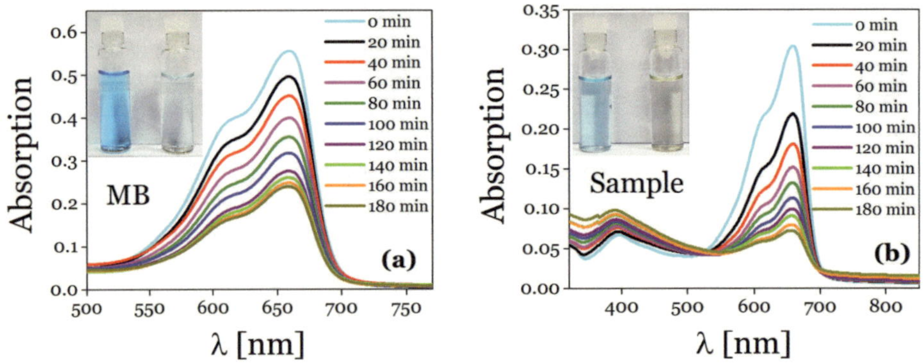

Figure 16.
Degradation of MB. (a) Absorption spectra of pure MB sample illuminated by a white light lamp taken at fixed time intervals. (b) Absorption spectra of sample containing small concentration of NCs under the same conditions as in (a) [6].

The MB absorbance decreases progressively faster than for pure MB. Dye degradation is easily identified by color change in the solution, from deep blue to faint light blue after exposure to white light (insets in the Figures). Degradation efficiency amounts to 80%, while for pure MB is only 55%.

3. Conclusion

FLASiS was used as a "green" method for synthesizing Ag, Ni and Fe metal nanocolloids in liquid media.

Comparison of Ag nanocolloids synthesized by FLASiS and chemical route in aqueous solutions of TSC and st has been discussed. Micro-Raman spectroscopy indicated that stabilizer molecules are adsorbed on the NPs surface and inhibit agglomeration, even up to one year in TSC. FLASiS synthesized NPs seem to stabilize faster that those generated by chemical route, as evidenced by the evolution of their optical extinction spectra.

Magnetic NPs like Ni and Fe colloids were generated in water and *n*-heptane. Spherical NPs shape is almost fully dominant, with a bimodal log-normal size distribution centered at roughly 4 nm and 9 nm radii. Different structures, like hollow type NPs and NiO species were observed. Nanocolloids in *n*-heptane exhibit a greater magnetic response than in water, both showing superparamagnetic behavior.

Fe NPs lack the characteristic optical plasmonic resonance exhibited by noble metals. Raman spectra of the Fe nanocolloids in water and ethanol show Raman peaks of magnetite, hematite or mixtures of them. For the case of water, maghemite signals were also detected.

Few atoms (2–20) Ag NCs can also be synthesized by FLASiS followed by several step centrifugation processes. Fluorescence spectra yield band structures in the range 250 nm to 625 nm, in agreement with the jellium model which predicts HOMO-LUMO type transitions. HRTEM analysis show clusters in different aggregation stages, with coexistence of 1 nm radius NPs together with few atoms NCs. Photocatalytic activity of Ag NCs was assessed against degradation of freshly prepared pure methylene blue. It was found that this efficiency rises 25% in the mixture of Ag NCs and MB.

Acknowledgements

We want to thank Dr. M.B. Fernandez Van Raap and Dr. D. Coral from Instituto de Física de La Plata (IFLP - CONICET) for the VSM measures. Besides, we acknowledge Dr. D. Muraca from Instituto de Física "Gleb Wataghin" (IFGW), Universidade Estadual de Campinas, Brasil and Dr. A. Caneiro from Y-TEC S.A., Argentina, for TEM analysis. We gratefully acknowledge their commitment and dedication.

These works were granted by PIPs 0394, 0280 and 0720 of CONICET, PME 2006-00018, PICTs 2012-1817 and 2016-3205 (ANPCyT), grants 11/I151 and 11/I197 (Facultad de Ingeniería), and 11/X651 and 11/X680 (Facultad de Ciencias Exactas) of Universidad Nacional de La Plata (UNLP), Argentina. We thank C2NANO-Brazilian Nanotechnology National Laboratory (LNNano) at Centro Nacional de Pesquisa em Energia e Materiais (CNPEM)/MCT (#14825, #14827, #16976, #18425, #19927 and #22345) and Research proposal TEM-16976 for the use of TEM. VSM was carried out at IFLP. Synthesis of nanocolloids by FLASiS, and OES, MRS, fluorescence and photocatalysis studies were performed at CIOp (CONICET - CIC - UNLP), La Plata. We acknowledge Y-TEC S.A. for the use of TEM FEI TALOS F200X.

D.C.S. and V.B.A. are researchers of CIC. L.B.S., J.M.J.S. and D.M.A. are researchers of CONICET, Argentina.

Author details

Jesica María José Santillán[1*], David Muñetón Arboleda[1*], Valeria Beatriz Arce[1], Lucía Beatriz Scaffardi[1] and Daniel Carlos Schinca[1,2]

1 Optical Research Center (CIOp, CONICET - CIC - UNLP), La Plata, Buenos Aires, Argentina

2 Engineering Faculty, National University of La Plata, La Plata, Buenos Aires, Argentina

*Address all correspondence to: jesicas@ciop.unlp.edu.ar
and davidm@ciop.unlp.edu.ar

References

[1] Gazit E. Self-assembled peptide nanostructures: the design of molecular building blocks and their technological utilization. Chem Soc Rev. 2007;36:1263-9. DOI: 10.1039/B605536M

[2] Kinkhabwala A, Yu Z, Fan S, Avlasevich Y, Müllen K, Moerner WE. Large single-molecule fluorescence enhancements produced by a bowtie nanoantenna. Nat Photonics. 2009;3:654-7. DOI: 10.1038/nphoton.2009.187

[3] Gao J, Gu H, Xu B. Multifunctional Magnetic Nanoparticles: Design, Synthesis, and Biomedical Applications. Acc Chem Res. 2009;42:1097-107. DOI: 10.1021/ar9000026

[4] Mafuné, F, Kohno J, Takeda Y, Kondow T, Sawabe H. Structure and Stability of Silver Nanoparticles in Aqueous Solution Produced by Laser Ablation. J Phys Chem B. 2000;104:8333-7. DOI: 10.1021/jp001803b

[5] Zhang D, Gökce B, Barcikowski S. Laser Synthesis and Processing of Colloids: Fundamentals and Applications. Chem Rev. 2017;117:3990-4103. DOI: 10.1021/acs.chemrev.6b00468

[6] Santillán JMJ, Muñetón Arboleda D, Muraca D, Schinca DC, Scaffardi LB. Highly fluorescent few atoms silver nanoclusters with strong photocatalytic activity synthesized by ultrashort light pulses. Sci Rep. 2020;10:8217. DOI: 10.1038/s41598-020-64773-z

[7] Pimpang, Pichitchai, Sutham, Withun, Mangkorntong, Nikorn, Mangkorntong, Pongsri, Choopun, Supab. Effect of Stabilizer on Preparation of Silver and Gold Nanoparticle Using Grinding Method. Chiang Mai J Sci. 2008;35:250-7.

[8] Amendola V, Scaramuzza S, Agnoli S, Granozzi G, Meneghetti M, Campo G, et al. Laser generation of iron-doped silver nanotruffles with magnetic and plasmonic properties. Nano Res. 2015;8:4007–DOI: 10.1007/s12274-015-0903-y

[9] Santillán JMJ, Fernández van Raap MB, Mendoza Zélis P, Coral D, Muraca D, Schinca DC. Ag nanoparticles formed by femtosecond pulse laser ablation in water: self-assembled fractal structures. J Nanopart Res. 2015;17:86. DOI: 10.1007/s11051-015-2894-8

[10] Zheng C, Wang H, Liu L, Zhang M, Liang J, Han H. Synthesis and Spectroscopic Characterization of Water-Soluble Fluorescent Ag Nanoclusters. J Anal Met Chem. 2013;2013:e261648. DOI: 10.1155/2013/261648

[11] Riboh JC, Haes AJ, McFarland AD, Ranjit Yonzon C, Van Duyne RP. A Nanoscale Optical Biosensor: Real-Time Immunoassay in Physiological Buffer Enabled by Improved Nanoparticle Adhesion. J Phys Chem B. 2003;107:1772-80. DOI: 10.1021/jp022130v

[12] Abreu AS, Oliveira M, de Sá A, Rodrigues RM, Cerqueira MA, Vicente AA, et al. Antimicrobial nanostructured starch based films for packaging. Carbohyd Polym. 2015;129:127-34. DOI: 10.1016/j.carbpol.2015.04.021

[13] Fernández JG, Almeida CA, Fernández-Baldo MA, Felici E, Raba J, Sanz MI. Development of nitrocellulose membrane filters impregnated with different biosynthesized silver nanoparticles applied to water purification. Talanta. 2016;146:237-43. https://doi.org/10.1016/j.talanta.2015.08.060.

[14] Lara HH, Ayala-Núñez NV, Ixtepan Turrent L del C, Rodríguez Padilla C. Bactericidal effect of silver nanoparticles against multidrug-resistant bacteria. World J Microbiol Biotechnol. 2010;26:615-21. DOI: 10.1007/s11274-009-0211-3

[15] Mori Y, Ono T, Miyahira Y, Nguyen VQ, Matsui T, Ishihara M. Antiviral activity of silver nanoparticle/chitosan composites against H1N1 influenza A virus. Nanoscale Res Lett. 2013;8:93. DOI: 10.1186/1556-276X-8-93

[16] Arboleda DM, Santillán JMJ, Arce VB, Fernández van Raap MB, Muraca D, Fernández MA, et al. A simple and "green" technique to synthesize long-term stability colloidal Ag nanoparticles: Fs laser ablation in a biocompatible aqueous medium. Mater Charact. 2018;140:320-32. DOI: 10.1016/j.matchar.2018.04.021

[17] Bohren CF, Absorption HDR. Absorption and Scattering of Light by Small Particles. John Wiley & Sons; 2008.

[18] Muñetón Arboleda D, Santillán JMJ, Mendoza Herrera LJ, van Raap MBF, Mendoza Zélis P, Muraca D, et al. Synthesis of Ni Nanoparticles by Femtosecond Laser Ablation in Liquids: Structure and Sizing. J Phys Chem C. 2015;119:13184-93. DOI: 10.1021/acs.jpcc.5b03124

[19] Schinca DC, Scaffardi LB, Videla FA, Torchia GA, Moreno P, Roso L. Silver–silver oxide core–shell nanoparticles by femtosecond laser ablation: core and shell sizing by extinction spectroscopy. J Phys D: Appl Phys. 2009;42:215102. DOI: 10.1088/0022-3727/42/21/215102

[20] Santillán JMJ, Scaffardi LB, Schinca DC. Quantitative optical extinction-based parametric method for sizing a single core–shell Ag–Ag_2O nanoparticle. J Phys D: Appl Phys. 2011;44:105104. DOI: 10.1088/0022-3727/44/10/105104

[21] Santillán JMJ, Videla FA, Fernández van Raap MB, Schinca DC, Scaffardi LB. Analysis of the structure, configuration, and sizing of Cu and Cu oxide nanoparticles generated by fs laser ablation of solid target in liquids. J App Phys. 2013;113:134305. DOI: 10.1063/1.4798387

[22] Santillán JMJ, Videla FA, Fernández van Raap MB, Muraca D, Scaffardi LB, Schinca DC. Influence of size-corrected bound-electron contribution on nanometric silver dielectric function. Sizing through optical extinction spectroscopy. J Phys D: Appl Phys. 2013;46:435301. DOI: 10.1088/0022-3727/46/43/435301

[23] Arce VB, Santillán JMJ, Muñetón Arboleda D, Muraca D, Scaffardi LB, Schinca DC. Characterization and Stability of Silver Nanoparticles in Starch Solution Obtained by Femtosecond Laser Ablation and Salt Reduction. J Phys Chem C. 2017;121:10501-13. DOI: 10.1021/acs.jpcc.6b12384

[24] Martina I, Wiesinger R, Jembrih-Simbürger D, Schreiner M. Micro-Raman characterisation of silver corrosion products: instrumental set up and reference database. E-Preserv Sci. 2012;9:1-8.

[25] Vinogradova E, Tlahuice-Flores A, Velazquez-Salazar JJ, Larios-Rodriguez E, Jose-Yacaman M. Surface-enhanced Raman scattering of N-acetylneuraminic acid on silver nanoparticle surface. J Raman Spectrosc. 2014;45:730-5. DOI: 10.1002/jrs.4544

[26] Kizil R, Irudayaraj J, Seetharaman K. Characterization of Irradiated Starches by Using FT-Raman and FTIR Spectroscopy. J Agric Food

Chem. 2002;50:3912-8. DOI: 10.1021/jf011652p

[27] Cael SJ, Koenig JL, Blackwell J. Infrared and raman spectroscopy of carbohydrates: Part III: raman spectra of the polymorphic forms of amylose. Carbohyd Res. 1973;29:123-34. DOI: 10.1016/S0008-6215(00)82075-3

[28] Landázuri N, Tong S, Suo J, Joseph G, Weiss D, Sutcliffe DJ, et al. Magnetic Targeting of Human Mesenchymal Stem Cells with Internalized Superparamagnetic Iron Oxide Nanoparticles. Small. 2013;9:4017-26. DOI: 10.1002/smll.201300570

[29] Orozco-Henao JM, Coral DF, Muraca D, Moscoso-Londoño O, Mendoza Zélis P, Fernandez van Raap MB, et al. Effects of Nanostructure and Dipolar Interactions on Magnetohyperthermia in Iron Oxide Nanoparticles. J Phys Chem C. 2016;120:12796-809. DOI: 10.1021/acs.jpcc.6b00900

[30] Wang Y-XJ. Superparamagnetic iron oxide based MRI contrast agents: Current status of clinical application. Quant Imaging Med Surg. 2011;1:35-40. DOI: 10.3978/j.issn.2223-4292.2011.08.03

[31] Veiseh O, Gunn JW, Zhang M. Design and fabrication of magnetic nanoparticles for targeted drug delivery and imaging. Adv Drug Deliv Rev. 2010;62:284-304. DOI: 10.1016/j.addr.2009.11.002

[32] Park J, Kang E, Son SU, Park HM, Lee MK, Kim J, et al. Monodisperse Nanoparticles of Ni and NiO: Synthesis, Characterization, Self-Assembled Superlattices, and Catalytic Applications in the Suzuki Coupling Reaction. Adv Mater. 2005;17:429-34. DOI: 10.1002/adma.200400611

[33] Xu R, Xie T, Zhao Y, Li Y. Quasi-homogeneous catalytic hydrogenation over monodisperse nickel and cobalt nanoparticles. Nanotech. 2007;18:055602. DOI: 10.1088/0957-4484/18/5/055602

[34] Ramírez-Meneses E, Betancourt I, Morales F, Montiel-Palma V, Villanueva-Alvarado CC, Hernández-RojasME.Superparamagnetic nickel nanoparticles obtained by an organometallic approach. J Nanopart Res. 2011;13:365-74. DOI: 10.1007/s11051-010-0039-7

[35] Rodríguez-Llamazares S, Merchán J, Olmedo I, Marambio HP, Muñoz JP, Jara P, et al. Ni/Ni oxides nanoparticles with potential biomedical applications obtained by displacement of a nickel-organometallic complex. J Nanosci Nanotechnol. 2008;8:3820-7. DOI: 10.1166/jnn.2008.199

[36] Solans C, Izquierdo P, Nolla J, Azemar N, Garcia-Celma MJ. Nano-emulsions. Current Opinion in Col Int Sci. 2005;10:102-10. DOI: 10.1016/j.cocis.2005.06.004

[37] Tartaj P, Morales MP, Veintemillas-Verdaguer S, Gonzalez-Carreño T, Serna CJ. chapter 5 Synthesis, Properties and Biomedical Applications of Magnetic Nanoparticles. In: Buschow KHJ, editor. Handbook of Magnetic Materials, vol. 16, Elsevier; 2006, p. 403-82. DOI: 10.1016/S1567-2719(05)16005-3

[38] Zhang D, Tong Z, Li S, Zhang X, Ying A. Fabrication and characterization of hollow Fe3O4 nanospheres in a microemulsion. Mater Lett. 2008;62:4053-5. DOI: 10.1016/j.matlet.2008.05.023

[39] Santillán JMJ, Muñetón Arboleda D, Coral DF, Fernández van Raap MB, Muraca D, Schinca DC, et al. Optical and Magnetic Properties of Fe Nanoparticles Fabricated by Femtosecond Laser Ablation in Organic and Inorganic Solvents.

ChemPhysChem. 2017;18:1192-209. DOI: 10.1002/cphc.201601279

[40] Amendola V, Riello P, Meneghetti M. Magnetic Nanoparticles of Iron Carbide, Iron Oxide, Iron@Iron Oxide, and Metal Iron Synthesized by Laser Ablation in Organic Solvents. J Phys Chem C. 2011;115:5140-6. DOI: 10.1021/jp109371m

[41] Crangle J, Goodman GM, Sucksmith W. The magnetization of pure iron and nickel. Proceedings of the Royal Society of London A Mathematical and Physical Sciences. 1971;321:477-91. DOI: 10.1098/rspa.1971.0044

[42] Maneeratanasarn P, Khai TV, Choi BG, Shim KB. The effect of laser energy on the preparation of iron oxide by a pulsed laser ablation in ethanol. J Korean Crystal Growth and Crystal Technology. 2012;22:134-8. DOI: 10.6111/JKCGCT.2012.22.3.134

[43] Chen W, Chen S. Oxygen Electroreduction Catalyzed by Gold Nanoclusters: Strong Core Size Effects. Angewandte Chemie. 2009;121:4450-3. DOI: 10.1002/ange.200901185

[44] Xu H, Suslick KS. Sonochemical Synthesis of Highly Fluorescent Ag Nanoclusters. ACS Nano. 2010;4:3209-14. DOI: 10.1021/nn100987k

[45] Ostuni E, Chen CS, Ingber DE, Whitesides GM. Selective Deposition of Proteins and Cells in Arrays of Microwells. Langmuir. 2001;17:2828-34. DOI: 10.1021/la001372o

[46] Santiago González B, Rodríguez MJ, Blanco C, Rivas J, López-Quintela MA, Gaspar Martinho JM. One step synthesis of the smallest photoluminescent and paramagnetic PVP-protected gold atomic clusters. Nano Lett. 2010;10:4217-21. DOI: 10.1021/nl1026716

[47] Buceta D, Piñeiro Y, Vázquez-Vázquez C, Rivas J, López-Quintela MA. Metallic Clusters: Theoretical Background, Properties and Synthesis in Microemulsions. Catalysts. 2014;4:356-74. DOI: 10.3390/catal4040356

[48] Donkers RL, Lee D, Murray RW. Synthesis and Isolation of the Molecule-like Cluster Au38(PhCH2CH2S)24. Langmuir. 2008;24:5976-5976. DOI: 10.1021/la801163t

[49] Wu, Zeng, Schelly ZA. Growth of Uncapped, Subnanometer Size Gold Clusters Prepared via Electroporation of Vesicles. J Phys Chem B. 2005;109:18715-8. DOI: 10.1021/jp0543476

[50] Yuan Z, Chen Y-C, Li H-W, Chang H-T. Fluorescent silver nanoclusters stabilized by DNA scaffolds. Chem Commun. 2014;50:9800-15. DOI: 10.1039/C4CC02981J

[51] Haberland H, editor. Clusters of Atoms and Molecules: Theory, Experiment, and Clusters of Atoms. Berlin Heidelberg: Springer-Verlag; 1994. DOI: 10.1007/978-3-642-84329-7

[52] Kreibig U, Vollmer M. Optical Properties of Metal Clusters. Berlin Heidelberg: Springer-Verlag; 1995. DOI: 10.1007/978-3-662-09109-8

Chapter 4

Gemini Imidazolinium Surfactants: A Versatile Class of Molecules

Kajol Bhati, Divya Bajpai Tripathy and Anjali Gupta

Abstract

Gemini imidazolinium surfactants fascinated the researchers and many industries towards it due to their distinct molecular structure. It belongs to the cationic surfactant group. The variation in the physicochemical properties of the gemini surfactant can be achieved by changing the characteristics in the structure. There are several applications of imidazolinium such as antistatic agents, fabric softener that makes it a demanding surfactant in detergent industries as well as in the laundry industries due to the immense number of properties like dispersibility, viscosity, desirable storage stability, emulsification, critical micelle concentration and fabric conditioning etc. This book chapter discussed about the Gemini imidazolinium surfactants and its various properties, synthesis methods and applications in various fields.

Keywords: gemini, surfactants, imidazoline, imidazolinium, emulsification

1. Introduction

The use of surfactants has been increasing due to their enormous applications in the field of chemistry. They belong to the organic compound group used in oil recovery, pharmaceuticals, nanoscience, biological activity, fabric softener, antibacterial and anti-foaming agents, and other technologies [1, 2]. Gemini is categorized as a surfactant and was first used by Menger in 1990 [3]. The word gemini means dimeric, an amphiphilic molecule, earlier it used to be synthesized by joining the two discrete surfactant molecules by a rigid spacer. It contains two terminal hydrocarbon tails (short or long); two polar head groups (cationic, anionic, or nonionic); and a spacer (short or long, flexible or rigid) [4]. The gemini surfactant has an efficiency of self- assembling at low concentration. In assessment with other surfactants, the gemini surfactant shows better surface activity. The presence of two polar groups and two terminal tails also made it more hydrophobic and hydrophilic as compared to monomeric surfactant systems. The substantial qualities of gemini viz., economic efficiency, flexibility, and functionality lead to its speedy demand in the field of research as well as in industry for examination and the use of it in various products. They also have other enhanced properties like low critical micelle concentration, wetting properties, efficient for high adsorption process, low surface tension, vesicle formation, helps in reduction of interfacial tension, and have the quality for aggregation [5]. The variation in the physicochemical properties of the gemini surfactant can be achieved by changing the characteristics in the structure. The cationic gemini surfactant has a wide range of purposes in the synthesis of nanorods, nanoparticles, construction of porous material, the formation of skincare products, drug development, gene therapy, and in antimicrobial process. Some examples of cationic gemini surfactants are piperidinum, pyridinium, imidazolium, imidazolinium, amino acid, and pyrrolidinum [6] (**Figure 1**).

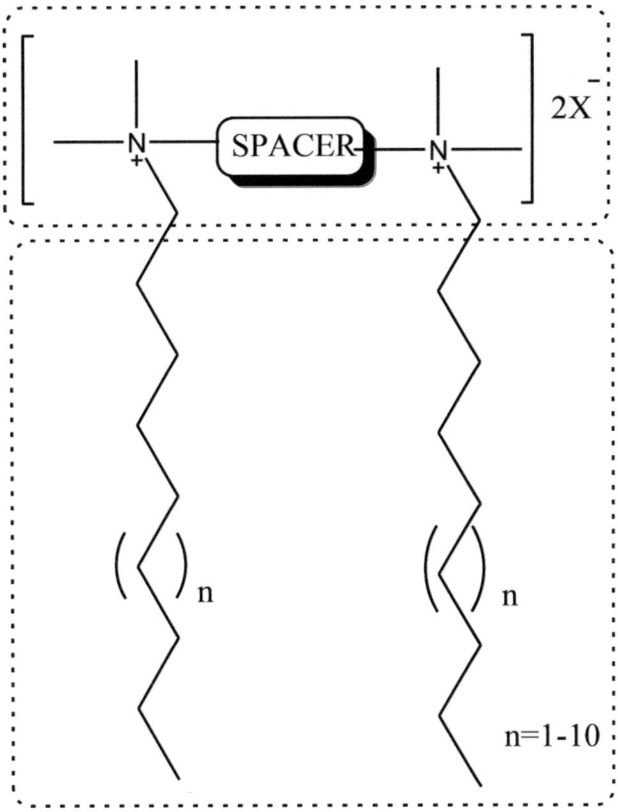

Figure 1.
Gemini surfactant [4].

As per the conducted studies, the spacer in the gemini surfactant has played a significant role in aggregation property. Examination conducted by Wanger et al. on cationic Gemini surfactants showed that spacer group has effect on the aggregation properties in aqueous solution. The use of hydrophilic compound and flexible spacer group helped in the formation of closely packed micelle structure as compared to the surfactant with rigid spacer group and hydrophobic compound. The micelle formation leads to decrease in surface tension of gemini surfactant which helps in increasing the surface area of surfactant. Therefore, use of hydrophilic compound and flexible spacer is in favorable condition for a better version of gemini surfactants. On gemini quaternary ammonium surfactants Zana et al. observed the behavior of association due to the spacer group in aqueous solution. Studies were conducted by Grosmaire et al. [7] on gemini surfactant spacer group to check the importance of carbon number on the micellization enthalpy for alkanediyl-α, ω-bis (dimethyl alkyl ammonium bromide) surfactants showed that the values of ΔH^{o}_{m} were strongly dependent on the spacer carbon number. In the transmission electron microscopy, the carbon position of C_{12}-C-C_{12} reflected the thread-like formation of micelle when concentration was less than 2% wt. and when the solution contained C_{12}-3-C_{12} with 7% wt.; the micelles shape was elongated. To cover better surface area, the formation of micelle concentration is crucial in surfactant as it lowers the surface tensions and hence, the minimum amount of gemini surfactant can be used in formulation process and for the applications.

1.1 Imidazolium gemini surfactant

The imidazolium is one of the varieties of cationic gemini Surfactant hence named gemini Imidazolium Surfactants. The nature of imidazolium is inherent and has greater potential than any other conventional surfactant. It has a self-aggregation tendency because of the high polarization nature of its head group. Researchers are focusing on imidazolium for advanced applications and for generating an enhanced variety [8]. Studies were conducted by Bhadani et al. for the synthesis of gemini surfactants taken from cardanol oil. They synthesized two sequences of imidazolium and pyridinium based upon phenoxy ring. The hydroxyl substituted pyridinium gemini surfactants with inconstant tail length and the other sequence with a variable length of the spacer group containing hydroxyl groups in their hydrophobic carbon chains for synthesis process [9, 10]. The synthesis of the gemini surfactant with variable length of the spacer comprising hydroxyl groups in their hydrophobic carbon chains Gemini surfactant has reported by P. Patial et al. [11]. They further assessed the surface properties of the synthesized surfactants [7, 12]. This carbon chain length is useful factor in the efficiency of surfactant, shorter the length of carbon chain, higher the suppressive efficiency of gemini surfactant [13].

1.2 Imidazolinium gemini surfactants

Gemini imidazoline surfactant fascinated the researchers and many industries towards it due to their distinct molecular structure. The bonding groups involvement is the crucial aspect in gemini surfactant for the modification of structure which affects the interface and solution properties [13, 14]. Conventional imidazolinium surfactant used to form with a polar imidazolinium head group and a long hydrocarbon tails, it used to be single chain structures whereas the Gemini imidazolinium surfactants are made up of two polar imidazolinium head group and two tails of hydrocarbon in which head groups are linked by a spacer [15]. It has enhanced surface-active properties than the conventional surfactants like corrosion inhibition, dispersibility, low critical micelle concentration, and hold better qualities as a softening agent [13, 16–18]. The Gemini surfactants are formed by adding two monomer surfactants with a binding group where the length of the monomer end chain can vary in length. It can be anionic, non-anionic, or cationic whereas the binding group varies in length and can be inflexible, soft, aromatic, or aliphatic. Other distinct chemicals and physical properties of gemini imidazoline surfactants are lower kraft point, the ability of self-assembling, high density, compatibility, inimitable rheological properties, etc. It has some other applications in drug delivery, nanoscience, and nanotechnology, molecular biology, in porous constituents, biological activity, etc.

The applications of imidazolinium such as antistatic agents, fabric softener make it a demanding surfactant in detergent industries as well as the laundry industries due to the immense number of properties like dispersibility, viscosity, desirable storage stability and fabric conditioning [16] (**Figure 2**).

As per the researches, the adaptation of various methods for synthesis and designing has been adopted which leads to the variation and enhancement in the synthesis methods and a better product. Some of the researchers prepared the gemini imidazolinium surfactant by microwave synthesis process that enhanced synthesis efficiency and also studied their surface properties. The microwave synthesized the surfactant in 5–10 minutes with a better yield of 80–91% as compared to conventional method i.e. thermal condensation which produced 75–80% [19–22]. The comparative study

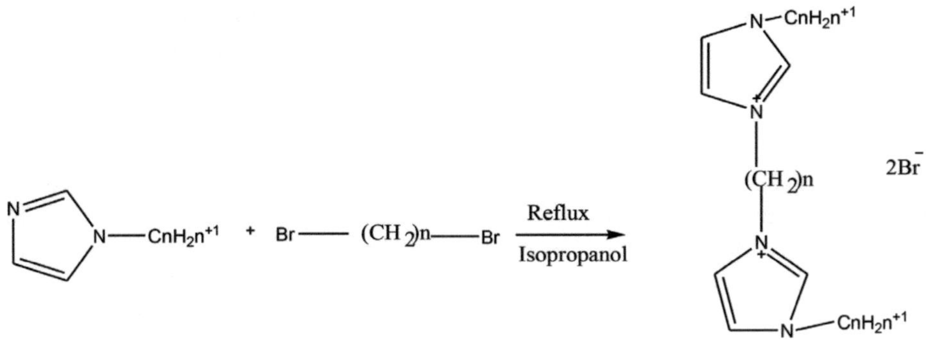

Figure 2.
Gemini imidazoline surfactant [16].

has also been done by Jianbin Huang on gemini and Bola, and the role of cephalic groups was found to be important as they alter their properties. A. Migahed et al. studied about the ligand length and its effect on the sustained release performance. The analysis showed that the shorter the length of the ligand, the better would be the effect of the sustained release [17]. Huaivu Yang et al. studied the relationship between sustained release and temperature and concluded that sustained release is directly proportional to the temperature [23]. A study done on emulsifier properties of gemini surfactants and applied it to emulsified asphalt and concluded that the minute quantity of gemini surfactant can be used in emulsified asphalt. Zhang Guanghua's et al. confirmed the sustained release effect on tinplate by synthesizing it. Yangiiang et al. studied the anti-corrosion properties of gemini surfactants and found that the surfactant works better during slow compound release [24]. All these studies analyzed the aspects of sustained release, surface properties, and length of the interval. Few of the methods used for synthesis helped in increasing the instability [25].

The gemini surfactants were compared with other single chain molecules, in which the gemini molecule structure was ambiguous. This was further studied to check the uniqueness and sustained release performance of gemini imidazoline surfactant. The comparison has been done between single- chain molecule and gemini imidazoline surfactant to observed the variations more precisely. The outcomes showed the great inhibitve efficiency in HCl solution as compared to single chain molecule. It also showed the better inhibitive effect on copper in HCl solution. The different concentration of gemini surfactant were taken to check the effect of concentration on inhibition property and it results that the inhibitive efficiency was higher when the concentration of HCl solution was less whereas for better inhibitive effect the concentration of gemini imidazoline surfactant must be more [25].

2. Classification

Gemini imidazoline surfactant can by classified on the basis physicochemical characterisitics.

2.1 On the basis of charge

Gemini imidazolinium belongs to the cationic gemini group but they are also amphoteric in nature. The imidazolinium gemini surfactant are dimeric molecules formed by two head groups associated with the spacer group and two hydrophobic tails [26] (**Figure 3**).

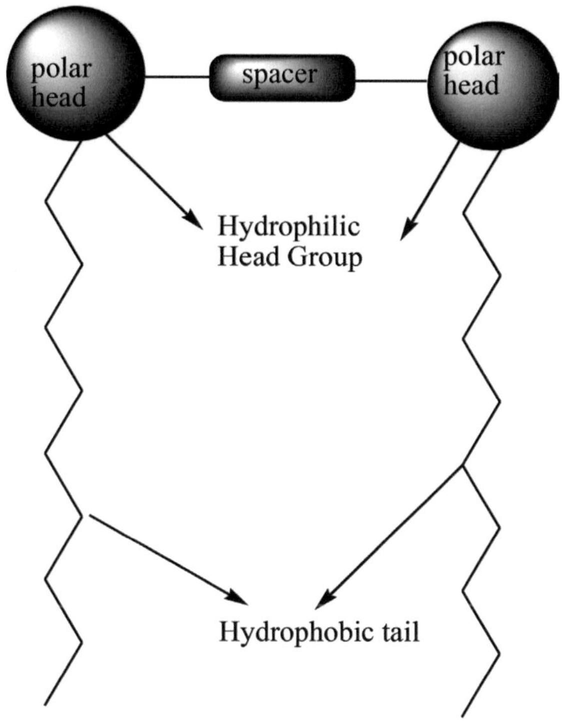

Figure 3.
Gemini imdazoline surfactant.

2.1.1 Cationic charge

The head group in the gemini imidazolinium surfactants contains the positive charge. It possesses long alkyl group, two polar imidazolinium head group and two tails of hydrocarbon in which head groups are linked by a spacer. They are soluble in organic solvents and dispense in water, eg., imidazoline-based dissymmetric bis-quaternary ammonium Gemini surfactant [26–28].

2.1.2 Amphoteric charge

The head group carries both negative and positive charge groups. They maintained the stability at acidic and basic pH of the solution, soluble in water, e.g. bisalkyl-bisimidazoline surfactant [26, 27, 29].

2.2 On the basis of spacer

The rigidity, length, polarity of the spacer can vary and leads to the variation in the structure of the gemini imidazoline surfactant as well as the in the surface properties [30, 31] (**Table 1**).

2.3 On the basis of hydrophobic tail

The variation in the hydrocarbon chain/tail can change the properties of surfactant. The chain can have two identical hydrophobic tails and two non-identical hydrophobic tails [30].

S.NO	Spacer properties	Characteristics	Sub-division
1.	Length	The surface activity of the surfactant varies with the number of carbon atoms present in the spacer group. Shorter the length of the surfactant better is the surface activity of gemini imidazoline surfactant.	• Shorter spacer chain. • Longer spacer chain.
2.	Rigidity	Shows higher surface activity. This can effect the aggregation of surfactant.	• Flexible spacer • Non-rigid spacer
3.	Polarity	Polarity affects the Critical micelle concentration of the surfactant	• Polar spacer

Table 1.
Classification of gemini imidazolinium surfactant on the basis of spacer group.

3. Synthesis

3.1 Synthesis of Gemini imidazole compound

Li et al. (1997) synthesized the Gemini imidazoline by a commercial process by taking 46.7 g of triethylene tetramine hydrate (0.25 mol), 146.6 g of oleic acid (0.52 mol) and added them in toludende (100 cm3) (DIAGRAM). With continuous stirring of the solution, it was heated until it reached to the toluene boiling point (120–130 C) almost for 3 hours. After that toluene azeotrope or water was collected from the reaction and temperature has been raised to 160–170 C at 12-16th hour of reaction by simultaneous removal of toluene from the Barrett distilling receiver. Thin layer chromatography was done to check the compounds in the reaction by taking chloroform/methanol (80: 20) as solvent and spot visualized using iodine. The completion of condensation reaction was checked by the TLC plate as the spot disappearance corresponding to the monoamides and diamides resulted in its completion. The reaction further continued till 16 hours for the collection of all byproduct water and yielded 96% of product which further recrystallized for the identification and structure confirmation from chloroform [32].

The two compounds Diethylenetriamine and lauric (molar ratio of 1:1.2) mixed with the xylene, which behaved as a solvent and kept at 140°C. Zinc powder has been added as a catalyzer. The Diethylenetriamine added into the flask at a slow rate and left for the reaction for 2 hours. Later, this flask was connected with the water separator at 200°C for 8 hours. A specific volume of water has been evaporated as per the theoretical calculation after the solvent were removed using the distillation method with the temperature maintained at 140°C to obtained the imidazoline intermediate is oily in nature and yellow in color. The distillation process was performed again to obtain the Single-chain imidazoline quaternary ammonium salt. The temperature was controlled around 80–90°C for 5 hours with the addition of dimethyl- carbonate in it with the same molar quantity as imidazoline intermediate. In this product, some amount of 1,3-dibromopropane has been added as half molar volume compared to intermediate and left for 8 hours for the reactivity, resulted in a sticky liquid of red-brown. Decreased the temperature to 50°C, after that added the less quantity of acetone to it, filtrated out to extract the zinc powder. Solution kept untouched until the crystal appeared, and lifted with sucking filtration and washed with acetone for numerous times. The obtained product was the Gemini imidazolinium surfactant that is solid khaki. (LG is 1,3-di(1-methyl-1-ethylamino-2-n-undecyl-4,5-dihydro-imidazoline) propane Gemini which is the Gemini cationic imidazoline surfactants based on lauric acid and LM is

1-enthylamino-2-n-undecyl-4,5-dihydro-imidazoline which is the cationic monomeric surfactants based on lauric acid [25] (**Figure 4**).

Migahed et al. (2018) synthesized the Gemini imidazole compound by taking the 0.2 mol dicarboxylic acid (aspartic or glutamic) in 100 ml of xylene, which was further mixed with the 0.4 ml of diethylenetriamine followed by refluxed process kept for 3 hours with the addition of catalyst PTSA in 0.1% amount at 140 C. Xylene has been removed from the solution after the required amount of water (0.2 mol) was obtained in the Dean-Stark tube. The resultant product was processed with diethyl ether for required amide compounds. For next 6 hours, the temperature of the reaction has been raised to 200 C. cool down the solution when collected water reached to the desired amount to obtain the Gemini compounds [17].

He further processed the obtained Gemini compounds for the synthesis of Gemini di-quaternary ammonium compound. The 1 mol compound was refluxed with 2 mol of dodecyl chloride in ethanol for 3 days. After that, ethanol was distilled off from the solution. Di- quaternary ammonium chloride has been obtained as resultant which was washed with the diethyl ether. The characterization and structure confirmation was done by FTIR (ATI Mattson series FTIRTM) revealed the functional groups of the compound [17].

3.2 Synthesis of Gemini imidazoline amphoteric surfactant

3.2.1 Synthesis of Bisalkyl-bis-imidazoline intermediate

A 250-ml four-necked flask was taken in which lauric acid with the twofold amount (in mol) and triethylenetetramine were charged and temperature maintained at 160 C for reaction under atmospheric nitrogen for 1.5 hours. Temperature has been raised to 200 C and heated for 1 hour and again raised to 250 C and heated for 1.5 hours. This process resulted in the formation of a light-yellow solid. This product further washed with ethanol, petroleum ether, and ethyl acetate (1:1:1 ratio) and resulted in bisalkyl-bisimidazoline [21] (**Figure 5**).

3.2.2 Quaternization of the Bisalkyl-bis-imidazoline intermediate

The intermediate of imidazoline was taken in a 250 ml four-necked flask and mixed with the two times the theoretical amount of sodium 2-chloroethanesulfonic acid.

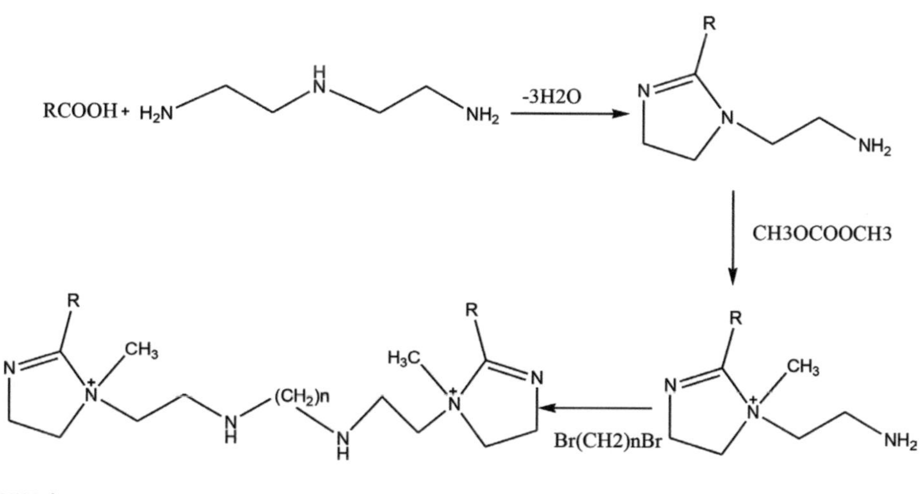

Figure 4.
Synthesis of LG [25].

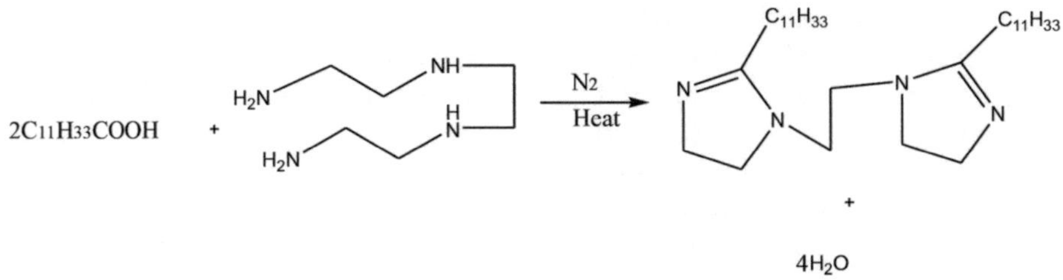

Figure 5.
Synthesis of Bisalkyl-bis-imidazoline intermediate [21].

Isopropyl alcohol or water mixture was added to it with continuation for 8 hours at 80–90 C. simultaneously, the pH of the solution has been around 8–9 by mixing the 10 wt.% NaOH solution. The supernatant liquid of the obtained product was taken in a beaker and mixed with 95% of alcohol for recrystallization. With the use of anhydrous alcohol, unreacted salts were removed from the solution. Thus, it resulted in the final product by removing the amide using chloroform [21] (**Figure 6**).

3.3 Synthesis of cationic gemini imidazoline surfactants

3.3.1 Synthesis of imidazolines

Tripathy et al. (2016) has synthesized the imidazoline using microwave synthesizer. Different compounds were taken in the beaker such as diethylenetriamine (20 mmol), fatty acids (Myristic acid, stearic acid, Lauric acid, Palmitic acid and Oleic acid- 40 mmol) and calcium oxide (20 g). The reaction was occurred at appropriate temperature as set in microwave. Later on, the mixture was cool down at room temperature. Then it further processed by adding the 80–100 ml of ethyl acetate and boiled, the remaining mixture was filtrated out and dried using vacuum. The optimization has been done and product was yielded at its maximum. The characterization and purity of the product was analyzed by spectrum [33].

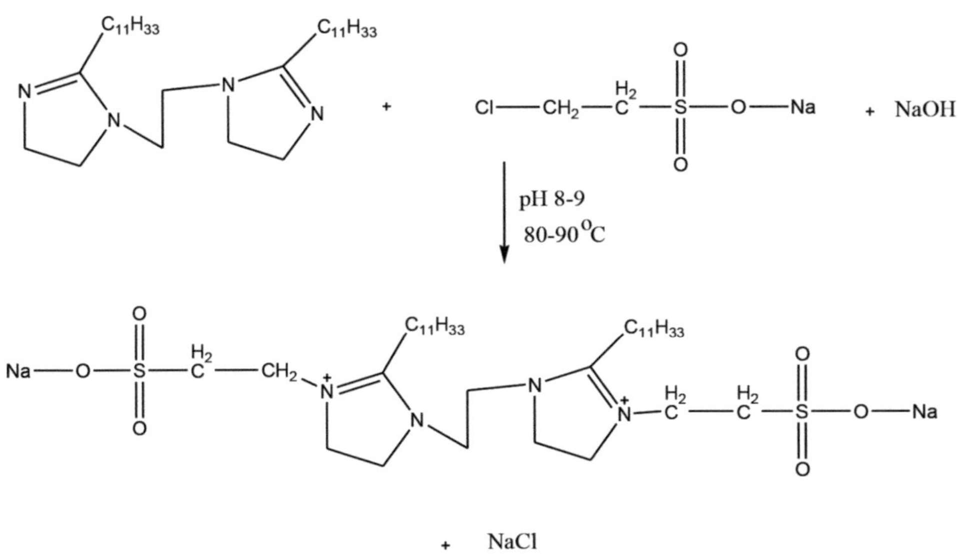

Figure 6.
Quaternization of the Bisalkyl-bis-imidazoline intermediate [21].

3.3.2 Synthesis of cationic Gemini imdazoline using waste frying oil

The synthesis of cost effective gemini imidazoline has also been reported by using waste frying oil like rapeseed oil, soyabean oil etc. The gemini imidazoline thus prepared have been analyzed for their surface active and performance properties. When compared to their monomeric counterparts, these geminis were found to be many folds superior to their corresponding monomeric counterparts [33].

3.3.3 Synthesis of cationic Gemini imidazoline surfactants with carbonate linkage

Tripathy et al. (2016) synthesized gemini imidazoline by the reaction of monomeric imidazoline and di iodoalkyl carbonated under microwave irradiation. Synthesis of di (iodoethyl) carbonate as quaternizing has been achieved by mixing the iodo alkanol and diphenyl carbonate (1:2 molar ratio) acted as carbonate exchange reaction in acetone at room temperature [22]. The resultant product was recrystallized by adding ethyl acetate and acetone (50:50) in it. The obtained product was Gemini imidazoline which further characterized by spectral analysis [33].

3.3.4 Synthesis of imidazoline-based dissymmetric bis-quaternary ammonium Gemini surfactant

Compound (I)N-(3-chloro-2-hydroxypropyl)-N, N-dimethyl alkylammonium chloride was formed by constantly mixing the N, N-dimethyl alkylamine hydrochloride, N, N-dimethyl alkylamine, and epichlorohydrin (1:1:1 molar ratio) for 14 hours in absolute ethanol. A light-yellow mixture has been obtained due to subsequently rotary evaporation of the ethanol. This product was further processed by repeated recrystallization from n-hexane and acetone to attained the white solid crude product. As per the study of this paper, compound (I) produced around 84.30% with the melting point of 47–48°C. The dissymmetric bis-quaternary ammonium salt with imidazoline ring (compound (II)) was synthesized using

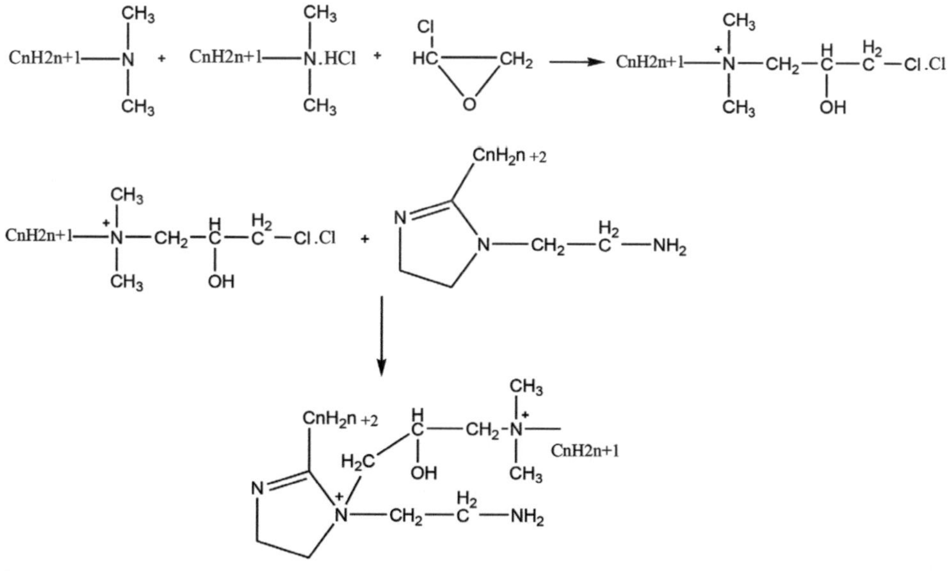

Figure 7.
Synthesis of imidazoline-based dissymmetric bis-quaternary ammonium Gemini surfactant [34].

alkyl imidazoline and N-dimethyl alkylammonium chloride, N-(3-chloro-2-hydroxypropyl)-N in isopropanol by keeping under the reflux for 10 hours and by putting the excess amount of alkyl imidazoline (10%). The resultant appeared as a waxy compound product. This product was further recrystallized with absolute acetone to obtain the desired compound in the form of a white solid with a melting point of 69–70°C. The characterization of the final product was performed to check the surfactant surfaces using Mass spectra and infrared spectrum, and 1H NMR (JNM ECP 600 MHz spectrometer) [34] (**Figure 7**).

4. Properties

4.1 Surface tension

Surface tension is the necessary exertion which is obligatory for the enhancement of the surface area of a liquid due to intermolecular force. Surfactant helps in reducing the surface tension of the liquid. An increase in the concentration of surfactants leads to a decrease in surface tension. Gemini imidazoline surfactant-containing stearic acid solution increased in the concentration from 0.1813 to 0.3626 g/l whereas the surface tension decreased from 42.8 to 28.6 mN/m. Again, the process repeated, the concentration of surfactant increased up to 0.6250 g/l leads to a decrease in the surface tension to 27.6 mN/m and a final increase in the concentration to 1.25 g/l and 2.5 g/l but the surface tension remained unchanged [22].

In the same study, the CMC values of 1% aqueous solution of gemini surfactants were found in the range from 0.0016 to 0.0032 mol/l that depends on the alkyl chain length (**Figure 8**). Greater the chain length of hydrophobic alkyl moiety, greater the CMC values of surfactant solutions [22].

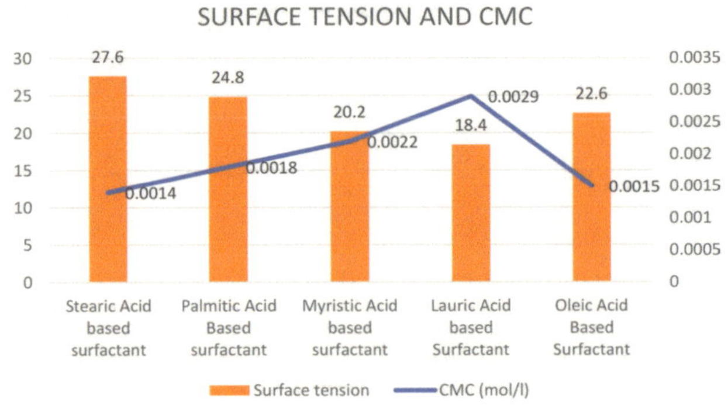

Figure 8.
Variation in surface tension and CMC values with hydrophobic chain [20].

4.2 Dispersion

An arrangement in which scattered particles of one substance dispersed to another substance is known as dispersion phenomena. Surfactants help in stabilizing the dispersion phenomena. Cationic gemini imidazoline surfactant showed stable and good dispersion capability. Gemini imidazoline surfactant-containing stearic acid had a cloudiness of about 4.8 ml that decreased with time. After 5 minutes, it was 3.6 ml and decreased to 3.0 ml after 10 minutes [20, 22]. The trend was found same when the gemini surfactants of different length of hydrophobic chain have been studied but in addition it was also revealed that decrease in the chain length of hydrophobic group increases the dispersibility of surfactants but the stability of dispersion was found to be decreased (**Figure 9**).

4.3 Softening

Cationic surfactants are used as fabric softening agents. The most common cationic surfactants which are used for softening agents are quaternary ammonium salts, imidazolinium salts, etc. [35]. The softening of surfactants decreases with the decrease in the length of the chain of the alkyl group. When compared with surfactants, it showed that surfactant-containing oleic acid had moderate softening behavior whereas fatty and highest alkyl chain has less softening. Fabrics treated with surfactants found to be soft as compared to untreated fabrics. Therefore, gemini imidazoline surfactants make the fabrics soft.

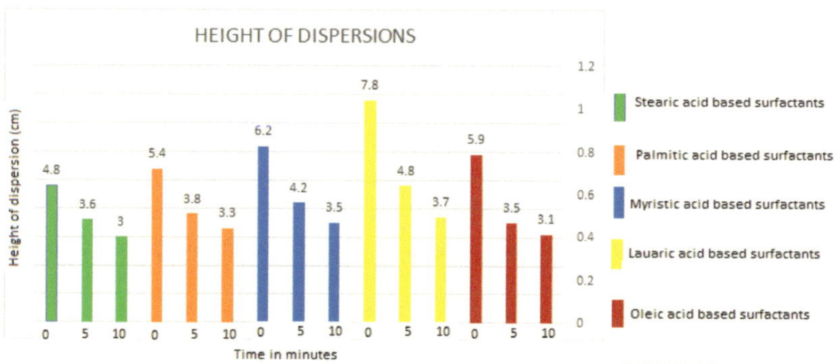

Figure 9.
Dispersibility of gemini surfactants based on different fatty acids.

4.4 Critical micelle concentration (CMC)

The surface-active agent present in solutions helps in the formation of micelles and this phenomenon is named as Micellization. This phenomenon occurs in critical micellar concentration or slender concentration. The surfactants can lower down the surface tension of the molecule which occurs due to the free monomer concentration and starts the micelle formation. This micelle formation in the solution is termed as critical micelle concentration. CMC is determined by the surface tension of the surfactant. An increase in concentration leads to a decrease in the surface tension until it reaches the critical micelle concentration. It was determined by plotting the graph of surface tension against the algorithm concentration of the surfactant. Critical micelle concentration and surface tension decreased with an increase in the length of the alkyl chain [10]. Zhang et al. research showed the CMC value of Gemini surfactant as 3.2_10_4 mol/L (194.9 mg/L) as per the graph reading in which surface tension plotted against log molar concentration of the surfactant and the breaking point revealed the mentioned value [6].

4.5 Wettability

The capability of a liquid to comes in contact with a surface of a solid and maintains it; this process is known as wettability. The adhesive and cohesive interaction helps in maintaining the interaction between liquid to solid and liquid to liquid. This property of gemini imidazoline surfactant increases with a decrease in the hydrophobic group length in the chain present in the surfactant. Whereas, the fall in the concentration of gemini imidazoline surfactant resulted in a decreased in the rewet ability and vice-versa [1, 20].

4.6 Inhibitor

Surfactants are used as corrosion inhibitors; they act as a protector on metal surfaces from corrosion. The gemini surfactants are amphiphilic, hence develops an affinity at metal/metal oxide–water interfaces for adsorption phenomena which leads to creating a barrier on metals and metal oxides surfaces and inhibits corrosion [36]. Imidazoline plays a vital role in corrosion inhibition as the imidazoline ring bond with the planar alignment to the surface of the metal. Further studies showed that the imidazoline inhibits the corrosion by blocking, activating, and using energy-related aspects together [37, 38].

4.7 Emulsification

A property of surfactants at the interface. Emulsification is a process that forms an emulsion between two immiscible liquids such as oil suspended in water [36]. Emulsification is an important property for gemini imidazolinium surfactants for the formation of the emulsion. The emulsification power of a surfactant varies with the length of the spacer group. It is directly proportional to the spacer chain length. Stable emulsions are required in drug formulation, cosmetics, solubilizations, etc. [5].

5. Applications

5.1 Biological activity

The antimicrobial activity of surfactants based upon the surface-active properties, hydrophobic chain length, and concentration. Gawali et al. (2019) synthesized the cationic Gemini surfactants and mentioned their application. The synthesized Gemini surfactants were assessed to check the biocidal activity contrary to a variety of bacteria such as Bacillus subtilis, *E. coli*, *Staphylococcus aureus*, and *P. aeruginosa*.

The R = C_7H_{15} [2-Octyl-1-diethylenediaminimidazoline based gemini surfactant (GSCTDH)] and R = $C_{11}H_{23}$ [2-dodecyl-1-diethylenediaminidazoline based gemini surfactant (GSLTDH)] compounds showed good microbial activity. The R = $C_{13}H_{27}$ [2-Tetradecyl-1-diethylenediaminimidazoline based gemini surfactant (GSMTDH)] surfactant was inactive for *Pseudomonas aeruginosa* but showed active behavior for Bacillus subtilis, *E. coli*, *Staphylococcus aureus*. The R = $C_{15}H_{31}$ [2-Hexadecyl-1-diethylenediaminimidazoline based gemini surfactant (GSPTDH)] compound showed biocidal activity against *E. coli*, *Staphylococcus aureus*. The GSCTDH surfactant compound containing the highest antimicrobial activity amongst the other Gemini synthesized compounds. As the compound showed the biocidal activity against the bacteria (gram-positive and negative), they can be considered to use as an antimicrobial agent in the form of surfactants. The test result showed the variation in the sensitivity of Gemini compound towards the gram-positive and gram-negative bacteria, they were more sensitive to gram-positive bacteria as compared to gram-negative bacteria. The reason behind that could be the behavior of the outer membrane of the bacteria which was less permeable in the case of gram-negative bacteria [39]. The surfactant having low critical micelle concentration values shows good corrosion inhibitor property due to their absorption quality at low concentration [40, 41]. Gawali et al. studied this property on carbon steel by making the surfactant with different concentrations at 30 c in 1 N H_2SO_4 [42].

5.2 Industrial applications

The gemini imidazoline surfactant with quaternary imidazolinium salts is used as a dispersant, emulsifiers, bleach agent, ant-static agent, and fabric softener as they show the better result as compared to the traditional surfactant and also have a mild effect on clothes, to eyes, and their biodegradability. Gemini imidazoline surfactant solubility is low in water [21]. They can be used for drug entrapment, oil recovery, also a probable vehicle for the transference of bioactive particles, for cleaning purpose, used as aerosol application [43].

5.2.1 Corrosion inhibitor

Zhuang et al. examined the corrosion inhibition property of gemini imidazoline surfactant on copper. They synthesized the imidazoline gemini surfactant using saturated fatty acids and studied the property by electrochemical method. Copper was taken in NaCl solution and according to the work, it showed that various factors like pH, surfactant concentrations, and length of carbon change affects the suppressive efficiency. The inhibition effect was better with the increase in gemini imidazoline surfactant concentration whereas suppressive efficiency is more when the length of the carbon chain is short [13].

Yang et al. synthesized the gemini imidazoline surfactant by adding oleic acid with triethylenetetramine (2:1). They studied the inhibition of carbon dioxide corrosion by linear polarization resistance in sparged beaker testing and concluded that the less concentration of gemini imidazoline was effective as corrosion inhibition when compared with traditional imidazoline and emulsion tendency was less than traditional imidazoline. The synthesized gemini imidazoline with oleic acid has improved film persistency, Higher surface activity, lower critical micelle concentration than traditional imidazoline. Therefore, this is a better surfactant as it is polluting the environment less than conventional imidazoline and also have better corrosion inhibition property [44]. Obot et al. also studied the corrosion inhibitors and concluded that the imidazoline based gemini surfactants are good corrosion inhibitors [37].

5.2.2 Laundry detergents

Gemini Imidazolinium surfactants have antistatic property, maintain the softness of fabrics. The studies showed that that the antistatic property of gemini quaternary ammonium salt surfactant helps in reducing the polyester fabric resistance to 107 Ω.S-HSJ-18and resulted in displaying the finest softening effect which leads to the low down the polyester fabric stiffness to be less than 2 mN·m. The pre-treatment process with KH560, cross-linking monomer N-hydroxymethyl acrylamide (MAM) and N, N-methylene bisacrylamide (MBA) of thermal setting fabric exhibited the greatest outcome for cross-linking that changed the wetting time of the fabric which could be more than 120 min and might stays the same after washing the fabric three-time i.e. 90 mins [45].

5.3 Nanoscience and nanotechnology

Li et al. developed the nanoparticles with a combination of silver and gemini imidazoline surfactants. The silver (Ag) nanoparticles well dispersed in nature were taken for the synthesis with novel imidazoline Gemini surfactant quaternary ammonium salt of di (2-heptadecyl-1-formyl aminoethyl imidazoline) hexane diamine at room temperature. Characterization of Ag nanoparticles was done by X-ray diffraction, Transmission electron microscopy, Fourier transform infrared, and UV- absorption spectra. The result revealed the enhanced micellized aggregation of gemini imidazoline surfactant in water as well as the Ag particles and the coordination and adsorption phenomena between the Ag nanoparticles and the imidazoline surfactant. It can act as environmentally friendly nanoparticles due to the modification in the surface of synthesized Ag nanoparticles. The developed Ag nano product act as a metal catalyst for methyl orange reduction reaction because of the active adsorption between methyl orange particles and Ag nanoparticles [46].

6. Conclusions

Gemini imidazolines are the surfactants that have been continuously getting explored for their varied application. At the one end where, the pendant structure of imidazoline surfactants make them easily absorbable on to the polar surface on the same tome the fatty alkyl chain imparts hydrophobicity thus defending so many polar surfaces by the adverse environmental impact and increase the life of surfaces. Presence of two surfactants molecules with in the molecule, drastically minimize their CMC values thus making the surfactants cost effective. Heterocyclic atom present in the ring of hydrophilic group imparts antimicrobial properties in the molecule thus making the surfactants suitable as antimicrobial agents. Furthermore, their existence in various forms like cationic, amphoteric, nonionic increase their applicability in various more industrial sectors.

Conflict of interest

The authors declare no conflict of interest.

Author details

Kajol Bhati[1], Divya Bajpai Tripathy[1*] and Anjali Gupta[2]

1 Department of Forensic Science, School of Basic and Applied Sciences, Galgotias University, Greater Noida, Uttar Pradesh, India

2 Department of Chemistry, School of Basic and Applied Sciences, Galgotias University, Greater Noida, Uttar Pradesh, India

*Address all correspondence to: divyabaj@gmail.com

References

[1] Tripathy DB, Mishra A, Clark J, Farmer T. Synthesis, chemistry, physicochemical properties and industrial applications of amino acid surfactants: A review. Comptes Rendus Chim 2018;21:112-30. https://doi.org/10.1016/j.crci.2017.11.005.

[2] Sar P, Ghosh A, Scarso A, Saha B. Surfactant for better tomorrow: applied aspect of surfactant aggregates from laboratory to industry. Res Chem Intermed 2019;45:6021-41. https://doi.org/10.1007/s11164-019-04017-6.

[3] Menger FM, Littau CA. Gemini Surfactants: A New Class of Self-Assembling Molecules. J Am Chem Soc 1993;115:10083-90. https://doi.org/10.1021/ja00075a025.

[4] Makhlouf A, Hajdu I, Badea I. Gemini surfactant-based systems for drug and gene delivery. Org. Mater. as Smart Nanocarriers Drug Deliv., Elsevier; 2018, p. 561-600. https://doi.org/10.1016/B978-0-12-813663-8.00013-0.

[5] Negm NA, El-Hashash MA, Mohamed DE, Marquis JM, Khowdiary MM. Gemini cationic surfactants: Synthesis and influence of chemical structure on the surface activity. J Surfactants Deterg 2013;16:733-8. https://doi.org/10.1007/s11743-013-1478-9.

[6] Pal N, Samanta K, Mandal A. A novel family of non-ionic gemini surfactants derived from sunflower oil: Synthesis, characterization and physicochemical evaluation. J Mol Liq 2019;275:638-53. https://doi.org/10.1016/j.molliq.2018.11.111.

[7] Zhao X, Liang W, An D, Ye Z. Synthesis and properties of tetrasiloxane Gemini imidazolium surfactants. Colloid Polym Sci 2016;294:491-500. https://doi.org/10.1007/s00396-015-3805-3.

[8] Ao M, Huang P, Xu G, Yang X, Wang Y. Aggregation and thermodynamic properties of ionic liquid-type gemini imidazolium surfactants with different spacer length. Colloid Polym Sci 2009;287:395-402. https://doi.org/10.1007/s00396-008-1976-x.

[9] Kamboj R, Singh S, Bhadani A, Kataria H, Kaur G. Gemini imidazolium surfactants: Synthesis and their biophysiochemical study. Langmuir 2012;28:11969-78. https://doi.org/10.1021/la300920p.

[10] Bhadani A, Singh S. Synthesis and properties of thioether spacer containing gemini imidazolium surfactants. Langmuir 2011;27:14033-44. https://doi.org/10.1021/la202201r.

[11] Patial P, Shaheen A, Ahmad I. Synthesis, characterization and evaluation of the surface active properties of novel cationic imidazolium gemini surfactants. J Surfactants Deterg 2014;17:253-60. https://doi.org/10.1007/s11743-013-1472-2.

[12] Shaheen A, Mir AW, Arif R, Wani AL. Synthesis, micellization behaviour and cytotoxic properties of imidazolium-based gemini surfactants. Colloids Interface Sci Commun 2020;36. https://doi.org/10.1016/j.colcom.2020.100257.

[13] Wenchang Z. Imidazoline Gemini Surfactants as Corrosion inhibitor for Copper in NaCl Solution. Int J Electrochem Sci 2020;15:8786-96. https://doi.org/10.20964/2020.09.43.

[14] Wang W, Han Y, Tian M, Fan Y, Tang Y, Gao M, et al. Cationic Gemini surfactant-assisted synthesis of hollow au nanostructures by stepwise reductions. ACS Appl Mater Interfaces 2013;5:5709-16. https://doi.org/10.1021/am4011226.

[15] Dahan E, Sundararajan PR. Solvent-dependent nanostructures of gels of a Gemini surfactant based on perylene diimide spacer and oligostyrene tails. Eur Polym J 2014;61:113-23. https://doi.org/10.1016/j.eurpolymj.2014.09.023.

[16] Shukla D, Tyagi VK. Cationic Gemini Surfactants: A Review. J Oleo Sci 2006;55:381-90. https://doi.org/10.5650/jos.55.381.

[17] Migahed MA, elgendy A, EL-Rabiei MM, Nady H, Zaki EG. Novel Gemini cationic surfactants as anti-corrosion for X-65 steel dissolution in oilfield produced water under sweet conditions: Combined experimental and computational investigations. J Mol Struct 2018;1159:10-22. https://doi.org/10.1016/j.molstruc.2018.01.033.

[18] Yin C, Kong M, Zhang J, Wang Y, Ma Q, Chen Q, et al. Influence of Hydroxyl Groups on the Inhibitive Corrosion of Gemini Surfactant for Carbon Steel. ACS Omega 2020. https://doi.org/10.1021/acsomega.9b02989.

[19] Bajpai D, Tyagi VK. Microwave synthesis of cationic fatty imidazolines and their characterization. J Surfactants Deterg 2008;11:79-87. https://doi.org/10.1007/s11743-007-1057-z.

[20] Tripathy DB, Mishra A. Convenient synthesis, characterization and surface active properties of novel cationic gemini surfactants with carbonate linkage based on C12[sbnd]C18 sat./unsat. fatty acids. J Appl Res Technol 2017;15:93-101. https://doi.org/10.1016/j.jart.2016.12.004.

[21] Xu Y, Xu H. Synthesis and Surface Active Properties of a Gemini Imidazoline Amphoteric Surfactant. J Surfactants Deterg 2016. https://doi.org/10.1007/s11743-016-1841-8.

[22] Tripathy DB, Mishra A, Wang XX, Zhuang W, Luo X, Zhang Y, et al. Convenient synthesis, characterization and surface active properties of novel cationic gemini surfactants with carbonate linkage based on C12[sbnd]C18 sat./unsat. fatty acids. J Appl Res Technol 2017;15:93-101. https://doi.org/10.1016/j.jart.2016.12.004.

[23] Lu T, Han F, Mao G, Lin G, Huang J, Huang X, et al. Effect of hydrocarbon parts of the polar headgroup on surfactant aggregates in gemini and bola surfactant solutions. Langmuir 2007;23:2932-6. https://doi.org/10.1021/la063435u.

[24] Zhang Y, Bansal N, Fujita Y, Datta-Gupta A, King MJ, Sankaran S. From streamlines to fast marching: Rapid simulation and performance assessment of shale-gas reservoirs by use of diffusive time of flight as a spatial coordinate. SPE J 2016;21:1883-98. https://doi.org/10.2118/168997-PA.

[25] Wang X, Zhuang W, Luo X, Zhang Y, Sun X, Li K. The inhibitive effects of Gemini imidazoline surfactants on copper in hydrochloric acid solution. Int J Electrochem Sci 2020;15:4338-51. https://doi.org/10.20964/2020.05.37.

[26] Khare U SP and KA. Applications of surfactants in pharmaceutical formulation development of conventional and advanced delivery systems. Int J Pharmacogn 2019;6:155-63.

[27] Sekhon BS. Surfactants: Pharmaceutical and Medicinal Aspects. J Pharm Technol Res Manag 2013;1:43-68. https://doi.org/10.15415/jptrm.2013.11004.

[28] Surfactants in pharmaceutical products and systems,. Encycl. Pharm. Technol. 3rd ed., Taylor and Francis; 2006.

[29] Sekhon BS. Gemini (dimeric) surfactants The Two-Faced Molecules. Resonance 2004:42-9. https://doi.org/10.1016/S1359-0294(96)80093-8.

[30] Mondal MH, Roy A, Malik S, Ghosh A, Saha B. Review on chemically bonded geminis with cationic heads: Second-generation interfactants. Res Chem Intermed 2016;42:1913-28. https://doi.org/10.1007/s11164-015-2125-z.

[31] Zhang Z, Zheng P, Guo Y, Yang Y, Chen Z, Wang X, et al. The effect of the spacer rigidity on the aggregation behavior of two ester-containing Gemini surfactants. J Colloid Interface Sci 2012;379:64-71. https://doi.org/10.1016/j.jcis.2012.04.052.

[32] Yang J, Gao L, Liu X, Qin W, Yin C, Zhang J. A highly effective corrosion inhibitor by use of gemini imidazoline. SPE J 2016;21:1743-6. https://doi.org/10.2118/173777-PA.

[33] Tripathy DB, Mishra A. MICROWAVE SYNTHESIS AND CHARACTERIZATION OF WASTE SOYBEAN OIL-BASED GEMINI IMIDAZOLINIUM SURFACTANTS WITH CARBONATE LINKAGE 2017;24:1-9. https://doi.org/10.1142/S0218625X17500627.

[34] Zhang J, Gong X, Song W, Jiang B, Du M. Synthesis of imidazoline-based dissymmetric bis-quaternary ammonium gemini surfactant and its inhibition mechanism on Q235 steel in hydrochloric acid medium 2011:1-10. https://doi.org/10.1002/maco.201006051.

[35] Puchta R. Cationic surfactants in laundry detergents and laundry aftertreatment aids. J Am Oil Chem Soc 1984;61:367-76. https://doi.org/10.1007/BF02678796.

[36] Zhu Y, Free ML, Woollam R, Durnie W. A review of surfactants as corrosion inhibitors and associated modeling. Prog Mater Sci 2017;90:159-223. https://doi.org/10.1016/j.pmatsci.2017.07.006.

[37] Obot IB, Solomon MM, Umoren SA, Suleiman R, Elanany M, Alanazi NM, et al. Progress in the development of sour corrosion inhibitors: Past, present, and future perspectives. J Ind Eng Chem 2019;79:1-18. https://doi.org/10.1016/j.jiec.2019.06.046.

[38] Ren C, Wang F, Zhang Z, Nie H, Li N, Cui M. Synthesis, surface activity and aggregation behavior of Gemini imidazolium surfactants 1,3-bis(3-alkylimidazolium-1-yl) propane bromide. Colloids Surfaces A Physicochem Eng Asp 2015;467:1-8. https://doi.org/10.1016/j.colsurfa.2014.11.031.

[39] Miyake M, Yamashita Y. Molecular structure and phase behavior of surfactants. Cosmet Sci Technol Theor Princ Appl 2017:389-414. https://doi.org/10.1016/B978-0-12-802005-0.00024-0.

[40] Branzoi V, Branzoi F, Baibarac M. Inhibition of the corrosion of Armco iron in HCl solutions in the presence of surfactants of the type of N-alkyl quaternary ammonium salts. Mater Chem Phys 2000;65:288-97. https://doi.org/10.1016/S0254-0584(00)00260-1.

[41] El-Tabei AS, Hegazy MA. Synthesis and Characterization of a Novel Nonionic Gemini Surfactant as Corrosion Inhibitor for Carbon Steel in Acidic Solution. Chem Eng Commun 2015;202:851-63. https://doi.org/10.1080/00986445.2013.867260.

[42] Gawali IT, Usmani GA. Synthesis, surface active properties and applications of cationic gemini surfactants from triethylenetetramine. J Dispers Sci Technol 2020;41:450-60. https://doi.org/10.1080/01932691.2019.1584112.

[43] Tyagi S, Tyagi VK. Novel cationic Gemini surfactants and methods for determination of their antimicrobial activity - Review. Tenside, Surfactants,

Deterg 2014;51:379-86. https://doi.org/10.3139/113.110319.

[44] Wang X, Yang H, Wang F. A cationic gemini-surfactant as effective inhibitor for mild steel in HCl solutions. Corros Sci 2010;52:1268-76. https://doi.org/10.1016/j.corsci.2009.12.018.

[45] FU Wei LL -s. -b. -f. Synthesis of gemini quaternary ammonium salt surfactants and their application performance on fabric. China Surfactant Deterg Cosmet 2009.

[46] Li W, Sun C, Hou B, Zhou X. Room Temperature Synthesis and Catalytic Properties of Surfactant-Modified Ag Nanoparticles 2012;2012. https://doi.org/10.1155/2012/638692.

Section 2

Structure, Dynamics and Colloidal Stability

Chapter 5

Aerogels Utilization in Electrochemical Capacitors

Ranganatha Sudhakar

Abstract

Supercapacitors are the integral part of electrochemical energy conversion and storage media. Energy storage mechanism is different in supercapacitors compared to batteries and results in exhibition of excellent power density. The supercapacitor performance is sensitive to material used as electrode, nature of electrolyte, etc. and the very significant is electrode surface nature. Based on the type of energy storage mechanism, supercapacitors are divided as electrochemical double-layer capacitors and pseudocapacitors. There is a practice to have both kind of these materials as electrode materials to achieve high electrochemical performance. Aerogels with inherent characteristics such as large pores, very high surface area, and superior mechanical stability make them superior candidates for electrode materials for high performance electrochemical supercapacitors. In this chapter, aerogels derived from different sources, their suitability and performance in view of electrochemical supercapacitors are discussed.

Keywords: supercapacitors, porous, mesoporous, specific capacitance, carbon materials

1. Introduction

Rapid growth of human society and subsequent need of energy is driving the manipulation of non-renewable sources in nature leading to depletion of the same. This is simultaneous with the increasing threats such as global warming, energy shortage, air pollution etc. Standardization of our life style and drastic change in dependence on electricity, demanding urgent need for high efficiency energy conversion and storage. Batteries and supercapacitors are excellent means of electrical energy conversion and storage including solar cells [1, 2].

Conventional condensers or capacitors utilize dielectric materials, e.g., ceramics, polymers which are non-conducting in nature, exhibit the capacitance in the range of pico to microfarad. Typically, anodic metal oxides mostly of Ta, Al, Nb are used in electrochemical capacitors which widen the capacity from micro to millifarad level. Recently, supercapacitors are devised involving energy mechanisms; electric double-layer capacitance (EDLC) and pseudocapacitance. Charge separation at electrode/electrolyte interface results in EDLC and fast, reversible reactions occurring on solid electrode surface leads generation of pseudocapacitance. RuO_2 and IrO_2 noble metals exhibit superior specific capacitance value of about 750 F/g but at the same time hazardous and non-economical. For this reason, oxides of transition metals, e.g., CoO_x, MnO_x, and NiO_x are extensively being studied as supercapacitor electrodes [2, 3].

2. Fundamentals of charge storage in electrochemical supercapacitors

Three categories of supercapacitors are made with respect to the mechanism involved in energy storage in them. The very first category is electrical double-layer capacitor (EDLC), wherein electrostatic charge gathered at the electrode electrolyte interface results in capacitance. Here capacitance is directly proportional to surface area accessed by the ions from electrolyte. Pseudocapacitor is the second category, where in reversible redox reactions by electroactive compounds are considered. In third type, combination of both of these kinds are made use to extract electrical energy. The type of material chosen plays a crucial role in the energy delivery. The challenges posed by supercapacitors are low energy density, low workable potential window, economy and self-discharge. The strategy to overcome these limitations is to design newer energy materials. Popularly, designing hybrid of a carbon material, pseudocapacitive metal compounds and conducting polymers. Synergistic effects, high surface area from carbon materials, high specific capacitance, redox processes from pseudocapacitive materials contribute to achieve high capacitance with good rate capability [1–4].

The first category supercapacitors, which are dependent on the formation of electric double layer, can be fabricated with the help of two carbon related electrode materials, an electrolyte and a separator. There will be no electron exchanges between electrolyte and electrode being non-faradaic. There occurs population of charges on electrodes when voltage is applied which drives the ion diffusion in the electrolytic solution. An electric double layer with oppositely charged ions on the electrode surface is formed to skip ionic recombination. Because of this mechanism, charge take up will be very fast and energy delivery too. Also, electrodes are benefitted with no swelling during charge and discharge cycles similar to the batteries [1, 5–8].

Considering pseudocapacitive materials as electrodes in supercapacitors, redox reactions take place between electrode and electrolyte which stores the charge. Oxidation or reduction reactions occur when a required voltage is applied, on the electrodes involving charge passage across double layer generating faradaic current. Metallic compounds and conducting polymers are best examples for this class of electrodes which suffer with lack of stability during cycling leading to lowering in power density [1]. As these faradaic reactions involve redox reactions, are slower which makes them to exhibit lower power density and poor cycling stability compared to EDLC type [1, 6–9].

There are few criteria to design a high performance supercapacitor electrode materials. To list out, high specific capacitance stands first. To achieve this, electrode material should pose a very high specific surface area which will eventually store high energy per unit mass and volume. So nanomaterials and porous materials can be expected to satisfy this criterion. Large rate capability and high cycle stability are major characteristic to be possessed by an ideal electrode material, which signifies capacitance retention at high scan rate and/or current density. Additionally economically viable and non-toxic materials are preferred.

Majorly the factors that determine the characteristics of high specific capacitance, rate capability and cycle stability are Surface area of the electrode, electronic/ionic conductivity and mechanical/chemical stability. As the charge storage mechanism involves the adsorption and desorption of the ions on the electrode surface, more the surface area higher will be energy storage. Specific capacitance and rate capability are highly relative to electronic and ionic conductivity. Higher values of these can maintain a rectangular shape of cyclic voltammogram which is typical for an ideal capacitor and symmetric profiles in galvanostatic charge discharge cycles.

3. Aerogels as supercapacitor electrode materials

Carbonaceous aerogels extended to three-dimensional structures are very much potential for high performance electrode materials in supercapacitor because of superior characteristics such as vast surface area and porous nature, facilitating uninterrupted paths for ionic movement by shortening diffusion pathways.

Based on the source, the carbonaceous gels are typically classified as (1) Aerogels derived from polymers (2) aerogels derived from carbon nanotubes, (3) Aerogels derived from graphene, and (4) aerogels sourced from biomass.

4. Polymer-derived aerogels for supercapacitors

Carbon aerogels (CAs) derived from polymers belong to wide category of carbonaceous aerogels. Pekala and Kong in 1989 firstly recorded the synthesis of CAs by inert atmosphere pyrolysis of resorcinol–formaldehyde (RF) organic aerogel [10, 11]. The organic aerogel from RF can be synthesized by aqueous mediated sol-gel-based resorcinol and formaldehyde monomeric poly-condensation followed by supercritical drying [12]. Alternatively, polymers of sol–gel source too utilized as starting materials for synthesizing carbon aerogels, such as phenol–melamine–formaldehyde gel [13], poly-benzoxazine gel [14], cresol, resorcinol and formaldehyde gel [15], resorcinol and pyrocatechol gel [16], cresol and formaldehyde gel [17], gel of resorcinol–methanol [18] and poly(vinyl chloride) gel [19]. Additionally, microwave drying [20] air drying [21, 22] and freeze-drying [23] too are preferred as the synthetic procedures for dried organic gels. The process of pyrolysis can also be employed to convert dried organic gel into CAs. Precursor configuration and conditions of pyrolysis process are very crucial parameters to design the desired structure of CAs.

CAs derived from polymers have been studied exclusively as supercapacitor electrode. First, Pekala in 1994 demonstrated capacity of CAs electrodes in supercapacitors [24]. The 3D CAs consist of nanoparticle assembly associated with variable sized pores. CAs possess the merits like of good electrical conductivity, high porosity, high surface area and tunable pore sizes [24, 25]. These unique structure and superior properties, make CAs potential candidates for the application as electrode materials in electrochemical supercapacitors.

An aerogel derived from pyrolysis of polybenoxazine by Katanyoota and coworkers with surface area 368 m^2/g with 2–5 nm pore size exhibits 56 F/g of specific capacitance in 3 M sulfuric acid [14]. CAs modified with pseudocapacitive materials such as metal oxides or conducting polymers can be expected to offer higher performances compared to their pristine counterparts. Resorcinol-methanol-derived CA modified with Mn_3O_4 doping shows 503 F/g in 0.5 M Na_2SO_4 [18]. Interestingly it possesses 577 m^2/g of surface area associated with 18 nm pore size. CA from RF precursor developed by Chien et al., [26] shows an excellent electrochemical behavior with 1700 F/g in 1 M potassium hydroxide. Though surface area was not so high, i.e., 206 m^2/g, the doping of pseudocapacitive material $NiCo_2O_4$ contributes towards high capacity. In an alternate report MnO_2 doped RF-derived CA offers 515 F/g in neutral 1 M Na_2SO_4 and corresponding surface area was 120 m^2/g [27]. Conducting polymer, polyaniline being a doping material in a CA for which RF were the precursors, shows of 710 F/g in acidic electrolyte 1 M H_2SO_4 [28]. There are some research efforts wherein secondary materials were used to modify the carbon aerogel derived from resorcinol-formaldehyde gel. Wang and co-workers doped nickel oxide particles to enhance the activity which resulted in exhibiting 356 F/g at 1 A/g in 6 M KOH medium [29]. In an alternate piece of work, carbon nanotubes were used as dopants to activate the CAs and delivered 141 F/g at 5 mV/s in 30% potassium hydroxide solution [30]. Also there are some reports wherein CAs are activated by CO_2 and KOH to enhance the electrochemical behavior [31]. These activated CAs possess hierarchical porous network structures with microporous, mesoporous and large pores with <2 nm, 2–4 nm and >30 nm correspondingly. These CAs deliver 250 F/g after KOH activation and 8.4 Wh/kg at 0.5 A/g of current density in 6 M potassium hydroxide as electrolyte solution. Doping with metal also found to influence the performance of CAs. Lee et al., [22] doped a series of CAs with different metals. They found metal doped CAs with higher capacitance comparing to pristine ones. Mn doping showed higher capacitance compared to those of Cu, Fe. The metal compounds doped CAs are also studied including Mn_3O_4 [18], $NiCo_2O_4$ [26], ZnO [32], FeO_x [33], MnO_2 [27], SnO_2 [34], NiO [29] and RuO_2 [35].

5. Carbon nanotube (CNT)-based aerogels for supercapacitors

Porous interconnects in 3 dimensions with carbon nanotubes as skeleton constitute this category. Different processes are employed to synthesize carbon nanotube-based aerogels, such as chemical vapor deposition (CVD), [36] freeze-drying, [37, 38] and critical-point-drying [39–41].

The CNT-based aerogels possess benefits of the carbon nanotubes, like excellent electrical conductivities, good mechanical resilience and superior thermal conductivity, and show the special characteristics of aerogels, too; 3D network with pores, less density, porous nature and high specific surface areas. These attractive characters direct carbon nanotube aerogels for applications as supercapacitor electrodes. Also, as these 3D networks possess number of pores facilitate substrate for holding other active materials like metal oxides, carbon, and polymers with conducting nature, improving storage capacities.

Most extensively used method to fabricate CNT-based aerogels is CVD. Bordjiba and coworkers synthesized CNT aerogels by CVD method with surface area 1059 m^2/g and modified with microfibrous carbon offered 524 F/g in 5 M KOH [36]. Polyaniline were made use to modify CNT aerogel by Zhong et al., to improvise the specific capacitance with the contribution from pseudocapacitive conducting polymer. It offered 189 F/g in in 1 M H_2SO_4 medium [42]. In an attempt Fang et al., Ni microfiber supported CNT aerogels were designed which showed 348 F/g in 5 M KOH electrolyte [43]. Preparing carbon nanotube layers on other three-dimensional networks using CVD process also results in the synthesis of CNT-based composite aerogels which can be expected to perform better. Bordjiba et al., targeted this kind of composite of CNT aerogel with carbon aerogel by CVD method [44]. This material with around 700 m^2/g of surface area delivered 524 F/g of specific capacitance in 5 M KOH electrolyte. Freeze-drying method was employed to design CNT-based aerogel using wet-gel precursors [45]. The aerogels prepared so, exhibit superior properties which include mechanical, thermal, etc. Authors of these report the suitability of cellulose-CNT hybrid aerogels for sensor applications for gases and other volatile organic compounds. Sun et al., demonstrated [46] direct freeze-drying synthesis of CNT-graphene composite aerogels which exhibit superior thermal property, electrical conductivity, and good adsorption characteristic, etc., which advocate suitability for electrodes in supercapacitors. Li and co-workers prepared

CNT aerogel using CVD process which successfully demonstrated capacitive behavior [47]. Under 50% of compressive strain, capacity retention was about 90% and it was 70% under 80% of strain. This demonstrates the quality of electrode as compressible and deformation sustainable electrodes which is of unique quality. A concept of decorating CNT aerogel with a conducting polymer was successful for superior quality supercapacitor electrodes. Lee et al., [48] were successful in doing so as CNT aerogel coating by polyl(3, 4-ethylenedioxythiophene) which successful by delivering a volumetric capacitance of about 40 F/cm^3 at 100 V/s and high volumetric energy 70 Wh/cm^3 along with added superior properties including high mechanical flexibility and strength. It will be high interest if a carbon material is composed with CNT aerogels which can eventually improve the storage capacity. There are reports which advocate this fact by fabricating composite materials of CNT aerogels with cellulose nanofibers [49] and mesoporous carbon [50] which successfully showed enhanced properties. Though CNT-based aerogels pose superior electrochemical properties, the high production cost hinders their application.

6. Graphene-based hydrogels and aerogels for supercapacitors

2D carbon material with one atom thick graphene has been very popular among scientific community because of its unique properties like superior thermal and electrical conductivity, appreciable flexibility and high mobility of charge carriers, very high specific surface area, mechanical and chemical stability make it potential for charge storage applications.

Graphene layers are basic skeleton for aerogels of graphene. The aerogels of graphene not only possess the merits of graphene but carry the inherited characters of aerogels. If water is replaced by air from graphene hydrogels, can result in three-dimensional cross-link, graphene aerogels. Mostly used method to prepare graphene aerogel is supercritical freeze-drying of the wet gels of graphene. There are few recent reports on the fabrication of graphene aerogels and their proven ability to behave as superior supercapacitor electrode materials. Liu and coworkers utilized freeze-drying method to prepare graphene aerogel which studied for electrochemical properties. It exhibited 172 F/g of capacitance when utilized as supercapacitor electrode in 1 M H_2SO_4 [51]. Supercritical-drying also adapted as synthetic route for graphene aerogels by Wu and Si, in a two different studies. The performance of these electrodes were fair enough which exhibited 153 F/g and 279 F/g respectively, where in corresponding electrolytes were ionic liquid and 1 M H_2SO_4 [52, 53]. Wu et al., [54] fabricated metal oxide composite with graphene aerogel which delivers a specific capacitance of 226 F/g by synergistic contribution from pseudocapacitive material, in 1 M H_2SO_4. Graphene synthesized via freeze-drying modified with L-ascorbic acid by Zhang et al., measure to be 512 m^2/g of surface area. Up on using it as supercapacitor electrodes, it exhibits 128 F/g as a full cell in 6 M KOH electrolyte [55]. Aerogels with modification of nitrogen and some atoms also attempted by scientific community with a hope that to have improved capacitance. Wu et al., were successful in doping nitrogen and Boron which eventually delivers 62 F/g at 5 mV/s in sulfuric acid-PVA medium. The material had a surface area of 249 m^2/g [56]. Carbohydrate modified graphene aerogel in neutral medium, i.e., Na_2SO_4 shows 162 F/g at 0.5 A/g [57]. However, this doped graphene had a surface area of about 365 m^2/g. Pyrolization was employed to fabricate graphene aerogel aiming to be used as electrode material for supercapacitor. Carbon modified such a graphene aerogel exhibited 122 F/g at 0.05 A/g in 6 M potassium hydroxide which was having a surface area in the range of 361–763 m^2/g [58]. He et al., designed aerogel of polypyrrole graphene with 3D hierarchical applied as supercapacitor electrodes [59]. It showed a considerably high 418 F/g at 0.5 A/g with an appreciable cyclability with 74% capacity retention in 1 M KOH. It can be noticed some literatures on rGO-based aerogel and their hybrids. In an attempt related to this, Boota's research group, utilized 2,5-dimethoxy-1,4-benoquinone and rGO to synthesize an electrode material which showed up to 650 F/g of specific capacitance at 5 mV/s in an acidic environment and interestingly 99% of initial capacity retained even after 25,000 cycles [60].

7. Biomass-derived aerogels for supercapacitors

Several merits of carbonaceous materials including low cost, easy accessibility and eco friendliness attracted appreciable interest for different applications. Especially, hydrogels and aerogels of biomass source consisting of 3D solid networks and porous structures carry excellent properties which make them being utilized as supercapacitor electrodes. Key factor to consider biomass-derived aerogels for supercapacitor application is its low production cost. Along with this, considerably high surface area unique structure in addition to greater mechanical behavior add to list.

- Carbonaceous gels were aimed by X L Wu from watermelon as a crude biomass source [61]. This showed interconnected network with an average 46 nm pore diameter. The Fe_2O_3 composite of this material showed a great electrochemical behavior with 333 F/g of capacity. In another effort, Lee and group used bacterial cellulose as carbon source to fabricate carbon fibers [62]. This nanocarbon electrode delivers 42 F/g specific capacitance and area normalized capacitance was 1617 F/cm^2. Graphene also made use to design composites with biomass-derived aerogels. Hybrid aerogels consisting of cellulose nanofibers and rGO designed by Gao et al., showed 207 F/g when used as supercapacitor electrode material [63]. CNTs also found place in the composite with aerogel derived from biomass. Cellulose nanofiber-multi walled CNT aerogels synthesized by Kang et al., which showed 178 F/g of specific capacitance [49]. Bacterial cellulose with lignin-resorcinol-formaldehyde carbon aerogel synthesized aiming towards efficient supercapacitor electrodes [64]. It performed well and showed 124 F/g at 0.5 A/g with 62.2 μF/cm^2 of aerial capacitance. Conducting polymer is frequently used to modify the aerogels to enhance the performance. Zhao and co-workers designed a 3D porous pectin/polyaniline aerogel in which functional groups of pectin such as carboxylic acid and hydroxyl groups ascribed to have hydrogen bonding with polymer leading to cross linking network [65]. This aerogel exhibits 184 F/g at 0.5 A/g and 71% of initial capacity retention. Cheng et al., prepared a cotton-derived carbon fiber aerogel and tested its electrochemical performance [66]. This carbon fiber aerogels were having 2307 m^2/g of surface area and possessed tubular morphology which facilitate conductive pathways for electron transport. This advocates shortening the ion transport lengths which eventually result in higher electrochemical performance with 283 F/g at 1 A/g and 224 F/g at 100 A/g.

8. Conclusion

In this chapter, fundamentals of supercapacitors and the utilization of carbonaceous aerogels in the fabrication of electrode materials of supercapacitors are acquainted along with their evaluations are successfully presented. The carbon aerogels sourced from polymers have rich resources, have highly tunable pores which make them to deliver high specific capacitance. But, poor mechanical stability restrict them from finding deployed in flexible supercapacitors. CNT and graphene-derived aerogels can overtime the shortcoming from having found application based on flexibility. Additionally, these aerogels exhibit properties such as excellent conductivity, superior mechanical properties, good flexibility and high surface area and stand potential candidates for electrodes for high performance electrochemical supercapacitors.

Author details

Ranganatha Sudhakar
Department of Chemistry, School of Engineering, Presidency University, Bengaluru, India

*Address all correspondence to: kamath.ranganath@gmail.com

References

[1] Conway BE. Electrochemical Supercapacitors: Scientific Fundamentals and Technological Applications. New York: Kluwer Academic/Plenum Publishers; 1999

[2] Arico AS, Bruce PG, Scrosati B, Tarascon JM, Schalkwijk WV. Nature Materials. 2005;**4**:366-377

[3] Simon P, Gogotsi Y. Nature Materials. 2008;7:845-854

[4] Hall PJ, Mirzaeian M, Fletcher SI, Sillars FB, Rennie AJR, Shitta GOB, et al. Energy. Environmental Sciences. 2010;**3**:1238-1251

[5] Wang G, Zhang L, Zhang J. Chemical Society Reviews. 2012;**41**:797-827

[6] Hou J, Shao Y, Ellis WM, Obert B, Moore, Yi B. Physical Chemistry Chemical Physics. 2012;**13**:15384-15402

[7] Patil S, Raut S, Gore R, Sankapal B. New Journal of Chemistry. 2015;**39**:9124-9131

[8] Ma X, Zhang W, Kong L, Luo Y, Kang L. New Journal of Chemistry. 2015;**39**:6207-6215

[9] Hu CC, Chang KH, Lin MC, Wu YT. Nano Letters. 2006;**6**:2690-2695

[10] Pekala RW, Kong FM. Polymer Preprints. 1989;**30**:221-223

[11] Wu D, Fu R, Zhang S, Dresselhaus MS, Dresselhaus G. Carbon. 2004;**42**:2033-2039

[12] Pekala RW. Journal of Materials Science. 1989;**24**:3221-3227

[13] Long D, Zhang J, Yang J, Hu Z, Cheng G, Liu X, et al. Carbon. 2008;**46**:1253-1269

[14] Katanyoota P, Chaisuwan T, Wongchaisuwat A, Wongkasemjit S. Mater. Sci. Eng. B. 2010;**167**:36-42

[15] Li W, Reichenauer G, Fricke J. Carbon. 2002;**40**:2955-2959

[16] Moreno-Castilla C, Dawidziuk MB, Carrasco-Mar'ın F, Zapata-Benabithe Z. Carbon. 2011;**49**:3808-3819

[17] Zhu Y, Hu H, Li WC, Zhang X. Journal of Power Sources. 2006;**162**:738-742

[18] Lin YH, Wei TY, Chien HC, Lu SY. Advanced Energy Materials. 2011;**1**:901-907

[19] Yamashita J, Ojima T, Shioya M, Hatori H, Yamada Y. Carbon. 2003;**41**:285-294

[20] Halama A, Szubzda B, Pasciak G. Electrochimica Acta. 2010;**55**:7501-7505

[21] Feng J, Zhang C, Feng J, Jiang Y, Zhao N. ACS Applied Materials & Interfaces. 2011;**3**:4796-4803

[22] Lee YJ, Jung JC, Yi J, Baeck SH, Yoon JR, Song IK. Current Applied Physics. 2010;**10**:682-686

[23] Tamon H, Ishizaka H, Yamamoto T, Suzuki T. Carbon. 1999;**37**:2049-2055

[24] Pekala RW, Mayer ST, Poco JF, Kaschmitier JL. Materials Research Society Symposium Proceedings. 1994;**349**:79-85

[25] Li J, Wang X, Wang Y, Huang Q, Dai C, Gamboa S, et al. Journal of Non-Crystalline Solids. 2008;**354**:19-24

[26] Chien HC, Cheng WY, Wang YH, Lu SY. Advanced Functional Materials. 2012;**22**:5038-5043

[27] Li GR, Feng ZP, Ou YN, Wu D, Fu R, Tong YX. Langmuir. 2010;**26**:2209-2213

[28] An H, Wang Y, Wang X, Li N, Zheng L. Journal of Solid State Electrochemistry. 2010;**14**:651-657

[29] Wang X, Wang X, Yi L, Liu L, Dai Y, Wu H. Journal of Power Sources. 2013;**224**:317-323

[30] Lv G, Wu D, Fu R, Zhang Z, Su Z. Journal of Non-Crystalline Solids. 2008;**354**:4567-4571

[31] Liu D, Shen J, Liu N, Yang H, Du A. Electrochimica Acta. 2013;**89**:571-576

[32] Kalpana D, Omkumar KS, Kumar SS, Renganathan NG. Electrochimica Acta. 2006;**52**:1309-1315

[33] Sassin MB, Mansour AN, Pettigrew KA, Rolison DR, Long JW. ACS Nano. 2010;**4**:4505-4514

[34] Hwang SW, Hyun SH. Journal of Power Sources. 2007;**172**:451-459

[35] Miller JM, Dunn B. Langmuir. 1999;**15**:799-806

[36] Bordjiba T, Mohamedi M, Dao LH. Advanced Materials. 2008;**20**:815-819

[37] Bryning MB, Milkie DE, Islam MF, Hough LA, Kikkawa JM, Yodh AG. Advanced Materials. 2007;**19**:661-664

[38] Zou J, Liu J, Karakoti AS, Kumar A, Joung D, Li Q, et al. ACS Nano. 2010;**4**:7293-7302

[39] Kim H, Oh Y, Islam MF. Advanced Functional Materials. 2013;**23**:377-383

[40] Kohlmeyer RR, Lor M, Deng J, Liu H, Chen J. Carbon. 2011;**49**:2352-2361

[41] Schiffres SN, Kim KH, Hu L, McGaughey AJH, Islam MF, Malen JA. Advanced Functional Materials. 2012;**22**:5251-5258

[42] Zhong J, Yang Z, Mukherjee R, Thomas AV, Zhu K, Sun P, et al. Nano Energy. 2013;**2**:1025-1030

[43] Fang Y, Jiang F, Liu H, Wu X, Lu Y. RSC Advances. 2012;**2**:6562-6569

[44] Bordjiba T, Mohamedi M. Journal of Solid State Electrochemistry. 2011;**15**:765-771

[45] Qi H, Mäder E, Liu J. Journal of Materials Chemistry A. 2013;**1**:9714-9720

[46] Sun H, Xu Z, Gao C. Advanced Materials. 2013;**25**:2554-2560

[47] Li P, Kong C, Shang Y, Shi E, Yu Y, Qian W, et al. Nanoscale. 2013;**5**:8472-8479

[48] Lee JA, Shin MK, Kim SH, Kim SJ, Spinks GM, Wallace GG, et al. ACS Nano. 2012;**6**:327-334

[49] Gao K, Shao Z, Wang X, Zhang Y, Wang W, Wang F. RSC Advances. 2013;**3**:15058-15064

[50] Tao T, Zhang L, Jiang H, Li C. New Journal of Chemistry. 2013;**37**:1294-1297

[51] Liu F, Song S, Xue D, Zhang H. Advanced Materials. 2012;**24**:1089-1094

[52] Wu X, Zhou J, Xing W, Wang G, Cui H, Zhuo S, et al. Journal of Materials Chemistry. 2012;**22**:23186-23193

[53] Si W, Wu X, Zhou J, Guo F, Zhuo S, Cui H, et al. Nanoscale Research Letters. 2013;**8**:247

[54] Wu ZS, Sun Y, Tan YZ, Yang S, Feng X, Müllen K. Journal of the American Chemical Society. 2012;**134**:19532-19535

[55] Zhang X, Sui Z, Xu B, Yue S, Luo Y, Zhan W, et al. Journal of Materials Chemistry. 2011;**21**:6494-6497

[56] Wu ZS, Winter A, Chen L, Sun Y, Turchanin A, Feng X, et al. Advanced Materials. 2012;**24**:5130-5135

[57] Ji CC, Xu MW, Bao SJ, Cai CJ, Lu ZJ, Chai H, et al. Journal of Colloid and Interface Science. 2013;**407**:416-424

[58] Meng F, Zhang X, Xu B, Yue S, Guo H, Luo Y. Journal of Materials Chemistry. 2011;**21**:18537-18539

[59] He Y, Bai Y, Yang X, Zhang J, Kang L, Xu H, et al. Power Sources. 2016;**317**:10-18

[60] Boota M, Chen C, Bécuwe M, Miao L, Gogotsi Y. Energy & Environmental Science. 2016;**9**:2586-2594

[61] Wu X-L, Xu A-W. Journal of Materials Chemistry A. 2014;**2**:4852-4864

[62] Lee KY, Qian H, Tay FH, Blaker JJ, Kazarian SG, Bismarck A. Journal of Materials Science. 2013;**48**:367-376

[63] Gao K, Shao Z, Li J, Wang X, Peng X, Wang W, et al. Journal of Materials Chemistry A. 2013;**1**:63-67

[64] Xu X, Zhou J, Nagaraju DH, Jiang L, Marinov VR, Lubineau G. Advanced Functional Materials. 2015;**25**:3193-3202

[65] Nyholm L, Nyström G, Mihranyan A, Strømme M. Advanced Materials. 2011;**23**:3751-3769

[66] Cheng P, Li T, Yu H, Zhi L, Liu Z, Lei Z. Journal of Physical Chemistry C. 2016;**120**:2079-2086

Chapter 6

Structure and Dynamics of Aqueous Dispersions

Tatiana Yakhno and Vladimir Yakhno

Abstract

The content of the chapter summarizes the long-term studies of the authors of the structure and dynamics of aqueous dispersions, including model protein-salt solutions, blood serum of healthy and sick people, food microdisperse systems. The data are considered in the context of the existing scientific literature on the stated problem. One of the important issues is the phase transitions of the protein when the ionic strength of the solution changes. To observe these processes, a drying droplet model, optical and atomic force microscopy are used. Based on the observations, a model of a cascade of phase transitions of a protein, from micelles to gel, was created. Parallels are drawn with the violation of these processes based on the results of the study of blood serum in seriously ill people. It is also shown that in the free volume of microdisperse systems there are near-hour oscillatory processes associated with the aggregation - disaggregation of the microdispersed phase. The surface tension of solutions, the mechanical properties of liquids, and the morphological characteristics of dried drops also fluctuate simultaneously. A model of self-oscillatory processes in such systems is presented, consistently describing the observed phenomena.

Keywords: aqueous microdisperse systems, drying drops model, protein phase transitions, ionic strength, free and bound water, nearhourly fluctuations

1. Introduction

The entire world around us, including the "starry sky above us," is an innumerable set of different-scale dispersed systems. The main feature of such systems is a large specific interfacial area, which provides them with excess energy. It is this excess energy that is the driving force that determines the structure and dynamics of dispersed systems. A dispersed system usually consists of two or more phases that practically do not mix and do not chemically react with each other. In the typical case of a two-phase system, the first of the substances (dispersed phase) is distributed in the second (dispersion medium). The result of the physical interaction of the lyophilic elements of the dispersed phase with the liquid dispersion medium is the formation of solvation shells around the dispersed particles, which is accompanied by a decrease in interfacial tension and an increase in the aggregate stability of such systems. In the case of aqueous dispersions, as a result of the physical interaction of hydrophilic particles with the environment, the particles acquire hydration shells from ordered layers of water molecules with a special spatial structure that differs from the structure of the bulk phase of water. In this paper, we will focus on the microstructure of aqueous solutions and microheterogeneous dispersions, as well as spontaneous and induced dynamic processes in them.

2. Processes in sessile drying drops of aqueous dispersions

An evaporating droplet sitting on a solid wettable substrate is a convenient model for observing phase transformations in aqueous colloidal dispersions. The shape of a sitting drop in the form of a truncated ball ensures the appearance of temperature gradients on its surface and the development of flows of thermocapillary nature: centrifugal flow and Marangoni flow (**Figure 1**).

Since the thinnest layer of liquid is located at the three-phase boundary, solid-state microdeposits are formed there first of all, ensuring the attachment of the droplet to the substrate. Further evaporation is accompanied by flattening of the droplet dome while maintaining its area. In this case, capillary forces arise, which ensure the predominant evaporation of water through the edges of the drop [1–5]. It is shown that the direction of the flows depends on the ratio of the thermal conductivities of the liquid and the substrate material [6]: if this ratio does not exceed 1.57, then the Marangoni flow promotes the removal of the colloidal phase to the edge of the droplet. If it exceeds, then the direction of the flow changes sign, and the colloidal phase is "swept away" to the center [7]. The ratio of thermal conductivity of glass and water is ~0.6. Therefore, when a drop of hydrocolloid dries up, the remaining solid sediment on the glass has the shape of a saucer with a volumetric rim along the edge. The same phenomenon is associated with the "effect of a coffee drop" - the formation of a colored rim along the periphery of the drop (**Figure 2**).

More complex and interesting processes take place in drying droplets if the water-colloidal system contains salt (for example, human biological fluids). **Figure 3** shows fragments of the drying process of a drop of a protein - salt solution, in which the ratio of total protein to salt corresponds to that in human serum (7% in BSA in 0.9% NaCl). At the beginning of drying (**Figure 3a–c**) the visible dynamics of the process is the same as in the salt-free solution. However, further (**Figure 3d–f**), salt crystallization begins in the center of the drop, and the final picture of the drop acquires a specific image [8, 9].

Let us take a closer look at the structure of the light circle that appears on the inner side of the protein roller (**Figure 3f**). In **Figure 4** it can be seen that this ring is composed of individual micron-sized protein aggregates. Closer to the center of the drop, the layer of protein structures transforms into a homogeneous layer of the protein gel, inside which salt crystallizes.

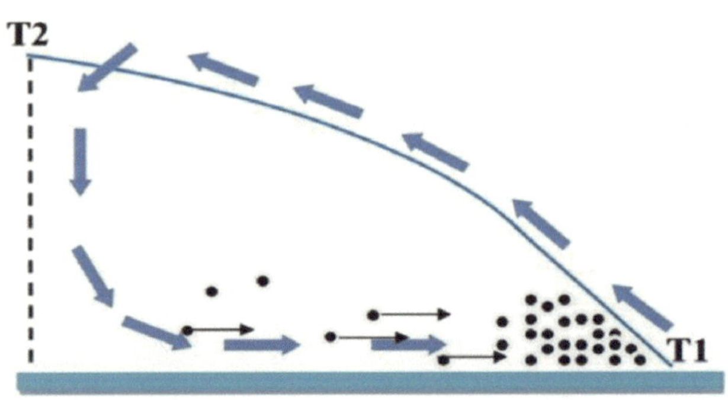

Figure 1.
Scheme of flows in a drying drop of an aqueous colloidal solution sitting on a glass substrate (a half of the 2D image). Capillary flow (black arrows) and Marangoni flow (blue arrows) promote the removal of the colloidal phase (black balls) to the periphery of the drop. T1 > T2.

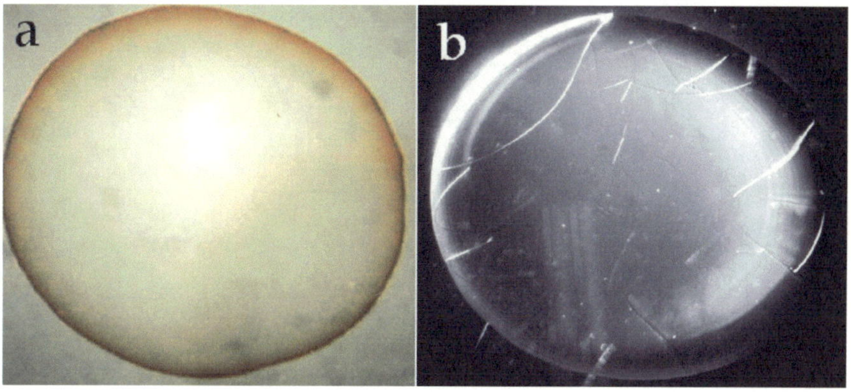

Figure 2.
Dried drops (5 μl volume) of an aqueous solution of coffee (a) and a 7% solution of bovine serum albumin (BSA) in distilled water (b). Formation of the roller of the colloidal phase along the droplet periphery.

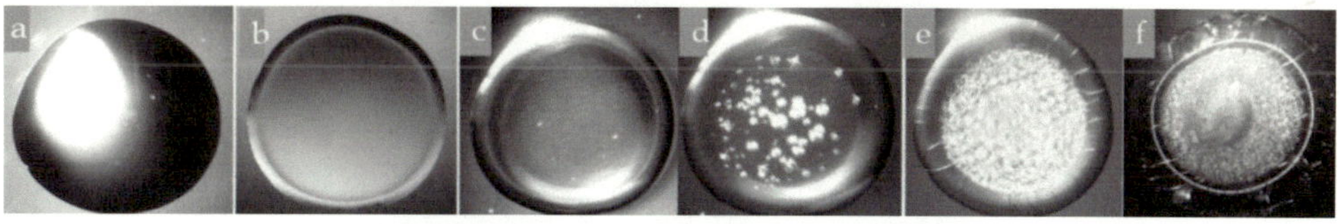

Figure 3.
Consecutive fragments of the drying process of drop of 7% BSA in a 0.9% NaCl physiological solution: (a-c) - redistribution of the colloidal phase and the formation of a protein roller; (d-f) - the process of salt crystallization in the protein matrix and the formation of the final image of the drop (f).

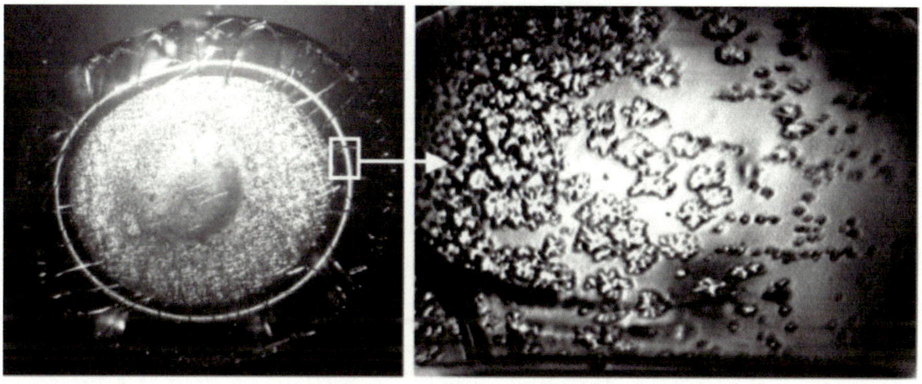

Figure 4.
Dried drop of protein-salt water solution (left) ×10. Light-diffusing circle is a place of protein structures formation. Right picture shows protein structures evolution from separated precipitates (right) to protein clusters that transform into gel. Magnification: ×280 [10].

With the help of physical modeling, it was found out how the ratios of the components in a protein-salt solution, drying in the form of a drop on a glass slide, change (**Figure 5**).

The protein roll formation time for a 3 μL drop is 4-6 minutes. During this time, the liquid part of the droplet loses ~30% of water (evaporates) and 70% of albumin (is carried out to the periphery of the droplet and becomes solid). As a result, the initial ratio of the components in the remaining solution changes, causing coacervation of albumin. The mechanism of these events is associated with the competition for hydration between colloidal particles and salt ions. The concentration of salt

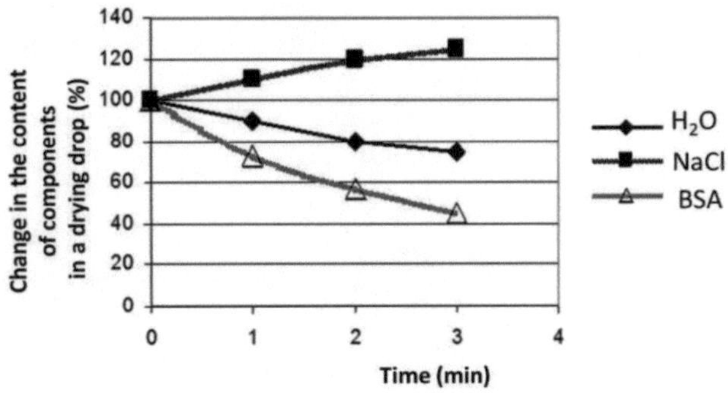

Figure 5.
Change in the relative content of components in the protein-salt solution when it dries in the form of a drop on a glass substrate (according to materials in [10]).

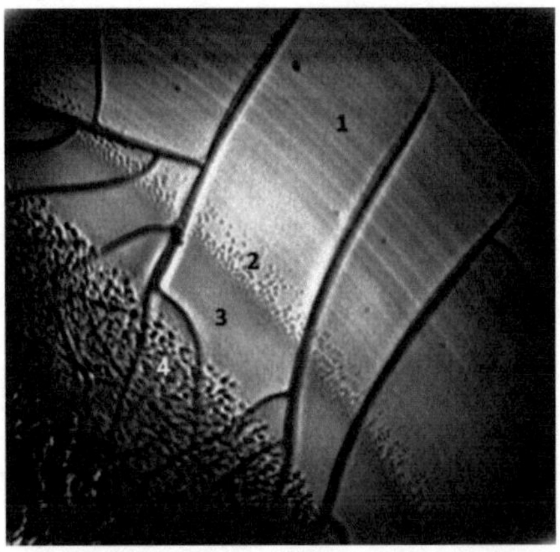

Figure 6.
Zones in dried drop of BSA-salt solution: 1 – homogenous protein film (colloidal glass); 2– zone of protein precipitates, from single ones to their clusters; 3 – gel; 4 - zone of salt structures in shrinking protein gel. Magnification: × 70.

per unit volume of protein increased greatly, which led to the loss of its aggregate stability and the beginning of the coagulation process. **Figure 6** shows the zones of different phase states of the protein, formed in the gradient of increasing salt concentration.

The cascade of protein phase transitions, according to the authors, can be represented as follows (**Figure** 7).

To examine the bottom adsorption layer in a dried drop of 7% BSA solution in saline NaCl, the top of the dried protein roll was carefully removed with a scalpel, as shown in **Figure 8**.

Investigation of the lower adsorption layer of a dried drop of a protein-salt solution using an atomic force microscope in zone 2 in **Figure 6** showed the presence of protein precipitates corresponding to the structures of the second order in **Figure** 7. Investigation of the lower adsorption layer of a dried drop of a protein-salt solution using an atomic force microscope in zone 2 in **Figure 6** showed the presence of protein precipitates corresponding to the structures of the second order in **Figure** 7 (**Figure 9**).

At a lower concentration of protein in physiological saline solution, the coacervation process begins earlier, which confirms the author's opinion about the nature of the observed phenomenon (**Figure 10**).

Thus, due to protein redistribution during drop drying, protein deposits on the drop edge and protein in the middle part of the drop are in different conditions, and

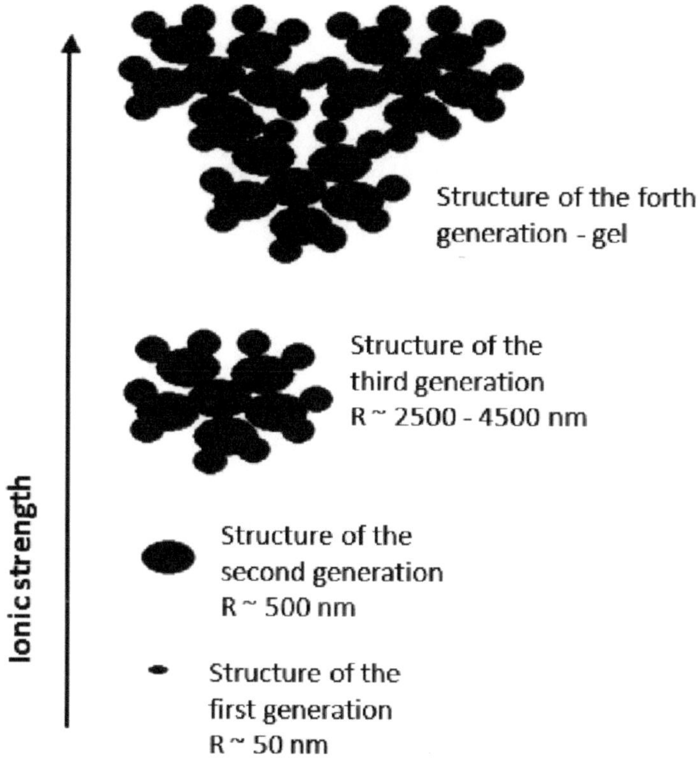

Figure 7.
Protein phase transitions in fluid part of a sessile drying drop of protein-salt-water solution. R is a radius of the structure [10].

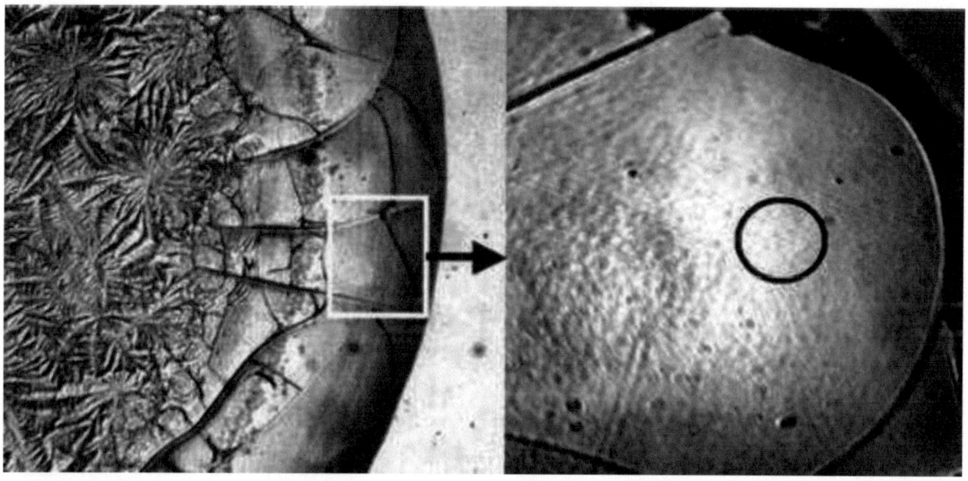

Figure 8.
Fragments of dried drops of 7% BSA in 0.9% NaCl water solution. Left – zone of protein ring (in a white rectangle) before removing; right – the same zone after removing the upper film. Black circle shows the area of bottom protein adsorption layer is used for AFM investigation.

form materials with different properties. The authors [11] argue that colloidal particles can form different structures: from colloidal glasses with very high volume fractions and low strength of interparticle attraction to colloidal gels with very low volume fractions and strong attraction between the particles (**Figure 11**). Before gelation, colloidal particles form fractal clusters, which turn into space-filling

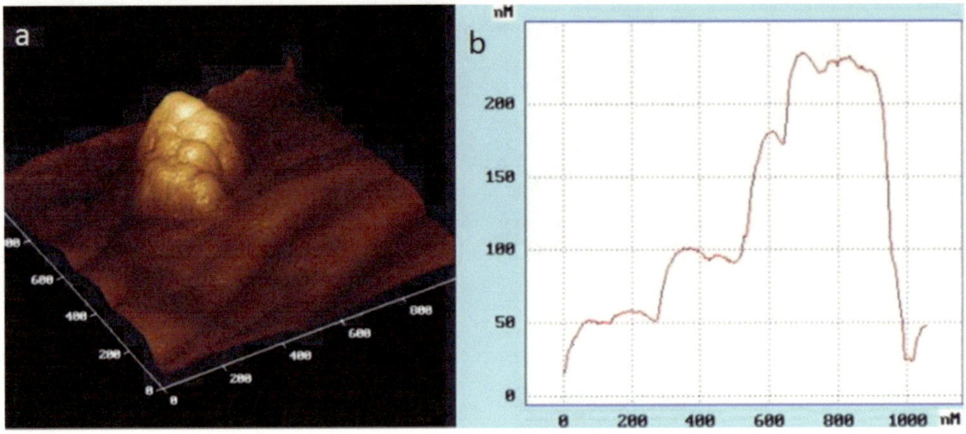

Figure 9.
AFM data: single protein precipitate (structure of the second generation) lying on the protein film in dried drop of BSA-salt – water solution. It consists of some subunits (structures of the first generation), which admittedly represents consolidated micells. (a) three-dimensional image; (b) precipitate profile.

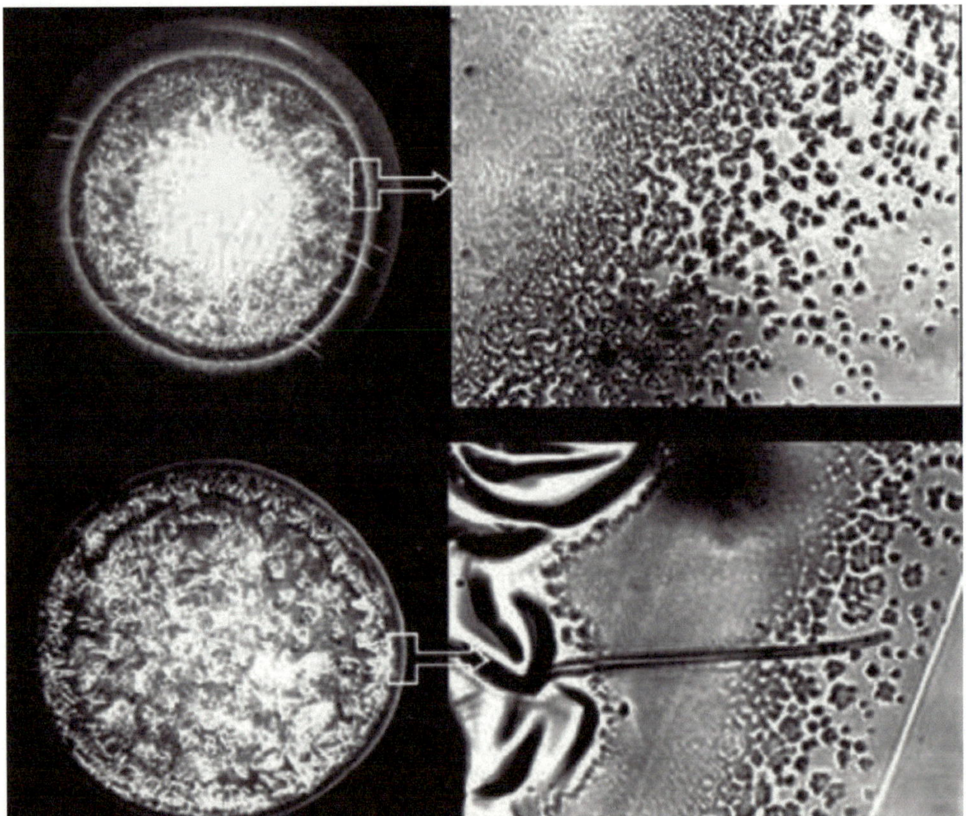

Figure 10.
Dried drops of protein-salt solutions: 7% BSA in 0.9% NaCl (above), and 2.5% BSA in 1.8% NaCl (bottom). Light-diffusing ring of protein structures has different positions (see the text). Magnification: left – ×10; right - ×70.

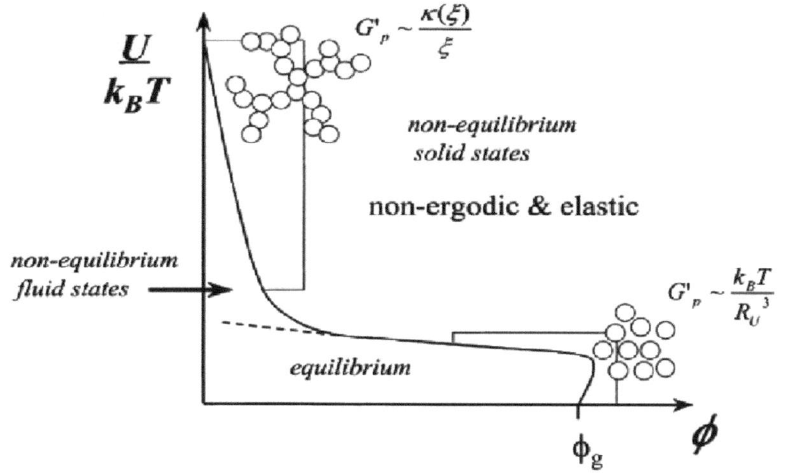

Figure 11.
Schematic state diagram of colloidal particles with short-range potentials, after V. Trappe, R. Sandkuhler [11]. Φ – volume fraction of colloid phase; U – strength of the interparticle attraction.

networks. Current investigations show that a drying drop of protein–salt aqueous solution is an excellent illustration of this dynamics. Taking into account hydrodynamic motion of the colloidal phase to the drop periphery and its rapid consolidation there, we suppose that this solid phase really represents the protein glass transition: it is transparent and extremely fragile. In contrast to the drop periphery, low protein volume fraction and high ionic strength in liquid residuals in the middle part of a drying drop stimulate liquid–liquid separation and further cascade of protein phase transitions leading to gel formation. Thus, protein gel probably forms only inside the protein glass ring of a drying drop.

More detailed information on the processes in drying drops of protein-salt solutions can be obtained in publications [12–22].

When working with biological fluids of healthy and sick people, it was noticed that in the case of severe diseases, the processes of protein structuring are disrupted, which is revealed in experiments with drying drops [23, 24]. **Figure 12** shows dried drops of blood serum of women after delivery at term and premature birth. Noteworthy is the significant expansion of the zone of protein structures and the formation of larger precipitates.

It is surprising that in some cases, in seriously ill people, regardless of the nature of the disease, micron-sized protein precipitates can be observed already in liquid blood serum (**Figure 13**).

The magnitude of the osmotic pressure created by the solution depends on the amount, and not on the chemical nature of the substances dissolved in it (or ions, if the molecules of the substance dissociate), that is, the osmotic pressure is a colligative property of the solution. The higher the concentration of a substance in a solution, the greater the osmotic pressure it creates. The volume and mass of a colloidal particle is much larger than the volume and mass of a molecule of low-molecular substances. At the same mass concentration of a substance, a unit volume of a sol contains significantly fewer particles than a unit volume of a true solution. Therefore, it is generally accepted that the osmotic pressure of colloidal solutions is negligible compared to that in true solutions. However, there is an opinion that a relatively small number of particles, much less than required by the “usual” colligation law, can create a high osmotic pressure if they have extensive hydrophilic surfaces [25]. Since, in addition to salts, blood serum contains other osmotically

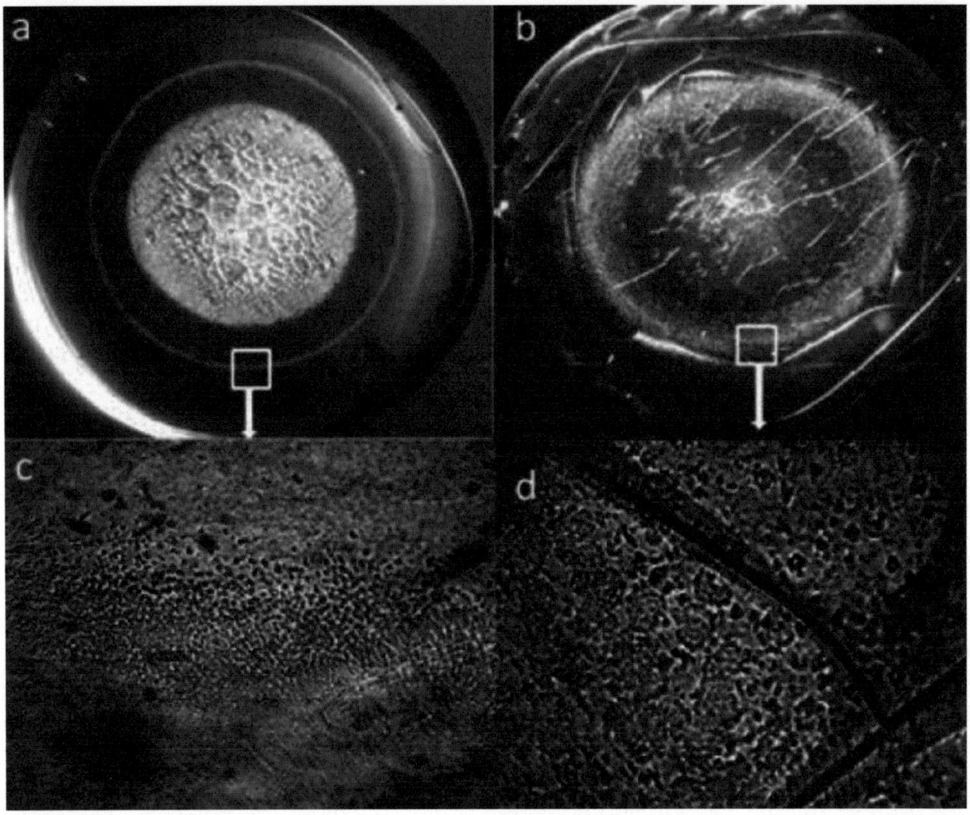

Figure 12.
Dried drops of serum of women in early afterbirth period: (a) is in-time birth (40 weeks); (b) is premature birth (34 weeks). Narrows show ring of micelles; (c) and (d) are selected by tetragon regions in more enlargement. Magnification: 15 × 40.

active components, it is customary in clinical practice to operate with the concept of "osmolarity," meaning the sum of the concentrations of cations, anions and non-electrolytes, that is, all kinetically active particles in 1 liter of plasma or serum. The osmolarity of biological fluids is a fairly strict indicator of homeostasis. So, the osmolarity of blood plasma in normal conditions can vary in the range of 280-300 mosm/l [26] (1).

$$\text{Osmolarity} = 195.1 + 0.74 \times \text{sodium} + 0.25 \times \text{urea nitrogen} + 0.03 \times \text{glucose} \quad (1)$$

where 195.1 is a free member; 0.74; 0.25; 0.03 - empirically found coefficients in the equation; sodium - in mmol/l, urea nitrogen and glucose - in mg%.

Calculations showed that in the examined patients with burn disease, the plasma osmolarity averaged 301.6 ± 6.56 mosm/l, fluctuating within the range of 286.16-320.01 mosm/l. However, in patients, the content of both total protein and albumin was decreased. That is, against the background of normal ionic strength of the solution, there was an average decrease in the mass fraction of albumin [27]. Violation of the protein-electrolyte balance led to coacervation of albumin in the liquid phase of the serum. In the course of successful treatment of the underlying disease, the disturbance of this balance began to decline, and coacervation in the liquid serum was not observed.

In this brief review, the phase transformations of protein in droplets of protein-salt solutions drying on glass were examined, and analyzed the cause-and-effect

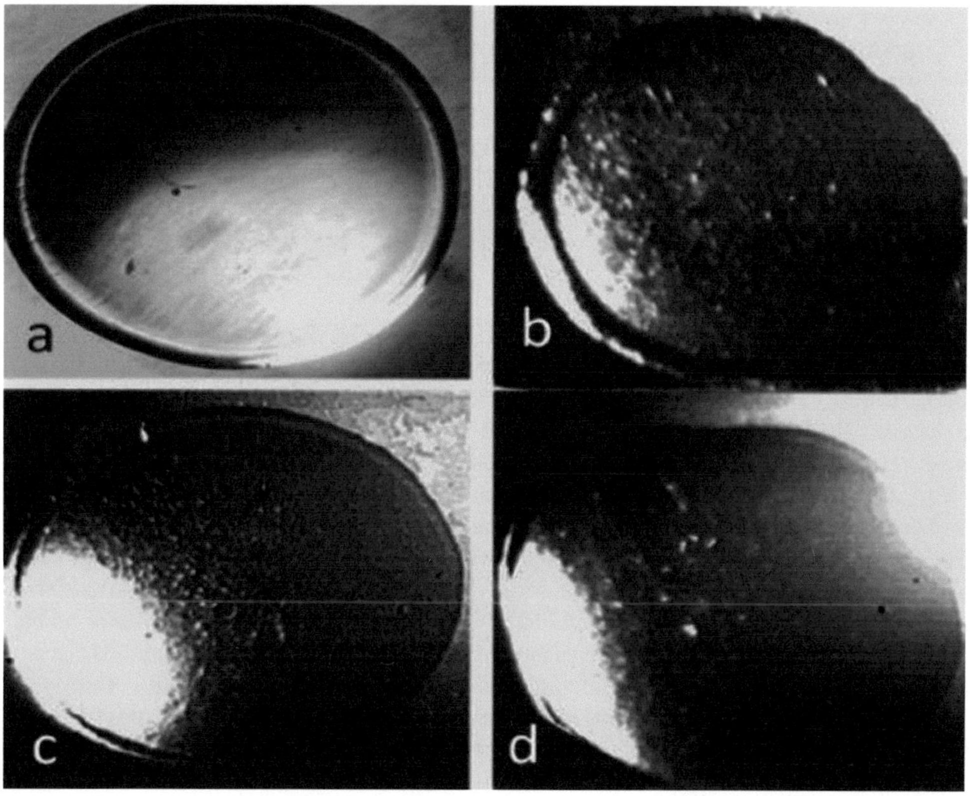

Figure 13.
The initial drying process of blood serum drops is the formation of a solid rim along the periphery of the drops: (a) - norm, (b) - chronic hepatitis B + C; (c) - burn disease; (d) - coxarthrosis. In the liquid phase of the patient's serum, micron-sized protein precipitates are visible. The initial volume of each drop is 3 μl.

relationships of these transformations. Also similar processes occurred in real human biological fluids was found. In this regard, it would like to note the amazing results of the successful treatment of hundreds of cases of serious diseases of a different nature, obtained by the Soviet doctor A.S. Samokhotskiy, which were carried out according to the method developed by him [28, 29]. Here is the conclusion of one of his articles [30]:

1. The concentration of electrolytes (sodium, potassium, calcium, magnesium) in the blood serum changes in a wide variety of diseases, but the ratios of these electrolytes can be similar in different diseases and different in the same disease in different people, as well as in one person for different stages of the painful process.

2. The use of medicinal compositions containing electrolytes, the concentration of which in the blood serum is relatively low, naturally increased their content and improved the patient's condition.

3. The use of medicinal compositions containing electrolytes, the concentration of which in the blood serum is relatively high, naturally increased their content (increased the gap in the ratios) and worsened the patient's condition. The deterioration usually was significant.

4. Very small doses of these elements are useful to normalize the ratios of sodium, potassium, calcium and magnesium in the blood serum and improve the patient's condition.

Unfortunately, after the death of the author in 1986, his work was not continued. Current investigation shows that continued research in this direction is very promising.

2.1 Materials and methods used when working with drying drops

7% w bovine serum albumin solution (BSA, 68 kDa, Sigma, USA) in distilled water or in physiological salt solution (0.15 M NaCl, chemically pure, "Reactiv, Inc.," Russia) were used. All solutions were prepared without buffering, a day prior to experimentation, refrigerated overnight and allowed to come to room temperature before testing. The samples under study were placed, using micropipette, onto clean glasses in the form of drops of volume of 3 μl (6-8 drops for each sample), and let for drying at room conditions. Morphological observations were carried out during drying, and 2-3 days after placing on the glasses, using LUMAM-I-3 microscope and video camera – computer setup. Dried drops also were investigated by means of atom force microscope (AFM) "Smena" NT-MDT (Russia), Russia, using a sensor CSG11. Samples of blood plasma and serum were obtained from 30 clinically healthy donors (the material supplied by Hemotransfusion Station, Nizhny Novgorod); 18 patients with viral hepatitis B and C in acute stage (the material supplied by the Hepatological Center, Nizhny Novgorod); 30 patients with burn disease, and 8 patients with diseases of articulations of inflammatory and degenerative character (supplied by the Federal Burn Treatment Center, Nizhny Novgorod Research Institute of Traumatology and Orthopedics); 40 women after normal or premature (second- and third-trimester) childbirth (supplied by the maternity and child-welfare services of Nizhny Novgorod).

3. Structure and dynamics of water microdispersed systems

Due to the thermodynamic instability of colloidal solutions, aggregation and disaggregation processes continuously occur in them, leading to a change in the number of osmotically active particles per unit volume, and, consequently, in the osmotic pressure. With an increase in the average radius of the particles of the system, as a result of their coagulation and the formation of aggregates, the osmotic pressure should drop very strongly. On the contrary, with the disintegration of aggregates into primary particles, the osmotic pressure should increase strongly. Since the phenomena of aggregation and disaggregation in colloidal systems very easily occur under the influence of sometimes even very weak external influences, the variability of the osmotic pressure of lyosols and their dependence on the prehistory of the solution becomes clear [31]. In recent years, evidence has appeared in the literature about the inhomogeneity of water and aqueous solutions at the micro level. For the first time, as far as we know, giant (millimeter-sized) clusters in a thin layer of water were detected using IR spectroscopy [32]. An assumption was made about their liquid crystal nature. As a result of the study of solutions of NaCl, citric acid, glucose, urea, vinegar and ethanol using static and dynamic light scattering, it was concluded that the dissolved substances in liquid media are distributed unevenly: areas with low and high concentration provide a contrast in light scattering during experimental observation. There are separate domains, close to spherical in shape, with a high density relative to the surrounding fluid. Their size can reach hundreds of nanometers [33]. Investigation of suspensions of fluorescent polystyrene microparticles (d = 1 μm, C = 0.2% in highly purified water) using a confocal laser scanning microscope for several hours allowed the authors to observe the appearance and growth of "voids" inside the colloidal phase [34]. According to the authors, the reason for this is the attraction initiated by counterions between like-charged particles [35]. The ability to move a particle forcibly placed in the resulting voids was severely limited in comparison with particles located in adjacent areas with a high packing density [36]. The addition of salt to the solution reduced the distance between the colloidal particles, but after reaching a certain limit, the "colloidal crystal" melted.

The results of experiments on determining the size distribution of optical inhomogeneities (clusters) in bidistilled water by the method of small-angle light scattering are presented [37]. The measurements showed the presence of a spectrum of cluster sizes in the water in the range (1.5-6.0) μm. With the help of laser interferometry, the formation of supramolecular water complexes with linear dimensions of 30–100 μm, distributed in continuous water, was shown [38, 39]. A critical review of modern water purification methods [40] states that water is easily contaminated with chemicals, gases, vapors and ions that are washed out of pipelines and containers. These can include sodium and silica from glass, plasticizers and ions from pipes, microbial particles and their endotoxins, and contaminants. Soluble organic contaminants can even be introduced from deionizing resins used during processing, especially if inadequate resins are selected or the resins have previously been contaminated. No cleaning method is perfect.

In our previous works it was also shown that water and aqueous solutions are microdispersed systems [41, 42]. Upon evaporation of free water, structures ranging in size from ten to hundreds of micrometers remain on the substrate, which are aggregates of a microdispersed phase (**Figure 14**). The aggregates do not evaporate at room temperature, have a viscous consistency and "melt" when the osmotic pressure rises. The unit of the microdispersed phase is NaCl microcrystals surrounded by a thick hydration shell. The water of hydration shells evaporates at a temperature of > 200° C and accounts for ~20% of the dry sediment mass [43].

The hydration shells of hydrophilic microparticles are denser liquid crystalline water, which forms a zone around the particle, displacing all impurities, including ions, from its volume — Exclusion Zone (EZ) [45]. The microheterogeneous structure of water was investigated using a conventional light microscope in the

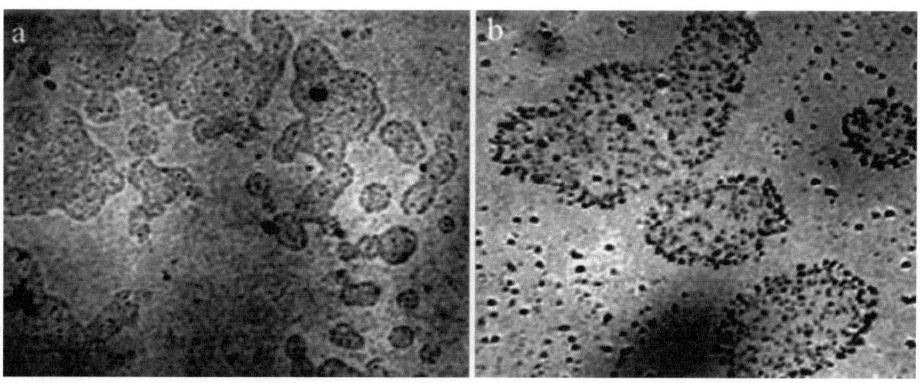

Figure 14.
Fragments of microstructure aggregates: (a) - dry white wine, in a thin layer of liquid (8 μm), frame width - 2.4 mm; (b) in the precipitate of a NaCl solution on a substrate after evaporation of free water, the frame width is 1 mm [44].

preparation between the slide and cover glass (layer thickness ~ 8 μm), as well as in a drop of water placed in a hole in a plastic plate 0.5 mm in diameter (**Figure 15**).

The dry residue mass after evaporation of free water from these liquids was 0.25%, 0.48% and 2.5% of the initial mass, respectively.

Now, when it became known that water and aqueous solutions are not homogeneous media, but are microheterogeneous dispersions with their characteristic dynamic processes, facts that previously did not have an adequate explanation become clear. For example, oscillatory processes in liquids revealed by different physical methods of analysis: determination of enzyme activity [46–48]; dynamic light scattering [49, 50], IR spectroscopy, Raman spectroscopy, UHF radiometry and NMR [51]. Continuous multi-hour studies of autonomous oscillatory processes in a number of beverages (tea, dry red wine [52], instant freeze-dried coffee [53, 54]), were conducted registering several parameters simultaneously: the dynamics of the complex mechanical properties of drying drops of these liquids, the dynamics of the surface tension of the solution, and the width of the edge roller for drops dried on glass. For periodic registration of complex mechanical characteristics of drying drops, a method developed by us earlier was used.

3.1 Methods used for studying dynamical processes in water microdispersed systems

3.1.1 Drying drop technology (DDT)

To monitor fluctuations of physicochemical properties of colloidal systems the DDT method was used based on acoustical impedancemetry developed in our laboratory earlier [55, 56]. Here only its main features would be explained. A coffee drop (volume of 3 μl) without any pretreatment dries on a polished end of a quartz plate. The quartz oscillates with a constant frequency of 60 kHz, which is equal to the resonance frequency of unloaded resonator. Acoustical – Mechanical Impedance (AMI) of the drop during drying is displayed as a curve on a screen (**Figure 16**). The parameter BS_2 reflects the dynamics of complex mechanical properties of the drying drop deposit (mechanical stress) and is calculated automatically by the software. In the same environment this parameter depends strongly on liquid composition and structure. The diagrams were built for temporal fluctuations of BS_2 parameter using Excel.

3.1.2 Optical investigation of coffee ring width fluctuations

The experiments were carried out using water solutions of the Nescafe Gold sublimated instant coffee bought in a store, with a concentration of 2.50 g/100 ml,

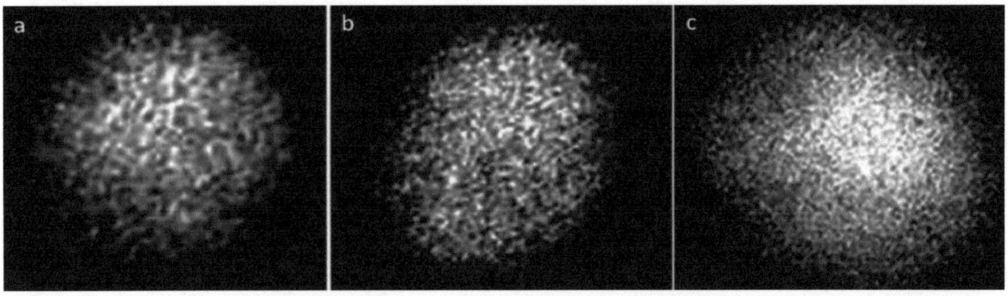

Figure 15.
Microstructure of liquids in the "hanging drop" preparation. Liquid placed in a hole in a plastic plate with a diameter of 0.5 mm: (a) - distilled water; (b) - tap water; (c) - water from the Black Sea. The drops were looked via microscope right through.

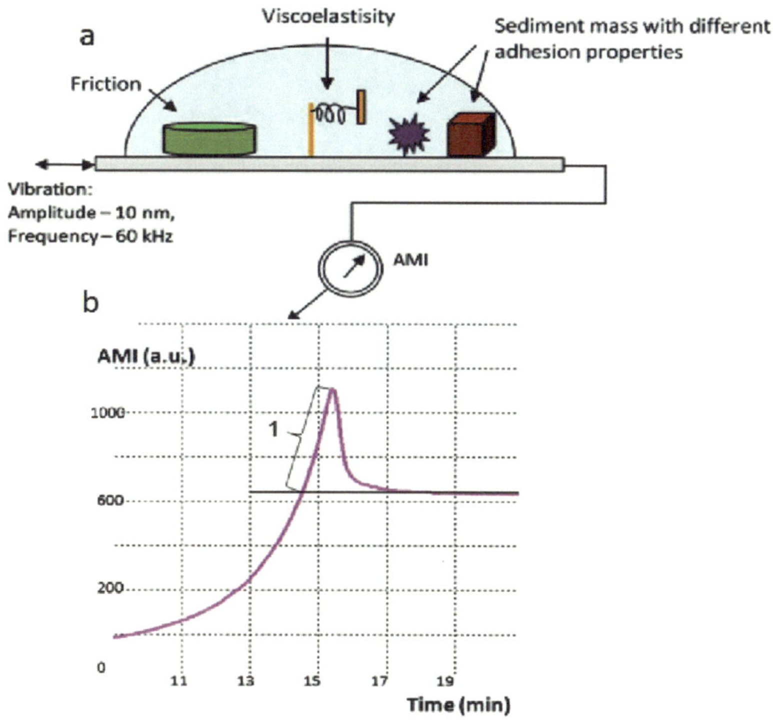

Figure 16.
DDT scheme: (a) – a drop as an object having a set of unique physical properties is drying on the surface of oscillating quartz resonator; (b) – typical AMI curve for drying drop of coffee: 1 – portion of the curve measured by total derivative (parameter BS_2).

at T = 22-23°C and H = 64-65%. A dry coffee sample was placed in a chemical glass, filled in with hot tap boiled water, and mixed by a glass stick until the coffee dissolved. Sampling was begun after cooling of solution to room temperature, at 9 o'clock Moscow time. Sampling was made each 30 minutes from the same glass of coffee solution standing on a table, using the zone equidistant from the center and edges of the glass, from the depth of about 2 cm by a microdispenser with removable tips. Such 30 minute intervals were stipulated by the duration of one test (20 min) and quartz treatment procedure. In some experiments we added to the colloidal system surfactant sodium dodecyl sulfate (SDS) with a concentration of 0.2% w. Ten repeated experiments were made for every type of investigation. The tests were carried out simultaneously with BS_2 parameter measurements. For every 30-minutecounting we took four drops having a volume of 3 μl: one drop for BS_2 parameter measurement, and 3 drops for coffee ring width measurements. Those 3 drops were placed on a new (without any treatment) microscope slides ApexLab, 7 countings of each slide (**Figure 17**), up to 22 countings.

The preparations were drying in horizontal position under room conditions, and were investigated the next day. Coffee ring width was measured using the

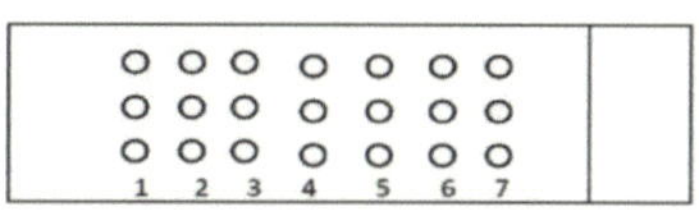

Figure 17.
Arrangement of drops on a glass for drying and microscopy.

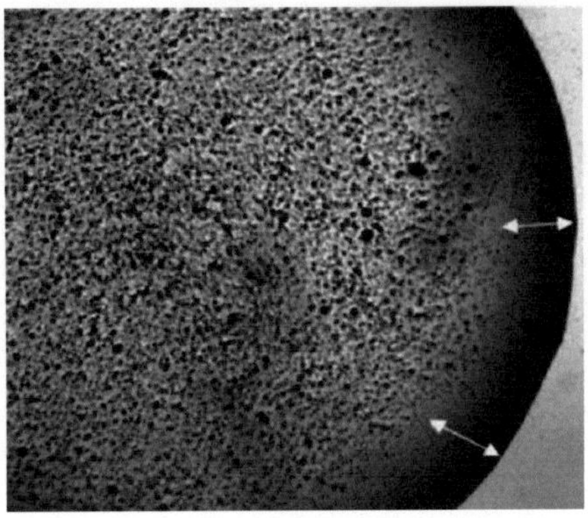

Figure 18.
Measurement of coffee ring width under microscope.

Levenhuk ToupView program in 3 positions into every drop (**Figure 18**), so for each 30-minute account 9 measurements were made. Arithmetic mean and standard deviation were calculated for further analysis.

3.1.3 Optical investigations of liquid samples

The microscopy investigation of coffee solution was carried out in freshly prepared samples by the method of "flattened drop." For this purpose a drop with a volume of 5 μl was placed on a new (without any treatment) microscope slide ApexLab (25.4 × 76.2 mm) then the drop was covered with a cover glass 24 × 24 mm in size (ApexLab), avoiding formation of air bubbles, and was studied under microscope Levenhuk with a digital camera connected to a computer. We made 10 pictures for every 30-minute step with the same magnitude and analyzed them later using the Levenhuk ToupView program. Morphometric measurements (diameters of avoids in the pictures) were made for every 30-minute step. Statistical analysis (calculation of mean and standard deviation) were made by Excel program. In some experiments, in parallel with the flattened drop, freshly prepared smears (without cover glass) were also examined under a microscope in polarized light.

3.1.4 Surface tension fluctuations detection

For detecting surface tension temporal changers we used a set of certified glass capillaries (10 μl Drummond Microdispenser, 100 Replacement Tubes, made in the USA by Drummond Scientific Company. Cat.# 3-000-210G). Each capillary was used once. A new dry capillary was submerged into liquid at regular intervals to a certain mark on a capillary and liquid raising level was measured. Simultaneously, the fluctuations of BS_2 were usually measured. Diagrams and calculations of the correlation coefficient were done by means of Excel program.

3.2 Results and discussion

3.2.1 Dynamic processes in liquid media

Figure 19 shows joint temporal fluctuations of parameter BS_2 and coffee ring width. Direct linear correlation between them at significant value $p = 0.005$ was 0.7 ± 0.16. This testified to a causal relationship between these parameters.

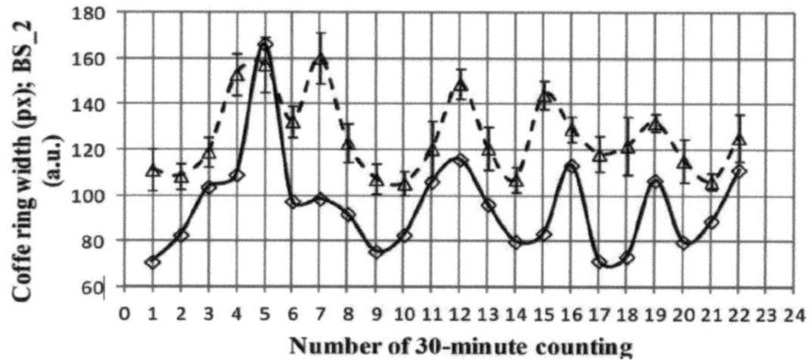

Figure 19.
Joint dynamics of parameter BS_2 (solid line) and coffee ring width (dashed line) in coffee water solution (2.5 g/100 ml).

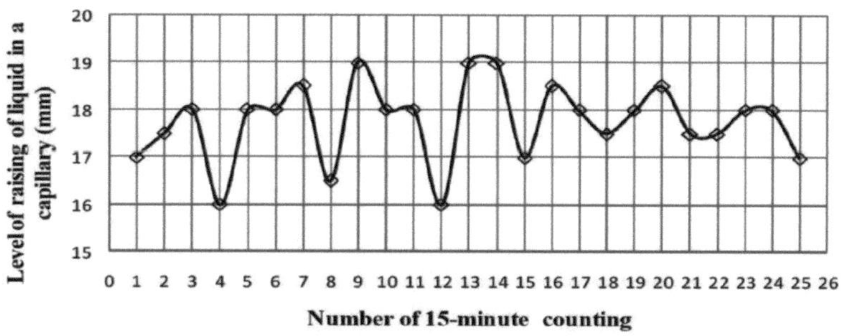

Figure 20.
Fluctuations of surface tension in coffee water solution (2.5 g/100 ml).

Fluctuations of surface tension in the same coffee solution could be measured more frequently. It was shown that one period took 30-40 minutes (**Figure 20**). Correlation coefficient between BS_2 and surface tension fluctuations in one and the same experiment was 0.8 ± 0.2 ($p = 0.01$) (**Figure 21**). It is interesting to note that fluctuations of these parameters did not disappear either on the third day of stay of this liquid in the same glass without cover on the table in laboratory. Despite a long

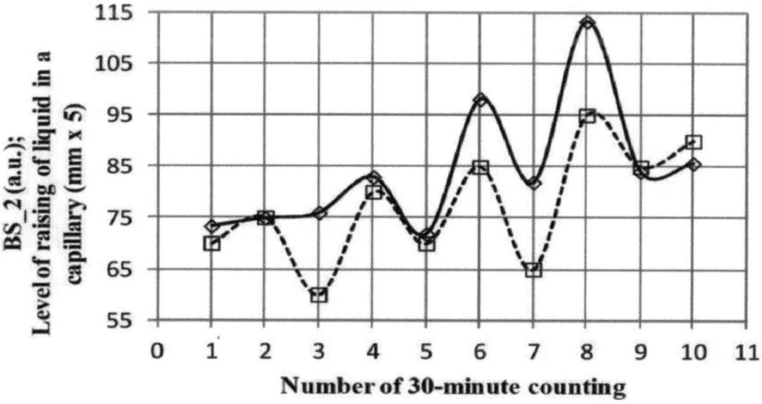

Figure 21.
Joint dynamics of parameter BS_2 (solid line) and level of rising of liquid in a capillary (dashed line) in coffee water solution (2.5 g/100 ml).

period of storage, fluctuations of parameters persist, and direct correlation link between them remains high ($r = 0.7 \pm 0.2$, $p = 0.01$).

It is important to note that when the concentration of coffee in the sample was halved, the amplitude was halved, and the period of oscillations doubled [53]. This makes it probable that oscillatory processes in liquids are associated with aggregation - disaggregation of the microdispersed phase.

A very important problem for the theory and practice is the development of methods for increasing the stability of colloidal systems. Such tasks can be decided in particular by means of adding surface modifying polymers [57] and a literature there]. In our research it was important to find out how addition of surfactant influences parameters of fluctuations and structurization of the drops drying on a glass support. According to **Figure 22**, SDS adding drastically reduced BS_2 value (mechanical stress during drop drying). It occurred due to decrease in interaction between colloidal particles as well as between the particles and quartz surface.

Diameter of the dried drops considerably increased, and the relief of their surfaces became smooth (**Figure 22**, drops 12, 13). Drops 5 and 6, corresponding to one of maxima of fluctuations of the BS_2 parameter before SDS adding were characterized by the relief coffee ring and presence of fragments of reticular structures on the surface. Drops 4 and 7, corresponding to the minimum BS_2 values,

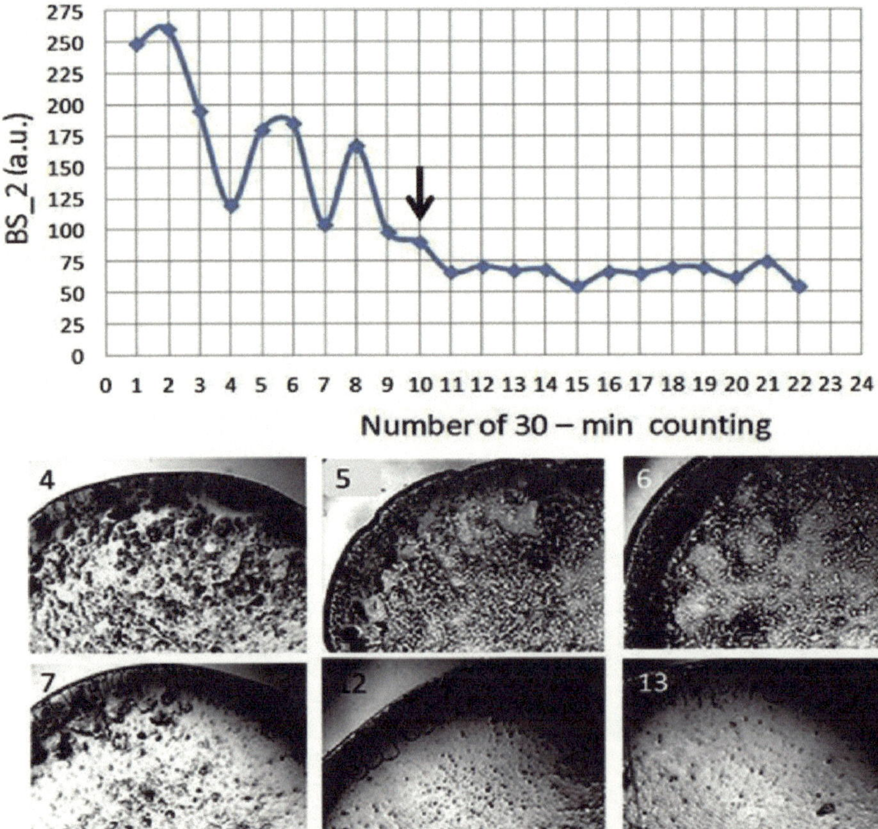

Figure 22.
Fluctuations of the BS_2 parameter in the drying drops of coffee solution (2.5 g/100 ml) before and after SDS addition (the moment of addition is specified by an arrow). From below - photos of the dried drops of coffee solution on glass support taken from total volume in different phases of the process (numbers of photos correspond to numbers of counting in above diagram).

had more flat coffee ring and did not contain the reticular structures. Instead of them separate clamps on a surface of the drops were observed. Our data agree with results of the work [58], showed that the interactions of colloids with (and at) liquid-solid and liquid-gas interfaces as well as bulk particle-particle interactions affect the morphology of the deposit. Now we can add that such interactions influence also the mechanical properties of dried materials from these colloids, which may be represented quantitatively. After surfactant addition the area of drops considerably increased, formation of the coffee ring has been complicated and structurization was suppressed, which corresponds to results of the research [59]. Thus, it can be stated that autonomous temporal fluctuations of mechanical properties of drying drops of colloidal suspensions revealed by us earlier [43, 52, 53], are also followed by coordinated fluctuations of surface tension. The authors will try to disclose the internal mechanism of these fluctuations looking directly into a liquid phase.

Observation of colloidal systems under optical microscope followed the same scheme: samples from one and the same volume of coffee solution were taken in certain periods of time and investigated them by the method of flattened drop. Perfectly shaped circles contoured by colloidal particles, sitting close to each other were observed everywhere (**Figure 23**). The circles were sitting on a glass substrate. Commonly it could be possible to find one central particle in each circle. Those round figures could associate, forming large – scale agglomerates [54].

It seems that growing "circles" pushed back colloidal particles, creating conditions for their convergence and coagulation. The size of the particles observed by us was not less than 1 μm so they did not participate actively in Brownian motion. Therefore during creation of spatial reticular structures their passive crowding due to the growing external structures seems to us more convincing than their active movement at the expense of the long-range attraction forces. **Figure 24** shows stages of temporal evolution of round structures in bulk, from small to big, and the remains of arches from colloidal particles after collapse of "round structures." Similar arches after collapse of round structures could be observed for some time in free floating (**Figure 25**).

The dynamics of growth and destruction of such round structures and their associates is shown in **Figure 26**. On the ascending and descending parts of the curve, size distribution of structures became bimodal due to the presence in the field of view, along with round structures, their large associates (see **Figure 23**, left). Nevertheless, our observations have revealed the rhythmic nature of formation and

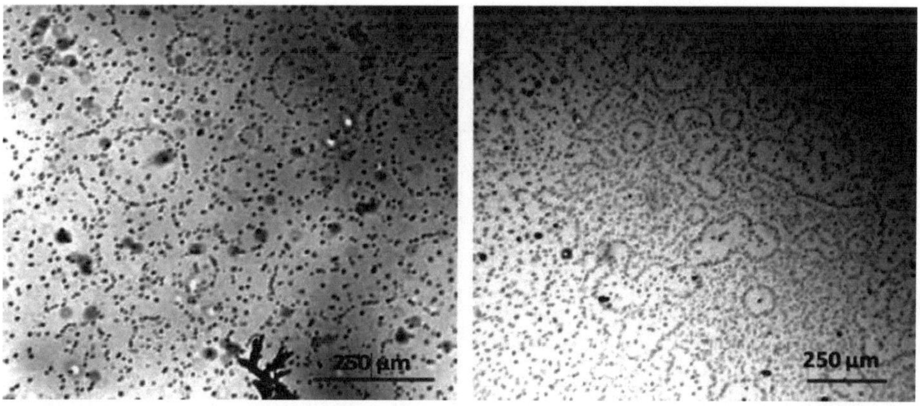

Figure 23.
Microphoto of water solution of coffee. Round figures and associates of round figures.

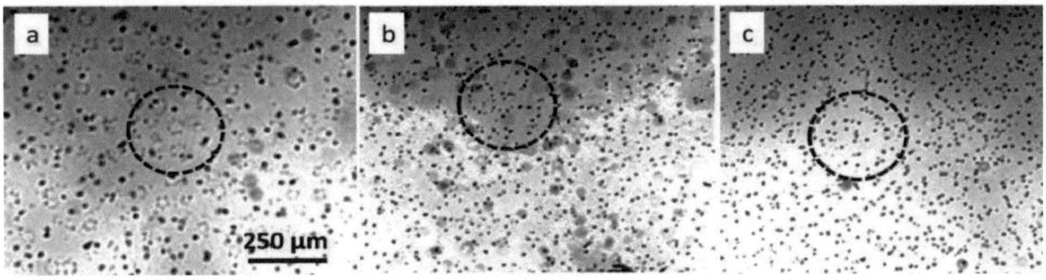

Figure 24.
Microphoto of water solution of coffee. Temporal evolution of round structures in bulk. Some structures in every picture are encircled (as a guide for eyes). Sampling time from the solution: (a) – 12:50, (b) – 13:30, (c) – 14:00.

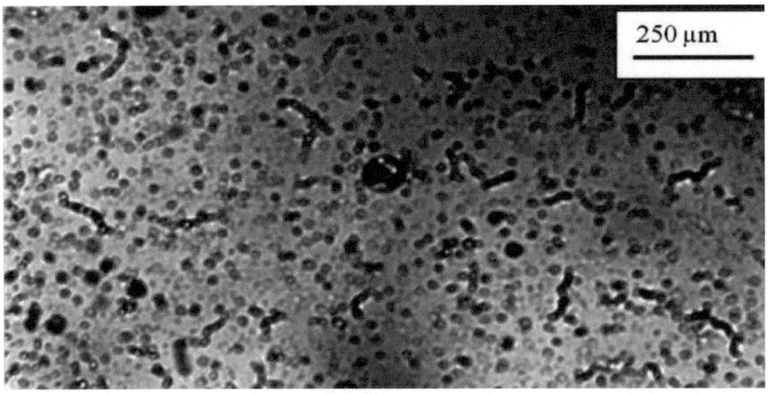

Figure 25.
Water immersion. Microphoto of water solution of coffee. Remainders of arches floating in solution after destruction of round structures.

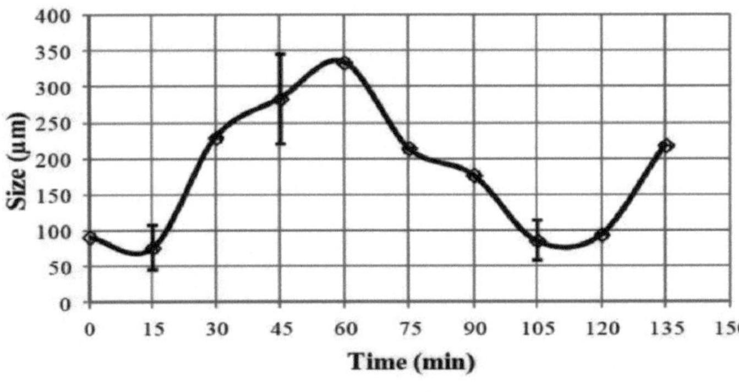

Figure 26.
Dynamics of growth and destruction of "round structures" in coffee solution (2.5 g/100 ml).

destruction of round structures, similar to a rhythm of fluctuations of physico-chemical parameters of this colloidal system. Our equipment allowed observing events only in two-dimensional option. Therefore, circular structures can be a projection of the balls on the plane. Data obtained by means of a laser scanning microscope manifestly showed spherical cavities in latex suspension (**Figure 2** in

[34]). Unfortunately, the authors did not pay attention to their shape. Those cavities looked empty, but now we believe that they were filled by transparent liquid crystal water. If so, then it is easy to explain the restricted movement of the particle placed in such media [16]. This assumption is confirmed by our observations of freshly prepared smears of coffee solution (**Figure 27c** and **d**). We could see real agglomerates of liquid crystal water. In a flat variant (between substrate and cover glasses), these agglomerates consist of round structures, which have visible borders due to adsorbing colloid particles (**Figure 27a** and **b**).

The mechanism of particle interaction in solution is currently actively discussed. Attraction of like charged gel beads with a diameter of 400-650 μm spaced several hundred micrometers apart in water was described in [60]. The authors measured the charge distribution around the beads with a pH sensitive dye and conjectured that the cause of the long-range attraction was a shell of multilayer structured water, formed around beads' hydrophilic surface. Here we can see the analogy with our experiment, where colloidal particles rather than gel beads interact. Moreover, their interaction is caused by spatial convergence which is due to the growing spheres of the liquid crystal water. In soft matter and nano-science, critical Casimir forces attract an increasing interest thanks to their capability of reversible particle assembly [61–67]. These forces are the thermodynamic analogue of the quantum mechanical Casimir force arising from the confinement of vacuum fluctuations of electromagnetic field. In its thermodynamic analogue, solvent fluctuations confined between suspended particles give rise to an attractive or repulsive force between them. Due to its unique temperature dependence, this effect allows in situ control of

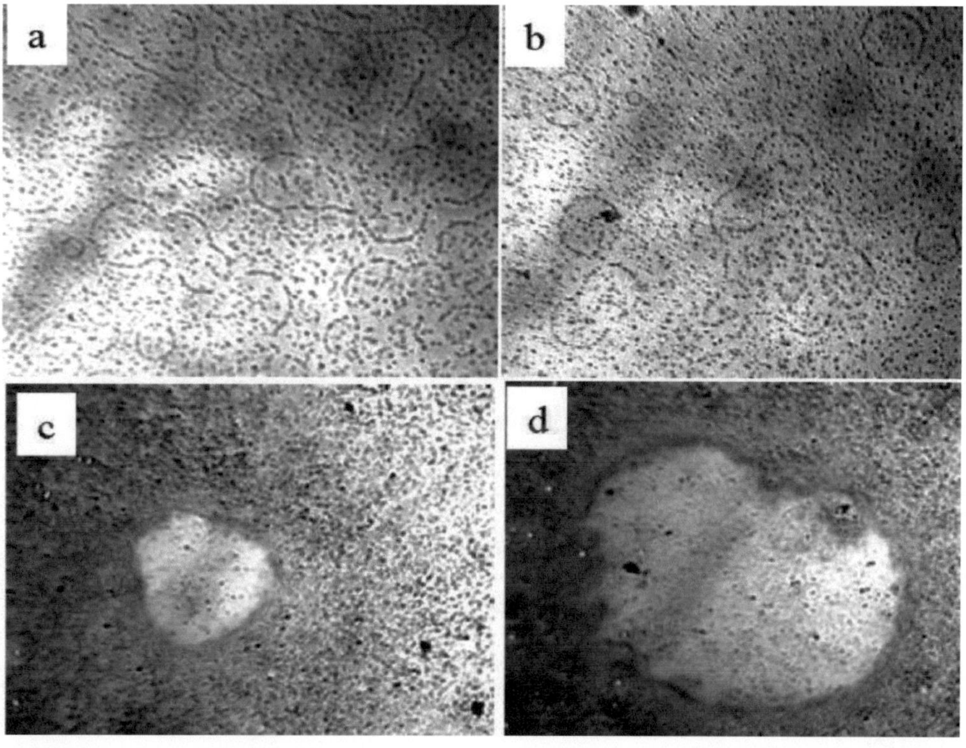

Figure 27.
Microphoto of water solution of coffee (a, b) prepared by the method of flattened drop; (c, d) – agglomerates of microdispersed particles with hydration shells of liquid crystal water (smears of the same coffee solution in polarized light). Each frame width is 1700 μm.

reversible assembly [63, 64]. The authors of [62] showed that in the system with negligible van der Waals forces a simple competition between repulsive screened Coulomb and attractive critical Casimir forces can account quantitatively for the reversible aggregation. Above the temperature Ta, the critical Casimir force drives aggregation of the particles into fractal clusters, while below Ta, the electrostatic repulsion between the particles breaks up the clusters, and the particles resuspend by thermal diffusion [62]. If the gap between the interacting surfaces is filled with a specially designed substance, the attraction between the surfaces can change their repulsion. If such interaction of surfaces with a dielectric constant Ɛ1 or Ɛ2, respectively, occurs in a medium with a dielectric constant Ɛ3, they will be attractive at (Ɛ1 - Ɛ3) (Ɛ2 - Ɛ3) < 0, and repulsive at (Ɛ1 - Ɛ3) (Ɛ2 - Ɛ3) > 0. These interactions are extremely sensitive to temperature, chemical composition of the medium and its physical characteristics [65, 66]. According to our data, the observed process is characterized by cyclic changes both in liquid solute concentration due to displacement of the ions and particles from Exclusion Zones (EZs) to the bulk, and in particle surface properties due to EZ shell growth around them. As these zones routinely generate protons in the water regions beyond, unequal proton concentrations in the respective areas may be responsible for creating both the pH and potential gradients, which may be ultimately responsible for the osmotic drive [30]. On the other hand, the surface water has different water activity and chemical potential to the bulk, leading to differences in osmotic pressure and other colligative properties [25]. When this increase in osmotic pressure next to the surface reaches a threshold, the mechanical instability of the system sharply growths, velocity of microstreams enhances, and aggregates of EZ spheres start to collapse. They break into small pieces and melt. Solute concentration and osmotic pressure decline. Free colloidal particles are distributed uniformly. Chains of particles coagulated on the surface of the water balls remain in solution. Growth of EZ balls begins again and the process recurs (**Figure 28**). As similar events (EZ growth) are registered for other polar liquids, we believe that the autonomous fluctuations based on rhythmic formation and destruction of liquid crystal spheres are the universal law of the nature. The considered processes have been used for creation of a phenomenological model showing a possibility of the existence of self-oscillatory modes in similar systems.

3.2.2 Model of dynamic processes in microdispersed water media

Let the volume of a colloidal system be a cube with edge length L. Let this cube house N_1 hydrophilic particles – seeds around which liquid crystal water spheres

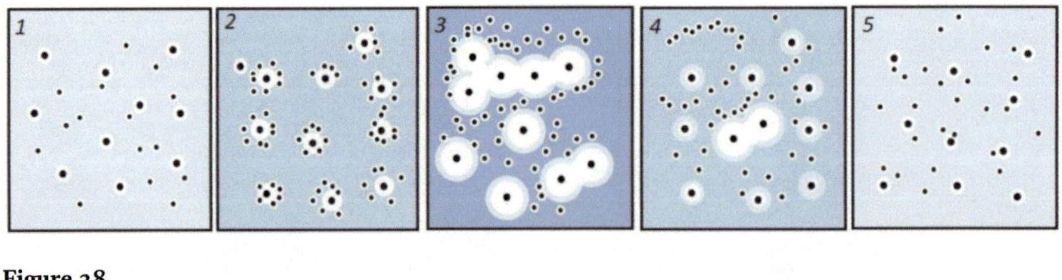

Figure 28.
Scheme of cyclic physicochemical transformations in colloidal system. Background coloring intensity corresponds to the concentration of ions and particles in dispersive phase 1 – Initial state, quasi-homogenous distribution of particles with small EZs; 2 – EZs growth around some particles is more intense than around others; 3 – giant EZ-balls aggregate and begin destruction due to high osmotic pressure; chains of particles coagulated on the surface of water balls remain in solution; 4 – destruction process continues, osmotic pressure decreases progressively; 5 – EZs are ready to grow again.

(EZ) are formed. Let n be the number of ions and colloidal particles determining osmotic pressure P at the interface between water spheres of radius r and dispersive medium. V is the amplitude of mean speed of microflows with a characteristic lateral dimension smaller than πr. This corresponds to the excited mode of mechanical instability for the sphere surface: $2\pi r/m$, where m = 2, 3, 4, 5… The estimated equations for integral processes in such a system can be written down in the following form: EZ growth near a seed particle can be described as (2).

$$dr/dt = l_0/\tau_{gr}(1 - V/V_{crit}) \tag{2}$$

where V is the average velocity of microstreams in bulk near the water balls; V_{crit} is the critical velocity of microstreams in bulk with sufficient energy for destruction of external borders of the water balls; l_0 is the increment of EZ shell thickness around a hydrophilic particle during time τ_{gr}. From the works [68, 69] and our own experiments we know that $l_0\ /\tau_{gr} \approx 1 - 10$ µm/sec. Formation of microstreams near the interface between the EZ shell and free water on achieving critical osmotic pressure P_{crit} can be described as (3)

$$dV/dt = -V/\tau_{visc} + 4\pi r^2 \cdot \gamma_m \cdot P \cdot F_{[x]} \cdot [P - P_{crit}] \tag{3}$$

where τ_{visc} is the characteristic time of reduction of microstreams velocity due to solution viscosity; $\tau_{visc} \approx$const; γ_m is the coefficient characterizing average change of destruction force depending on the created mode of spatial nonuniformity on the destroyed external border of EZ; $F_{[x]}$ is the step-type function equal to zero if $x = P - P_{crit}$ 0, and equal to 1 if $x = P - P_{crit} > 0$.

We assume that the speed of diffusion of ions and colloidal particles is much more than the growth rate of EZ shell and speeds of delay of microstreams. Then we can use Vant Hoff's law for stationary conditions (4)

$$P = n \cdot R \cdot T/V \tag{4}$$

where R is universal gas constant, T is absolute temperature.

The volume of colloidal liquid except for the volume of water spheres is found from equation (5)

$$V = L^3 \cdot \left[1 - 4/3\pi r^3 \cdot N_1/L^3\right] \tag{5}$$

Thus, the status of water spheres in the bulk of the remaining colloidal liquid can be described by Eqs. (2)–(5). We introduce the following notation:

$$\beta = 4\pi N_1/3L^3 \tag{6}$$

$$\alpha = 4\pi\gamma_m \cdot nRT/L^3 \cdot V_{crit} \tag{7}$$

$$\chi = V/V_{crit} \tag{8}$$

and rewrite the above equations correspondingly:

$$\begin{cases} dr/dt = l_0/\tau_{gr}(1 - \chi) & (9) \\ d\chi/dt = -\chi/\tau_{visc} + \alpha r^2/1 - \beta r^3 \cdot F_{[x]} \cdot [P - P_{crit}] & (10) \\ P = n \cdot R \cdot T/L^3 \cdot (1 - \beta r^3) = (V_{crit}/4\pi\gamma_m) \cdot \alpha/(1 - \beta r^3) & (11) \end{cases}$$

On the basis of this system of equations and understanding of the physics of the dynamical process we can distinguish 3 stages of the process:

a. EZ shell growth around hydrophilic colloidal particles to the size of huge liquid crystal water spheres; osmotic pressure growth in a bulk;

b. Osmotic pressure growth in bulk over critical value, forming conditions for the development of mechanical instability at the interface between water spheres and bulk (similar to the Rayleigh-Taylor instability [67]); microstream strengthening, causing erosion until complete destruction of water spheres.

c. Microstreams slowdown due to viscosity and transition of the system to stage 1.

Let us consider the dynamics of the process based on Eqs. (9)–(11) in simplified form.

Stage 1. EZ shell growth around hydrophilic colloidal particles (12)–(14):

$$\begin{cases} dr/dt = l_0/\tau_{gr}(1-\chi), & \chi > 1 & & (12) \\ d\chi/dt = -\chi/\tau_{visc}, & \chi \approx 0 & (\text{at t} = 0) & (13) \\ P = (V_{crit}/4\pi\gamma_m)\cdot\alpha/(1-\beta r^3) & < P_{crit} & & (14) \end{cases}$$

Stage 2. Development of instability and destruction of water spheres (15)–(17):

$$\begin{cases} P = P_{\text{crit}} & (15) \\ r^3 = r^3{}_{crit} = 1/\beta\,[1-(\alpha\cdot V_{crit}/4\,\pi\gamma_m\cdot P_{crit})] & (16) \\ d\,\chi/dt = (\chi_{max}-\chi)/\tau_{visc}, & (17) \end{cases}$$

where $\chi_{\max} = \alpha\bullet\tau_{visc}\bullet r^2{}_{crit}/\,(1 - \beta\cdot r^3{}_{crit})$. If $\chi \le 1$, then r continues to grow. According to our observations, $r_{crit} \approx 250$ μm.

Stage 3. Microflows slowdown.

Since turbulent flows are formed at this stage, τ_{gr} may depend on χ and P. However, for our simplified representation, we shall assume that τ_{visc} = const, as in the case of laminar flow (18)–(21):

$$\begin{cases} P < P_{crit}. & (18) \\ dr/dt = l_0/\tau_{gr}\,(1-\chi), \chi > 1 & (19) \\ d\chi/dt = -\chi/\tau_{visc} & (20) \\ P = (V_{crit}/4\,\pi\gamma_m)\bullet\alpha/(1-\beta r^3) < P_{crit} & (21) \end{cases}$$

This dynamics can also be represented on phase plane in χ and r coordinates (**Figure 29**).

4. Conclusions

In this part of work an instant coffee as a sample of complex colloidal system was used. Making long repeated measurements of the dynamics of complex mechanical properties of drying drops by DDT [55, 56] slow periodic fluctuations of the

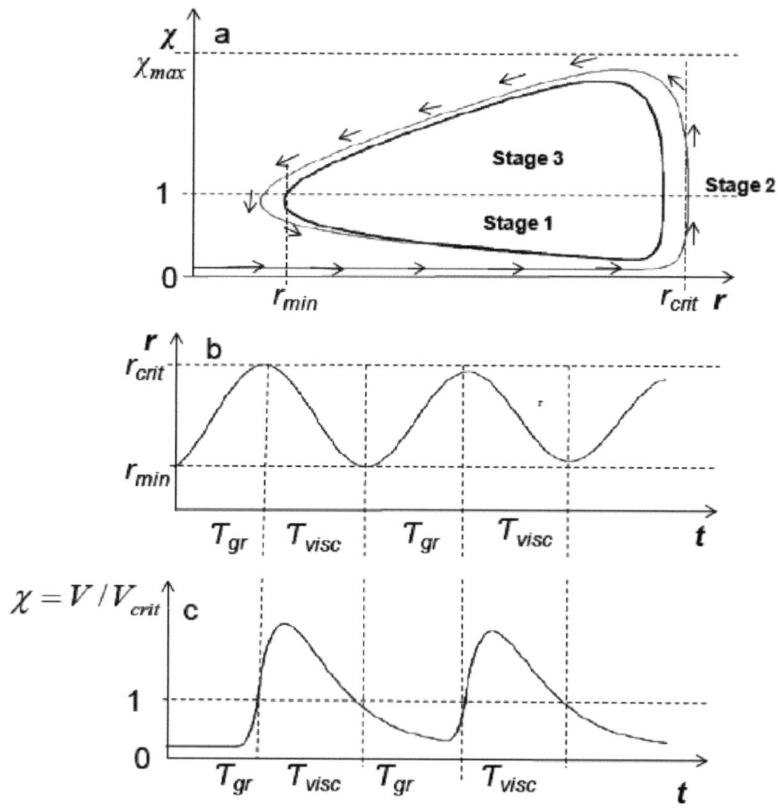

Figure 29.
Qualitative description of cyclic changes of variables in the course of growth of water spheres, their destruction and formation of conditions for their new growth: (a) - view of phase trajectories: process begins at the zero point (χ = 0; r = 0) and reaches limit cycle; (b) - temporary change of radius of water spheres; (c) - temporary change of velocity of microstreams.

measured parameter were noticed. Additional studies showed that, simultaneously with the above mentioned parameter, there occurred coordinated fluctuations of the surface tension and the character of structuring in drying drops. It had been found earlier that parameters of fluctuations depended on the extent of dilution and did not depend on additional hashing of solution and shielding from external electromagnetic fields [53, 54]. So, it was decided to elucidate the mechanism of such fluctuations in liquid phase. The basis for such a study was, on the one hand, information available on reversible formation of voids inside colloidal dispersions [34–36] and, on the other hand, description of slow fluctuations and inhomogeneity of fluids of different types obtained in the light scattering studies [33, 49, 50]. Moreover, it had been earlier described similar fluctuations of optical density in blood and plasma diluted with physiological saline solution in different proportions [27]. For the liquid phase study, the method of flattened drop and smears of freshly prepared coffee solution were used. The existence of spherical structures in the liquid phase of coffee, their periodical occurrence, growth, destruction and re-emergence, which agreed with fluctuations of physicochemical parameters of the system, was described for the first time. It had been showed morphologically that these spheres are liquid crystal water which forms EZs around hydrophilic colloidal particles. Agglomerates of such crystal water spheres look like voids inside colloidal dispersions under confocal scanning laser microscope. These findings are based on modern and last-century studies of wall layer of water mentioned in the book of Gerald Pollack [45]. We proposed a water-induced mechanism of self-oscillatory processes in colloidal systems and confirmed the principle possibility of its existence by a simple mathematical model. These data allow a fresh look at the process of aggregation - disaggregation of colloidal particles in the solution. The experiments showed that the growth of water spheres pushes colloidal particles to the borders of the spheres, promotes their crowding and coagulation. On the border of water spheres, the particles form chains (reticular structures) which exist some time in liquid phase after destruction of liquid crystal water shells. Thus, all complex dynamics is controlled by the phase transitions of water – from the free to the bound (liquid crystal) state and back. Osmotic pressure acts as the intermediary messenger and the synchronizer of these transformations in a whole volume of liquid. Actually, these processes are not very sensitive to temperature (unlike the models based on calculation of Casimir forces [61–66]), and do not notably depend on liquid disturbance by hashing. Hashing by a turning of a test tube initiates emergence of streams of liquid with a characteristic size of an order of the size of a test tube (1.5×9.0 cm [53]), when the most part of the brought energy is spent for movement of spheres in bulk. Destruction of water spheres happens at initiation of microstreams of 10-100 μm in size, which correspond to the sizes of the destroyed objects. Osmotic pressure P in accordance with Eq. (3) is proportional to the absolute temperature ($\approx$295 K). Possible changing the temperature at a few degrees is negligible to affect the considered processes. For obvious reasons parameters of fluctuations depend on concentration of the components. The described mechanism can explain some phenomena that were not clear before and deserves special research. The authors believe that we deal with a universal phenomenon of undoubted importance for fundamental and applied science. We are sure that our hypothesis will stimulate researchers with other ideas and other tool kits to join this direction of study.

5. Concluding remarks

In this chapter, the authors tried to present their ideas about the structure and dynamics of aqueous microdispersed media, which were developed by them over the course of last 20 years. The authors followed the phase transformations of the protein using the drying drop model; found similar changes in the blood serum of seriously ill people and came to the conclusion that they have a violation of the required proportion between the content of protein and salt components of blood serum; found confirmation of our conclusions in the practical activities of Dr. A.S. Samokhotskiy, who drew the attention to the importance of the problem posed and the prospects of its solution for the drug-free treatment of patients. It has been shown that water and aqueous solutions of salts are also microdispersions. There is no clean water in the world around us. Using instant coffee as an example, we made sure that aqueous dispersions live in their own rhythm, the parameters of which depend on the ratio of the volume of the dispersed phase and the ionic strength of the dispersion medium. These parameters do not depend on the volume and shape of the vessel, as well as on mechanical disturbances. Surfactants reduce the amplitude of oscillations without significantly affecting their phase. In phase with the fluctuations in the mechanical properties of dispersion media, their surface tension and the width of the edge ridge of the drops dried on glass change. The authors associated these processes with the phenomenon of aggregation - disaggregation of the dispersed phase, the elementary unit of which is a hydrophilic microparticle surrounded by a hydration shell of liquid crystalline water. A model of self-oscillatory processes in microdispersed systems, which consistently describes the observed phenomena, was proposed.

The authors regard this work as a stage on the path of knowledge, which must be refined and corrected on the endless path of the development of science.

Conflict of interest

The authors declare no conflict of interest.

Funding

This project was funded by the Ministry of Education and Science of Russia (project no 14.Y26.31.0022).

Author details

Tatiana Yakhno[1*] and Vladimir Yakhno[1,2]

1 Institute of Applied Physics of Russian Academy of Sciences, Russia

2 Lobachevsky University of Nizhny Novgorod, Russia

*Address all correspondence to: yakhta13@gmail.com

References

[1] Deegan RD, Bakajin O, Dupont TF, Huber G, Nagel SR, Witten TA. Contact line deposits in an evaporating drop. Phys. Rev. E. 2000;62(1):756-765. DOI: https://doi.org/10.1103/PhysRevE.62.756

[2] Deegan RD. Pattern formation in drying drops. Phys. Rev. E. 2000;61(1): 475-485. https://doi.org/10.1103/PhysRevE.61.475

[3] Popov Yu. Evaporative deposition patterns: spatial dimensions of the deposit. Phys. Rev. E. 2005;71:036313. DOI: 0.1103/PhysRevE.71.036313

[4] Nguen TAH, Nguen AV, Hampton MA, Xu ZP, Huang L, Rudolf V. Theoretical and experimental analysis of droplet evaporation on solid surfaces. Chemical Engineering Science. 2012;69(1):522-529. https://doi.org/10.1016/j.ces.2011.11.009

[5] Hu H, Larson RG, Analysis of the effects of Marangoni stresses on the microflow in an evaporating sessile droplet. Langmuir. 2005;21(9): 3972-3980. https://doi.org/10.1021/la0475270

[6] Ristenpart WD, Kim PG, Domingues C, Wan J, Stone HA. Influence of substrate conductivity on circulation reversal in evaporating drops. Phys. Rev. Lett. 2007;99:234502. https://doi.org/10.1103/PhysRevLett.99.234502

[7] Bhardwaj R, Fang X, Attinger D. Pattern formation during the evaporation of a colloidal nanoliter drop: a numerical and experimental study. New J. Phys. 2009;11:075020, Online at http://www.njp.org/doi:10.1088/1367-2630/11/7/075020

[8] Yakhno TA, Yakhno VG, Sanin AG, Sanina OA, Pelyushenko AS. Protein and salt: spatiotemporal dynamics of events in a drying drop. Technical Physics. The Russian Journal of Applied Physics. 2004;49(8):1055-1063

[9] Yakhno TA, Yakhno VG, Structural evolution of drying drops of biological fluids. Technical Physics. 2009;54(8): 1219-1227. https://doi.org/10.1134/S1063784209080210

[10] Yakhno T, Salt-induced protein phase transitions in drying drops. J. Colloid Interface Sci. 2008;318:225-230. http://dx.doi.org/10.1016/j.jcis.2007.10.020 .

[11] Trappe V, Sandkuhler R. Colloidal gels—low-density disordered solid-like states. COCIS. 2004;8: 494-500. http://dx.doi.org/10.1016/j.cocis.2004.01.002

[12] Pathak B, Christy J, Sefiane K, Gozuacik D. Complex pattern formation in solutions of protein and mixed salts using dehydrating sessile droplets. Langmuir. Just Accepted Manuscript. DOI: 10.1021/acs.langmuir.0c01122 • Publication Date (Web): 27 Jul 2020.

[13] Zang D, Tarafdar S, Tarasevich Y Y, Choudhury M D, Dutta, T. Evaporation of a Droplet: From physics to applications. Physics Reports. 2019;804:1-56. DOI: 10.1016/j.physrep.2019.01.008

[14] Thampi S P, Basavaraj M G. Beyond coffee rings: drying drops of colloidal dispersions on inclined substrates. ACS Omega. 2020;5(20):11262–11270. DOI: 10.1021/acsomega.9b04310

[15] Sett A, Ayushman M, Dasgupta S, DasGupta S. Analysis of the distinct pattern formation of globular proteins in the presence of micro-and nanoparticles. J. Phys. Chem. B. 2018; 122(38): 8972-8984. https://doi.org/10.1021/acs.jpcb.8b05325

[16] Ragoonanan V, Aksan A. Heterogenity in desiccated solutions:

implications for biostabilization. Biophys. J. 2008;94:2212-2227.

[17] Pauchard L. Patterns caused by buckle-driven delamination in desiccated colloidal gels. Europhys. Lett. 2006;749(1):188-192. DOI: 10.1209/epl/i2005-10493-3

[18] Tarasevich Yu Yu, Vodolazskaya IV, Isakova O. P. Desiccating colloidal sessile drop: dynamics of shape and concentration. Colloid and Polymer Science. 2011;289(9):1015-1023 DOI: 10.1007/s00396-011-2418-8

[19] Sobac D, Brutin D. Structural and evaporative evolutions in desiccating sessile drops of blood. Phys. Rev. E., 2011;84:011603. https://doi.org/10.1103/PhysRevE.84.011603

[20] Sobac D, Brutin D. Triple-line behavior and wettability controlled by nanocoated substrates: influence on sessile drop evaporation. Langmiur. 2011;27(24):14999-15007. https://doi.org/10.1021/la203681j

[21] Sobac D, Brutin D. Thermocapillary instabilities in an evaporating drop deposited onto a heated substrate. Phys. Fluids. 2012;24:032103. https://doi.org/10.1063/1.3692267

[22] Collins KD, Neilson GW, Enderby JE. Ions in water: Characterizing the forces that control chemical processes and biological structure. Biophys. Chem. 2007;128: 95-104 https://doi.org/10.1016/j.bpc.2007.03.009

[23] Yakhno TA, Sedova OA, Sanin AG, Pelyushenko AS. On the existence of regular structures in the liquid human blood serum (plasma) and phase transitions in the course of its drying. Technical Physics. 2003; 48(4):399-403. https://doi.org/10.1134/1.1568479

[24] Yakhno T. Protein phase instability developed in plasma of sick patients: clinical observations and model experiments. Natural Science. 2010;3: 220-227. DOI: 10.4236/ns.2010.23034; also available: http://www.scirp.org/journal/NS

[25] Chaplin M. Self-generation of colligative properties at hydrophilic surfaces, 2020. https://arxiv.org/ftp/arxiv/papers/1203/1203.0206.pdf

[26] Gubler E.V. Computational methods of analysis and recognition of pathological processes. L .: Medicine. 1978, 294 p. (In Russ.)

[27] Yakhno T. Blood as a polydisperse system. Lambert Academic Publishing GmbH & Co, Germany, 2011. ISBN: 978-3-8443-5767-7, 313 p. (in Russ.)

[28] Portraits: Alexander Svyatoslavovich Samokhotsky [Internet] (In Russ.) Available from: http://actualmed.ru/portrety-aleksandr-svyatoslavovich-samoxockij/ [Accessed: 2020-08-13]

[29] Samokhotskiy AS. Experience in determining therapeutic patterns. Dissertation, 1946, Faculty surgical clinic of the Odessa Medical Institute. (in Russ.) Available from: http://lib.ru/URIKOVA/METAMED/SAMOHOCKIJ/disstitulioglav.txt [Accessed: 2020-08-13]

[30] Samokhotskiy AS. About nervousness and its medical problem. (in Russ.) Available from: http://lib.ru/URIKOVA/METAMED/SAMOHOCKIJ/samah1.txt [Accessed: 2020-08-13]

[31] Voyutsky S.S. Colloidal chemistry course. 2nd ed. M .: Chemistry, 1976, 512 p. (in Russ.)

[32] Fesenko EE, Terpugov EL. On the unusual spectral properties of water in a thin layer. Biophysics. 1999; 44 (1):5-9.

[33] Sedlák M. Large-scale supramolecular structure in solutions of

low molar mass compounds and mixtures of liquids: I. Light scattering characterization. J. Phys. Chem. B. 2006;110(9):4329–4338 https://doi.org/10.1021/jp0569335

[34] Ito K, Yoshida H, Ise N. Void Structure in colloidal dispersions. Science. 1994;263(7):66-68 doi: 10.1126/science.263.5143.66

[35] Ise N. Like likes like: counterion-mediated attraction in macroionic and colloidal interaction. Phys. Chem. Chem. Phys. 2010;12:10279-10287 https://doi.org/10.1039/C000729C

[36] Yoshida H, Ise N, Hashimoto T. Restricted motion of a particle trapped inside a void in a colloidal dispertion. Langmuir. 1995;11:2853-2855 https://doi.org/10.1021/la00008a002

[37] Bukaty VI, Nesteruk PI. Study of optical inhomogeneities (clusters) in bidistilled water by the optical method of small angles. Polzunovsky Bulletin. 2011; 3/1: 106-108 (in Russ.)

[38] Smirnov AN, Syroeshkin AV. Supranadmolecular complexes of water. Russian Chem. J. 2004; XLVIII (2): 125-135 (in Russ.)

[39] Smirnov AN. New data on structures in liquid water. Applied Physics and Mathematics. 2017;2: 17-21 (in Russ.)

[40] Laboratory Water. Its Importance and Application. / National Institutions of Health, 2013, 22 p. Also available from: https://www.orf.od.nih.gov/TechnicalResources/Documents/DTR%20White%20Papers/Laboratory%20Water-Its%20Importance%20and%20Application-March-2013_508.pdf

[41] Yakhno T, Yakhno V. A study of structural organization of water and aqueous solutions by means of optical microscopy. Crystals. 2019;9(1): DOI: 10.3390/cryst9010052. http://www.mdpi.com/2073-4352/9/1/52

[42] Yakhno T, Drozdov M, Yakhno V. Giant Water Clusters: Where Are They From? Int. J. Mol. Sci. 2019;20:1582; doi: 10.3390/ijms20071582. https://www.mdpi.com/1422-0067/20/7/1582

[43] Yakhno TA, Yakhno VG, Zanozina VF. Phase transitions of water as a source of slow oscillatory processes in liquid media. Actual problems of biological physics and chemistry. BFFC-2017: materials of the XII international scientific and technical conference, Sevastopol, October 2-6, 2017; Sevastopol, p. 23-27 (in Russ.)

[44] Yakhno TA, Yakhno VG. Study of the role of microdispersed phase of water during its transition to activated state. Russian Journal of biological physics and chemistry. 2020, accepted (In Russ.)

[45] Pollack G. The fourth phase of water: beyond solid, liquid and vapor. Ebner & Sons publisher, Seattle WA, USA, 2013, 357 p. http://www.ivoviz.hu/files/GHP_thefourthphaseofwater.pdf

[46] Selkov, EE. Oscillations in biochemical systems. Experimental data, hypotheses and models. Proceedings of the All-Union Symposium on Oscillatory Processes in Biological and Chemical Systems. Pushchino-on-Oka, March 21-26, 1966 / Ed. G.M. Frank. M.: Nauka, 1967, 7-22

[47] Shnol SE. Conformational oscillations of macromolecules. Proceedings of the All-Union Symposium on Oscillatory Processes in Biological and Chemical Systems. Pushchino-on-Oka, March 21-26, 1966 / Ed. G.M. Frank. M.: Nauka, 1967, p. 22-41

[48] Shnol SE. Cosmophysical factors in random processes. Stockholm, 2009, 338 p.

[49] Chernikov FR. Fluctuations in the intensity of light scattering in aqueous

solutions of proteins. Biophysics. 1986; 31(4):695-699 (in Russ.) http://www.biophys.ru/archive/h2o-00005.pdf

[50] Chernikov FR. Ultra-slow oscillations of light scattering in liquids of different types. Biophysics. 1990;35 (5):711-715 (in Russ.) http://www.biophys.ru/archive/h2o-00006.pdf

[51] Drozdov AV, Nagorskaya TP. Quasiperiodic character of intermolecular interactions in water. Biophysics. 2014;59(6):1195-1208 (In Russ.) https://doi.org/10.1134/S0006350914060025

[52] Yakhno TA, Sanin AG. Fluctuations of the dynamic parameters of structurization of drying droplets periodically sampled from the total volume of multicomponent aqueous solutions. Collection of scientific papers of the VII International Congress "Weak and superweak fields and radiation in biology and medicine", September 7 - 11, 2015, St. Petersburg, v.7, pp. 114-115. www.biophys.ru/archive/congress2015.pdf

[53] Yakhno TA, Yakhno VG. The coffee drop phenomenon and its temporary fluctuations. Autonomous oscillatory processes in colloidal liquids. Technical Physics. 2017;7(3):323-330. DOI 10.21883 / JTF.2017.03.44233.1778 http://journals.ioffe.ru/articles/44233

[54] Yakhno TA, Yakhno VG. Water-induced self-oscillatory processes in colloidal systems by the example of instant coffee. JOBARI. 2017;20(2):70-83

[55] Yakhno T, Sanin A, Pelyushenko A, Kazakov V, Shaposhnikova O, Chernov A, Yakhno V, Vacca C, Falcone F, Johnson B. Uncoated quartz resonator as a universal biosensor. Biosensors and Bioelectronics. 2007;22 (9-10):2127-2131. https://doi.org/10.1016/j.bios.2006.09.029

[56] Yakhno TA, Sanin AG, Vacca CV, Falcione F, Sanina OA, Kazakov VV, Yakhno VG. A new technology for studying multicomponent liquids using a quartz crystal resonator: theory and applications. Technical Physics. 2009;54 (10):1423–1430. DOI: 10.1134/S1063784209100041

[57] Goicochea AG, Nahmad-Achar E, Perez E. Colloidal stability dependence on polymer adsorption through disjoining pressure Isotherms. Langmuir. 2009;25:3529-3537 https://doi.org/10.1021/la802585h

[58] Anyfantakis M, Baigl D. Manipulating the coffee-ring effect: Interactions at work. Chem Phys Chem. 2015;16:2726–2734. DOI: 10.1002/cphc.201500410

[59] Rechell FJ, Zhang W, Sang W, Steenhuis NS. Surfactant-mediated control of colloid pattern assembly and attachment strength in evaporating droplets. Langmuir. 2013;29(6):1831–1840. DOI: 10.1021/la304685b

[60] Nagornyak E, Yoo H, Pollack G. Mechanism of attraction between like-charged particles in aqueous solution. Soft Matter. 2009;5:3850-3857. Available: http://pubs.rsc.org/en/Content/ArticleLanding/2009/SM/b905080a#!divAbstract

[61] Nguyen VD, Dang MT, Nguyen TA, Schall P. Critical Casimir forces for colloidal assembly. Critical review. J Phys Condens Matter. 2016;28(4): 043001. DOI: 10.1088/0953-8984/28/4/043001

[62] Bonn D, Otwinowski J, Sacanna S, Guo H, Wegdam G, Schall P. Direct observation of colloidal aggregation by critical Casimir forces. Phys Rev Lett. 2009;103(15):156101 https://doi.org/10.1103/PhysRevLett.103.156101

[63] Nguyen VD, Faber S, Hu Z, Wegdam GH, Schall P. Controlling colloidal phase transitions with critical Casimir forces. Nat Commun. 2013;4:

1584. DOI: 10.1038/ncomms2597 Available: http://www.ncbi.nlm.nih.gov/pubmed/23481392

[64] Dang MT, Verde AV, Nguyen VD, Bolhuis PG, Schall P. Temperature-sensitive colloidal phase behavior induced by critical Casimir forces. J. Chem. Phys. 2013;139:094903. Available: http://scitation.aip.org/content/aip/journal/jcp/139/9/10.1063/1.4819896

[65] Kats EI. Van der Waals, Casimir, and Lifshitz forces in soft matter. Physics Uspekhi. 2015; 185(8):964–969 DOI: 10.3367/UFNe.0185.201509g.0964

[66] Schlesener F, Hanke A, Dietrich S. Critical Casimir forces in colloidal suspensions. J. of Statistical physics. 2003;110(3):981-1013. Available: http://link.springer.com/article/10.1023%2FA%3A1022184508016#/page-1

[67] Sharp DH. An overview of Rayleigh-Taylor instability. Physica D: Nonlinear Phenomena. 1984;12(1–3):3-10. https://doi.org/10.1016/0167-2789(84)90510-4

[68] Zheng J, Wei-Chun Chin WC, Khijniak E, Khijniak E. Jr, Pollack G. Surfaces and interfacial water: Evidence that hydrophilic surfaces have long-range impact. Advances in Colloid and Interface Science. 2006;127:19–27. http://courses.washington.edu/bioe555/Zheng.pdf

[69] Chai B, Pollack GH. Solute-free interfacial zones in polar liquids. J. Phys. Chem. B. 2010; 114(16), 5371–5375. https://doi.org/10.1021/jp100200y

Chapter 7

Colloidal Stability of Cellulose Suspensions

Marina Stygar Lopes

Abstract

Cellulose is the most abundant natural compound in nature and the main component of the cell wall of plants. It is a linear polymer with a high degree of polymerization, responsible for most of the properties of wood. Colloidal phenomena are often used in various industrial production processes. Suspended cellulose, used worldwide in the paper and cellulose industries, with regard to stability, has a high tendency to aggregate and form clots. The different interactions between the dispersed phase and the dispersion phase are one of the critical points in the study of the behavior and stability of colloids. Cellulose is no different, as several studies seek to improve the colloidal stability of cellulose in aqueous media by observing the specific characteristics of the colloid, such as its geometry, mass and area/volume ratio, and the possible interactions between particles that make up the cellulose dispersion in order to understand and control colloidal stability. Therefore, the objective of this chapter is to define the main characteristics of colloids, to classify them, to present the main methods of preparation, to address important aspects about colloid stability and the colloidal stability of cellulose.

Keywords: colloids, suspension, cellulose, colloidal stability, cellulose fibers, lignocellulosic material

1. Introduction

Colloids are heterogeneous mixtures of at least two distinct phases, with the material of one of the phases in a finely divided form (solid, liquid or gas), called dispersed phase, mixed with the continuous phase (solid, liquid or gas), called medium dispersion [1].

Understanding and controlling the stability of colloidal dispersions is essential for its satisfactory use. For both economic and environmental reasons, water is often required as a dispersing phase, even when the particles that need to be kept in suspension are hydrophobic, as is the case with cellulose [1].

Cellulose has been gaining importance in the industrial scenario due to the growing interest in sustainability and environmental protection, becoming a competitive material since it is renewable, abundant, low cost, non-petroleum and non-toxic [2].

Suspended cellulose has a tendency to aggregate. In this way, some strategies to avoid cellulose self-agglomeration in aqueous medium have been used in order to reduce the hydrophilic character of cellulose, avoiding the formation of additional hydrogen bonds between cellulose fibers [3].

Therefore, this chapter aims to contribute to the field of study of colloids and their characteristics, in addition to cellulose with regard to its characteristics and behavior of aqueous solutions of cellulose and alternatives sought to improve the colloidal stability of cellulose suspensions.

2. Colloids

Colloids are systems formed by macromolecules or particles dispersed in a medium, in which one or more components have at least one of their dimensions within the range of 1 nm to 1000 nm [4].

Colloidal systems have been used since the dawn of humanity. Ancient people used gels from natural products as food, clay dispersions for the manufacture of ceramic utensils and colloidal pigment dispersions to decorate cave walls [5].

Colloidal systems are present in our daily lives in several products and technologies, such as personal hygiene (shampoo, toothpaste, foam, shaving cream, makeup, cosmetics) and in food (milk, coffee, butter, vegetable creams, fruit jellies, beer, soda or ice cream). During a single day we are consuming several colloids [5]. Colloids are also present in several consumer goods production processes, including drinking water, in the separation processes in the biotechnology industries and in the treatment of the environment.

In addition, colloidal phenomena are frequently used in industrial processes for the production of polymers, detergents, paper, soil analysis, food products, fabrics, precipitation, chromatography, ion exchange, flotation and heterogeneous catalysis. In orthomolecular therapeutic medicine, knowledge of the properties of colloidal systems can assist in the elucidation of diseases, such as Alzheimer's and Parkinson's [5].

2.1 Colloid characteristics

The factors that most contribute to the characteristics of a colloid are:

- The particle dimensions;
- The shape and flexibility of the particles;
- Surface properties;
- Particle-particle interactions;
- Particle-solvent interactions.

Colloids have specific characteristics such as high mass, high particle area/volume ratio and are relatively large. On the separation surfaces (interfaces) between the dispersed phase and the dispersion medium, characteristic surface phenomena are manifested, such as adsorption and double electrical layer effects, phenomena of great importance in determining the physicochemical properties of the system as a whole [6].

Depending on the affinity between the particles of a dispersion and the medium in which they are dispersed, we can classify colloids in two ways: lyophilic and lyophobic colloids. Lyophilic colloids are those in which the particle surface has an affinity for the solvent, keeping the dispersion more stable and minimizing aggregation. Lyophobic colloids, on the other hand, are those in which the particles have greater interaction with each other, which ends up leading to a rapid aggregation process [7].

2.2 Colloid classification

Regarding the colloid classification, there are the following categories:

- Aerosol: consists of a solid or a liquid dissolved in a gas.
- Foam: consists of a gas dispersed in solid or liquid.
- Emulsion: are colloids formed by liquid dispersed in another liquid.
- Sol: are colloids formed by the dispersion of a solid in a liquid or solid.
- Gel: solid of gelatinous material formed from a colloidal dispersion, in which the dispersed is in the liquid state and the dispersant in the solid state.

The **Table 1** shows some different types of colloids according to the state of the continuous and dispersed phases, and examples found in everyday life.

		Scattered		
		Gas	Liquid	Solid
Dispersant	Gas	Does not exist. All gases are soluble with each other	Aerosol Liquid Examples: cloud, fog	Aerosol solid Examples: smoke, dust in suspension
	Liquid	Liquid foam Example: soap foam, shaving cream, whipped cream	Emulsion Examples: milk, honey, mayonnaise, creams	Sol Example: paints, colored glass
	Solid	Solid foam Example: pumice, expanded polystyrene	Gel Examples: gelatin, cheese, jam	Solid Sun Example: ruby and sapphire crystal, metal alloys

Table 1.
Classification of colloids according to the dispersed phase and dispersion medium.

Colloidal systems can be divided into three types: colloidal dispersions, true macromolecule solutions and association colloids [8].

Colloidal dispersions are heterogeneous systems composed of two or more phases, as shown in **Table 1**, and these systems are thermodynamically unstable, due to their high surface free energy. In a colloidal dispersion, the interfacial area of the dispersed phase is very large, which requires a lot of energy to keep it dispersed. In an attempt to minimize the free energy of the surface, the system tends to minimize the area, based on the aggregation of the dispersed phase [8].

True macromolecule solutions are thermodynamically stable colloidal systems, that is, they will not separate phase. Polymeric solutions are examples of this class of colloids. Association colloids, which are also thermodynamically stable, are formed by the association of surfactant molecules, that is, micellar aggregates [8].

2.3 Colloid preparation

For the production of colloids there are two groups with different production methods, they are: dispersion methods and condensation methods [3].

2.3.1 Methods of dispersion

Mechanical spraying: for the production of colloids a solid substance is used added to a liquid, using colodal mills, which is a method used in the industry for the production of colloidal pigments.

Electric spraying: produced using the Bredig method, in which two electrodes are immersed in water to generate an arc. Spraying results in a coarse suspension from the metal particles. In the suspension you get the hydrosol. Electric spraying is used in the production of metallic colloids. In the reaction of substances such as benzene and ethyl ether, alkaline and alkaline earth colloids are produced.

Spraying by Ultrasound: from mechanical vibrations, which can be produced under a piezoelectric quartz generator in an excitation process, the formation spray of colloidal solutions is generated.

Peptization: performed with peptizing agents that have the ability to disintegrate, with colloids as the final product. These materials are used, for example, in the food industry in the production of gelatins, gums, and agar from the use of hot water, which is a peptizing agent.

2.3.2 Condensation methods

The condensation method is a means of producing colloids carried out with the precipitation of an insoluble substance by means of a chemical transformation between solvent substances. During its chemical transformation, the insoluble product is in the molecular state, occurring after condensation.

2.4 Stability of colloids

The different interactions between the dispersed phase (particles) and the dispersion phase (continuous) constitute one of the critical points in the study of the behavior and stability of colloids. The interactions between the particles that make up a dispersion and the dispersing medium are fundamental to understand colloidal stability [9].

The stability of a dispersion can be thermodynamic or kinetic and one of the ways to understand the difference between them is in terms of the colloid stabilization time. While a thermodynamically stable colloid will remain unchanged for an infinite time, maintaining properties like temperature and concentration unchanged, kinetically stable colloids tend to aggregate over time. Therefore, the study of colloidal chemistry makes it possible to change the time in which the colloid remains kinetically stable [8].

When it comes to particles, the energy in van der Waals' interactions comes from integrating the potential of all the molecules that make it up [10]. Van der Waals interactions between two particles will always be attractive if the particles are made of the same material, no matter what medium they are in [11]. If the particles are different in nature, van der Waals interactions can be attractive or repulsive [12]. In the study of colloidal dispersions, the focus is mostly on the interaction of particles of the same nature, that is, they are attractive interactions [13].

To increase the stability of a colloidal dispersion the steric effect of macromolecules is used to prevent the particles from aggregating by adding a stabilizer that will adsorb on the surface of the particle [13]. If the adsorbed macromolecule is in a good solvent, its chains expand. When it encounters a chain from another particle, there is a restriction in the conformation of both chains in the volume between the two particles, causing a decrease in configurational entropy and an increase in free energy [14]. To minimize this effect, the chains of the macromolecules repel each other, causing a repulsive effect between the particles, preventing aggregation.

Regarding the stability of aqueous colloidal dispersions, they are sensitive to the presence of electrolytes and polyelectrolytes (charged polymers of high molecular mass), since the colloidal particles can irreversibly aggregate in the presence of electrolytes and result in large and compact aggregates (clots) by a process called coagulation, while in the presence of polyelectrolytes there may be the formation of less dense aggregates (floccules), which can be easily broken and dispersed by mechanical agitation [15].

Understanding and controlling the stability of colloidal dispersions is essential for its satisfactory use. Some specific applications require that such dispersions be maintained over a wide range of temperatures and chemical conditions [8].

For both economic and environmental reasons, water is often required as a dispersing phase, even when the particles that need to be kept in suspension are hydrophobic. Water is a highly structured material, due to the hydrogen bonds that connect the molecules to each other. In the vicinity of a hydrophobic surface, ruptures of the hydrogen bonds between water molecules occur, increasing the free energy in relation to the solution. As a consequence, water is expelled to regions more favorable to hydrogen bonding. The migration of water molecules results in a mutual attraction between hydrophobic surfaces that implies a reduction in the free energy of the system [11].

3. Cellulose

Cellulose has stood out in the last 20 years as a study material for several applications, as it is the most abundant, renewable and natural polymer on the face of the Earth [15], and can be found mainly in woody plants (wood), annuals and in grasses [16]. Cellulose is located mainly in the secondary cell wall, corresponding to approximately 40 to 45% of the wood mass [17].

Cellulose (C6H10O5) n is a polysaccharide, linear chain containing from hundreds to thousands of chemical bonds involving carbon, hydrogen and oxygen atoms (**Figure 1**) [18]. The cellulose chain is of high molecular weight, which tends to form hydrogen bonds between the molecules [19, 20]. The hydroxyl groups of cellulose molecules form hydrogen bonds that can be intramolecular or intermolecular,

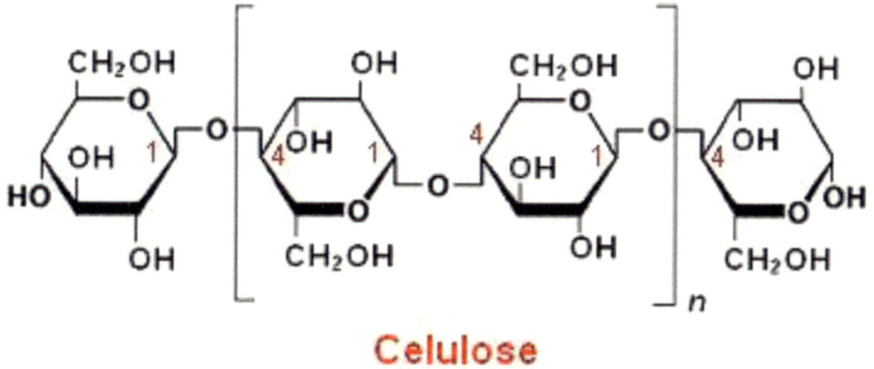

Figure 1.
Chemical structure of cellulose.

directing the crystalline packaging, and it is these bonds that make cellulose a stable polymer and appreciated as reinforcement in composites [21, 22].

Its organized structure is formed by cellulose microfibrils, which due to intermolecular bonds form the fibrils, which in turn are composed in an orderly fashion in order to form cellulosic fibers. Cellulose fibers are made up of two regions, the crystalline region, in which the microfibrils are presented in an extremely orderly manner, and the amorphous region, in which they are arranged in a less ordered manner [17], and for some lignocellulosic sources the amorphous regions can reach 50% of the structure [22].

Despite the hygroscopic nature of the individual cellulose molecules, the absorption of water molecules is only possible in the amorphous zones, since there is a lack of empty spaces in the crystalline structure. Hydroxy groups are the most abundant groups in the cellulose molecule, followed by the acetal bonds that form the ring of pyraneses [23].

In the crystalline regions of cellulose, we also have that the intra and intermolecular interactions can vary, giving rise to the various polymorphs [18]. The degree of polymerization and the crystallinity of cellulose vary according to the lignocellulosic source [1]. Due to the presence of crystalline and amorphous regions, cellulose can be classified as a semicrystalline fibrillar material [24].

Using cellulosic materials has several advantages, such as: its low cost, low density, high mechanical resistance and high elastic modulus. Due to the stable structure of their crystalline regions, cellulose fibrils have high mechanical properties along the longitudinal direction [24]. It is also possible to benefit from the high stiffness of the cellulose crystal which, when used on a nanometer scale for the production of composite materials, makes it possible to preserve the optical properties of the original material while improving the mechanical properties [25].

3.1 Colloidal stability of cellulose

Cellulosic pulp is a material whose characteristics and properties are determined by its origin. Cellulose modification methods are used when carrying out processes carried out in an aqueous medium, but cellulose is an amphiphilic polymer, that is, it presents a hydrophilic region that dissolves in water, and another hydrophobic region that does not dissolve in water, due to the presence of crystalline and amorphous regions.

The geometry, size and surface density of the particles are also properties that interfere with the processes of coagulation and flocculation. The polymers used with water retention agents increase the forces of colloidal attraction and induce flocculation through different mechanisms, based on different effects. We can mention: flocculation by bridge effect, flocculation by depletion effect and flocculation by reinforced bridge effect [1].

In the case of cellulose fibers, these properties are not well defined due to the variety in the size and shape of the fibers. However, it is known that cellulose fibers when dispersed in water have a pH of around 6, which indicates the acidic character of the surface, therefore a tendency to preferentially adsorb OH- group. In this way, the aqueous dispersions of cellulose fibers are influenced in their colloidal stability by the presence of a double electrical layer under their surface, resulting from the dissociation of different functional groups, such as carboxylics [26].

The pure cellulose fiber in suspension has a high tendency to aggregate and form clots by the action of gravity. However, studies show that through the addition of symmetrical or asymmetric electrolytes the tendency to coagulate the cellulose fiber suspension can be maximized or minimized depending on the final objective. The addition of cationic starch and calcium carbonate to the cellulose fiber suspension causes a change in the charge signal of the fiber surface, resulting in phenomena of fiber-fiber interaction that guarantees greater stability in relation to pure cellulose fiber [1].

3.2 Cellulose used in paper production

In the refining process, for example for the production of paper, cellulose fibers are immersed in water. The fibrils, which make up the cells, are composed of crystalline regions that, when immersed in water, absorb a quantity of this water across all exposed crystalline surfaces, causing their swelling and decreased attraction between the fibrils. The mechanical action of shearing the fibers through refiners speeds up this swelling, as it exposes the surfaces previously located inside the fibers, causing an increase in surface exposure, which promotes a greater number of contacts and connections between the fibers, resulting in this stronger paper [1].

The steps of converting cellulose to paper involve many surface chemical interactions, interactions between fibers and colloidal particles. Understanding these interactions is useful for product development and improving the resolution of operational problems.

4. Conclusions

This chapter sought to define the main characteristics of colloids, as well as their classification, methods of preparation and finally to address characteristics of colloid stability. Cellulose, the most abundant biopolymer in the world, is a colloid widely used in several industries. This colloid proves challenging for some segments due to its detailed characteristics throughout of the chapter. Studies continue to be carried out on this topic in order to bring solutions to improve the colloidal stability of cellulose.

Notes/Thanks/Other declarations

I thank IntechOpen for the opportunity and the Federal University of Paraná for my training from undergraduate to doctorate.

Author details

Marina Stygar Lopes
Federal University of Paraná, Curitiba, Brazil

*Address all correspondence to: marinastygar@gmail.com

References

[1] PANDOCHI, L. "Estudo do Comportamento Coloidal de Suspensão de Fibra de Celulose, Carbonato de Cálcio, Amido Catiônico: Variação da Força Iônica e do pH." [thesis]. Araraquara: Paulista State University; 2009.

[2] MILANEZ, D. H.; AMARAL, R. M.; FARIA, L. I. L.; GREGOLIN, J. A. R. Assessing nanocellulose developments using science and technology indicators. Materials Research, vol. 16, V. 3, pp. 635-641, 2013.

[3] BECK, S.; BOUCHARD, J.; BERRY, R. Dispersibility in water of dried nanocrystalline cellulose. Biomacromolecules, vol. 12, pp. 1486-1494, 2012.

[4] SHAW, D. J. Introdução à Química dos Colóides e de Superfícies; Edgard Blucher: São Paulo, SP, 1975.

[5] LEGRAND, C.P., LAFUMA, F., AUDEBERT, R., "Rheological behavior of colloidal dispersions of hydrophobic particles stabilised in water by amphiphilic polyelectrolytes", Colloids and Surfaces A, v. 152, pp. 251-261, 1999.

[6] ROZENBURG, I.M. Química Geral; Editora Edgard Blücher Ltda;1°edição; 2002; p. 431-432.

[7] MISAWA, T.; HASHIMOTO, K.; SHIMODAIRA, S. CORROS. Sci. 1974, 14 (2), 131-149.

[8] KLINE, S.R.; KALER, E.W., "Aggregation of Colloidal Silica by n-Alkyl Sulfates", Langmuir, v. 12, pp. 2402-2407, 1996.

[9] JAFELICCI JUNIOR, M.; VARANDA, L. C. O mundo dos colóides. Química Nova na Escola, n. 9, p. 9-13, 1999.

[10] MATIJEVIĆ, E.; SCHEINER, P. J. Colloid Interface Sci. 1978, 63 (3), 509-524.

[11] Colloid Science: Principles, methods and applications, 2nd ed.; Cosgrove, T., Ed.; John Wiley & Sons Ltd., 2010.

[12] ISRAELACHVILI, J. N. Intermolecular and Surface Forces, 3rd ed.; Elsevier Inc., 2011; Vol. 1.

[13] EVANS, D. F.; WENNERSTROM, H. The Colloidal Domain, 2nd ed.; Wiley-VCH, 1999.

[14] CORNELL, R. M.; SCHWERTMANN, U. The Iron Oxides, 2nd ed.; Wiley-VCH, 2003; Vol. 39.

[15] RODRIGUES, R. K.; DA SILVA, M. A.; SABADINI, E. LANGMUIR 2008, 24 (24), 13875-13879.

[16] DUFRESNE, A. (2013). Nanocellulose: A new ageless bionanomaterial. Materials Today, 16(6), 220-227. https://doi.org/10.1016/j.mattod.2013.06.004

[17] SJÖSTRÖM, E. Wood chemistry: fundamentals and applications. 2° edition. Elsevier: San Diego, 1993.

[18] BRINCHI, L. COTANA, F.; FORTUNATI, E.; KENNY, J. M. Production of nanocrystalline cellulose from lignocellulosic biomass: Technology and applications. Carbohydrate Polymers, vol. 94, pp. 154-169, 2013.

[19] MIMMS, A. Kraft pulping, a compilation of notes. 2° edição. Atlanta: TAPPI PRESS, 1993.

[20] KHALIL, H. P. S. A.; DAVOUDPOURA, Y.; NAZRUL ISLAM, M. D.; MUSTAPHA, A.; SUDESH, K.; DUNGANIA, R.; JAWAID, M.

Production and modification of nanofibrillated cellulose using various mechanical processes: A review. Carbohydrate Polymers, v. 99, p. 649-665, 2014.

[21] DAMASIO, R. A. P. Caracterização e aplicações de celuloses nanofibrilada (CNF) e nanocristalina (CNC) [thesis]. Viçosa: Federal University of Viçosa; 2015.

[22] SIQUEIRA, G.; BRAS, J.; DUFRESNE, A. Cellulosic bionanocomposites: a review of preparation, properties and applications. Polymers, vol. 2, pp. 728-765, 2010.

[23] MARIANO, M.; KISSI, N. E.; DUFRESNE, A. Cellulose Nanocrystals and Related Nanocomposites: Review of some Properties and Challenges. Journal of Polymer Science, part B: Polymer Physics, vol. 52, pp. 791-806, 2014.

[24] AGODA-TANDJAWA, G.; DURAND, S.; BEROT, S.; BLASSEL, C.; GAILLARD, C. ; GARNIER, C. ; DOUBLIER, J. L. Rheological characterization of microfibrillated cellulose suspensions after freezing. Carbohydrate Polymers, vol. 80, N. E, pp. 677-686, 2010.

[25] NAKATANI, H.; MIYAZAKI, K.; HΛMΛDΛTE, M.; TERANO, M. Syndiotactic Polypropylene/Microfibrous Cellulose Composites: Effect of Filler Size on Tensile Proporties. Journal of Applied Polymer Science, vol. 128, pp. 915-922, 2012.

[26] TANAKA, H. Paper. In: OSHIMA, H.; FURUSAWA, K. (Ed.) Eletrical phenomena at interfaces: fundamentals, measurements, and applications. 2th ed. New York: Marcel Dekker, 1998. Cap 19. (Surfactant Science, 76).

Section 3

Application of Colloidal Particles

Chapter 8

Removal of Copper Ions from Aqueous Solution Using Liquid Surfactant Membrane Technique

Huda M. Salman and Ahmed Abed Mohammed

Abstract

Liquid Surfactant Membrane (LSM) as an alternative extraction technique shows many advantages without altering the chemistry of the oil process in terms of efficiency, cost effectiveness and fast demulsification post extraction. Copper (Cu) extraction from aqueous solution using Liquid Membrane (LM) technology is more efficient than the sludge-forming precipitation process and has to be disposed of in landfills. In this chapter, a liquid surfactant membrane (LSM) was developed that uses kerosene oil as LSM 's key diluent to extract copper ions as a carrier from the aqueous waste solution through di-(2-ethylhexyl) phosphoric acid (D2EHPA). This technique has several benefits, including extracting one-stage extracts. The LSM process was used to transport Cu (II) ions from the feed phase to the stripping phase, which was prepared, using H_2SO_4. For LSM process, various parameters have been studied such as carrier concentration, treat ratio (TR), agitating speed and initial feed concentration. After finding the optimum parameters, it was possible to extract Cu up to 95% from the aqueous feed phase in a single stage extraction.

Keywords: copper, D2EHPA, extraction, surfactant, liquid membrane

1. Introduction

Increased use of metals and chemicals in process industries has led to the production of large volumes of effluent containing high levels of toxic heavy metals and their presence, due to their non-degradable and persistent existence, poses problems with disposal. World Health Organization (WHO)-based aluminum, cobalt, chromium, iron, cadmium, nickel, zinc, copper, lead and mercury are the most toxic metals [1–3].

Leather tanning, mining, electroplating, textile dyeing, coating operations, aluminum conversion, and pigments are the main industries that introduce water contamination by chromium. Owing to the decreasing availability of natural resources and the rising contamination in the atmosphere, the removal of ions from their effluents has taken on greater significance in the recent past [4–6]. For environmental purposes, the removal of copper (Cu) from aqueous solutions requires an effective method (toxic ions if they are beyond the WHO limits). The minimization of liquid effluents containing hazardous metals is a general concern.

The solvent extraction process is a conventional method to eliminate Cu from solutions. A well-established Cu extract, such as diketones or hydroxytoxic agents [7], should be used in this technique. Both are LIX acid (Cognis) and phosphoric acid (D2EHPA) di- (2-ethylhexyl) [8–11].

Liquid surfactant membrane (LSM) for the isolation of solvents, such as phenols, biochemical products and metal pollutants [2, 12–19], has been considered as an alternative to solvent extraction. LSM is a form of triple dispersion, where a primary emulsion (water/oil or oil/water) is dispersed to be processed in the feed process (E). The liquid membrane consists of three phases: I internal, ii) external and iii) organic. The organic phase includes a diluent, an emulsifier to stabilize the emulsion and, in the case of metal ion separation [10], an extractant. The solution is transferred through the membrane through the stripping phase droplets during the mixing between the feed phase (E) and the emulsion (organic + internal) and is concentrated [20]. After extraction, the emulsion is isolated from the raffinate process and the emulsion is typically demulsified by high voltage or heat application. There are several advantages to LSM, such as single-stage re-extraction, large specific surface area for extraction, concurrent extraction and the need for an expensive extractant in small quantities [10, 21, 22].

The aim of this research was to investigate the potential of a liquid surfactant membrane (LSM) to extract copper ions from the feed solution. Despite studies in this area, the study examined different experimental parameters, such as extractant concentration, ratio of treatment, rate of agitation, and initial feed concentration, to determine the best conditions that would give the LSM the greatest efficiency.

2. Experimental protocols

2.1 Reagents

The phosphoric acid di-(2-ethylhexyl) (D2EHPA) worked as a shuttle and the nonionic emulsifier was Sorbitan monooleate (Span 80 C24H44O6), both of which were supplied by Sigma-Aldrich (Merck, Darmstadt, Germany). The Southern Oil Company (SOC) (Al Basra-Iraq) supplied kerosene used as a diluent, while the removing agent was sulfuric acid (H_2SO_4) and was purchased from the acid and base factory (Babylon, Iraq). Copper solutions were prepared from nitrate of copper (Chemical, Company, Co., Ltd. Korea).

2.2 Procedure

The experimental work consists of four parts: emulsion preparation as a first step, stock solution preparation, extraction process execution, and emulsion demulsification. In this article, **Figure 1** shows the LSM process.

2.2.1 Emulsion preparation

Mixing those volumes of kerosene, Span80, and D2EHPA using SR30 digital Homogenizer, (model: 670/340 W, 10-2000 ml, 3000–27,000 rpm) at a speed of 17,500 rpm to reach the oil process. The sulfuric acid (H_2SO_4) solution was applied dropwise to the oil process as a stripping agent until the necessary volume ratio was obtained from the oil solution to the stripping solution. To achieve a stable Water/Oil LSM, the solution was continuously stirred for 10 minutes.

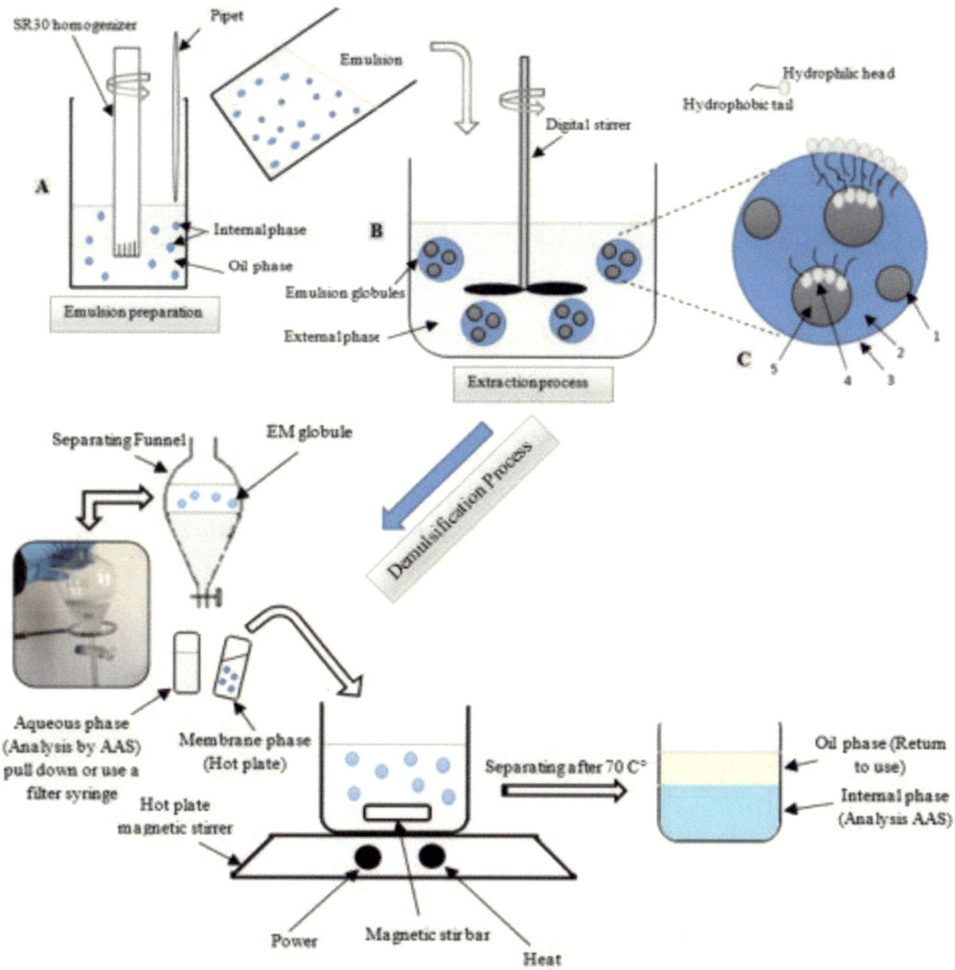

Figure 1.
LSM technique: (1) droplets, (2) organic phase, (3) globules, (4) emulsifier, (5) internal phase and Cu.

2.2.2 Feed phase preparation

This stage was prepared to obtain the necessary concentrations (200 ppm) of copper by adding distilled water (conductivity, 1 μs/m) to Cu $(NO_3)_2$ (solid form) and then adding some drops of sulfuric acid to pH 4.

2.2.3 Extraction

At a temperature of 25 ± 1°C, all experiments were performed. The prepared emulsion (2.2.1) has been added to a specific feed solution volume. The production of double emulsions of water / oil/water was obtained by stirring the contents with a digital stirrer (12,700 rpm) for 12 minutes. The external solution (E) was drawn from the syringe and filter syringe and then analyzed by AAS (atomic absorption spectrophotometry). The resulting solution was allowed to be separated by gravity into an emulsion (water/oil) and an external solution (E) in a 24-hour separation funnel. The external phase was drawn after two-phase separation and the concentration of Cu was analyzed using AAS (Atomic Absorption Spectrophotometer) in the internal phase. The Cu(II) ions remain in membrane process can be determined by mass balance. The extractant concentration, initial Cu concentration, treatment ratio (TR) and stirring speed were varied to observe their effects on Cu extraction in order to understand the important variables relating to the extraction of Cu.

2.2.4 Demulsification of the emulsion

After the extraction experiment, the loaded emulsion was broken into the internal Cu concentrated phase and the organic phase by means of a hot plate magnetic stirrer (70° C for 43 minute). The internal phase (I) was analyzed and the Cu concentration determined after that.

2.3 Extraction mechanism in the ELM system

The prepared emulsion (Section 2.2.1) containing a certain concentration of copper ions at pH 4 (adding some drops of 0.2 M H_2SO_4) was transferred to the external process. For 0–12 minutes, a robotic mixer was used to stir the solution. Eqs. 1 and 2 elucidate the extraction and stripping reactions of the copper ions.

Here, RH refers to an extractant's protonated form (D2EHPA in this paper) [23]. **Figure 2** [24, 25] reveals the D2EHPA structure.

Extraction reaction of the copper ions:

$$Cu^{+2}{}_{(S)} + 2(RH)_{(L)} \rightarrow PbR_2(RH)_2 + 2H^+ \tag{1}$$

Stripping reaction of the copper ions:

$$CuR_2(RH)_2 + 2H^+ \rightarrow Cu^{+2}{}_{(S)} + 2(RH)_{2(L)} \tag{2}$$

At the membrane (O)-external (E) interface, Eq. (1) denotes the reaction, whereas Eq. (2) shows the reaction where the copper ions are stripped at the oil (O)-internal (W) interface. **Figure 3** describes the movement of Cu (II) ions by an extractant from the external phase to the internal phase. Based on the Eq. (3), the extraction percentage (E percent) is found:

$$E\% = \frac{C_{in} - C_{out}}{C_{in}} \times 100\% \tag{3}$$

In the external phase, where C_{in} is the initial copper concentration, and C_{out} is the concentration of copper ions after the extraction phase.

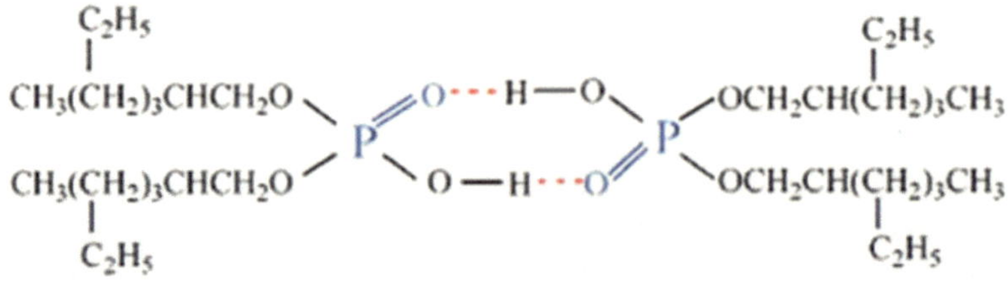

Figure 2.
Depicts the structure of D2EHPA.

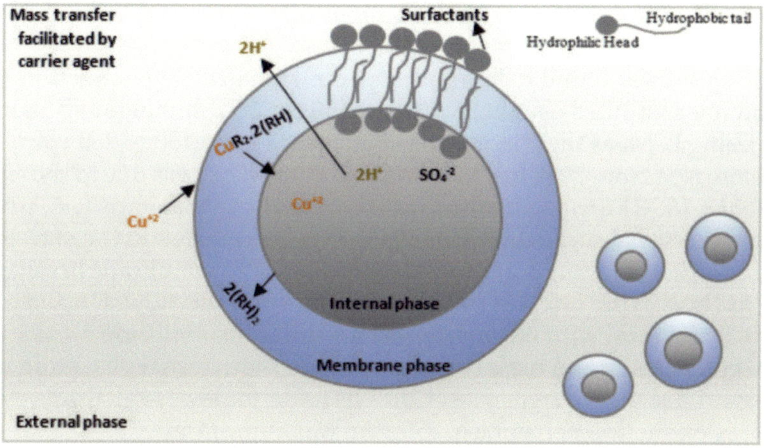

Figure 3.
Depicts the transfer mechanism of LSM.

3. Results and discussion

3.1 Effect of changes in carrier concentration on copper removal efficiency

As expected, this paragraph presented in **Figure 4**, as soon as the mixing began, the extraction efficiency increased in the first 0.5 minutes due to the efficacy of the carrier in carrying the copper ions and the increase of the shuttle D2EHPA concentration from 6–8% (v / v) provides only a 2% increase in the quantity extracted using LSM. At 10% D2EHPA, the E percentage decreased significantly. It should be noted that the D2EHPA concentration in the membrane process was observed to decrease the rate of copper extraction in the range of 2% (v / v) to 4% (v/v) under optimum conditions for copper extraction from nitrate solution, as observed by [2, 23]. An improvement of 2 percent from an economic point of view is very low, so 6 percent of D2EHPA is used in the experiments.

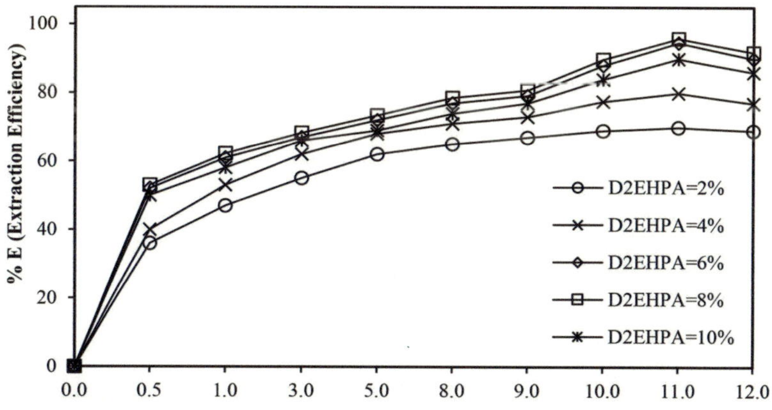

Figure 4.
Effect of D2EHPA concentration on the Cu extraction at optimal conditions using LSM. (O/I = 1/1, span 80 = 4 v/v%, H_2SO_4 = 0.5 M, feed concentration≈ 200 mg/L, pH = 4, TR = 1:10, mixing speed = 250 rpm).

3.2 Effect of changes of stirring speed on the copper removal efficiency

Stirring speed was found to be another parameter affecting extraction to a large extent, and it was studied using LSM1 in the 150 to 550 rpm range and shown in **Figure 5**. Using LSM, as the stirring speed increased from 150 to 250 rpm, copper removal increased from 82% to 94.7% in 11 minutes. This was due to the small size of the globules (SSG) formed by the shear force of the stirrer impellers, which provided more interfacial surface area for efficient mass transfer. In the external phase, no copper was detected for more than 11 minutes due to membrane breakage. However, as the stirring rate was increased to 300 rpm, the emulsion and external phase were introduced with more shear, which promotes emulsion breakage. The interfacial contact area and mass transfer between the external phase and the emulsion decreased due to the larger size of the emulsion for lower agitating velocity. For a satisfactory extraction percentage, 250 rpm was appropriate.

The proportion starts to decrease after 250-rpm extraction. A further increase in the mixing speed resulted in a breakdown of the liquid surfactant membranes, resulting in the outflow of extracted lead into the external phase. This is due to a higher mixer speed, which usually results in greater transport of water into the inner strip process beyond limits, causing the membrane to swell [26, 27]. 250 rpm was therefore chosen as the optimum speed of mixing for Cu (II) extraction.

3.3 Effect of changes of treat ratio (TR) on the copper removal efficiency

The ratio of the emulsion phase to the feeding phase in an LSM extraction is the treatment ratio. Generally, rising TR contributes to an improvement in the loading ability and extraction rate. This case occurred due to an increase in emulsion volume and an increase in D2EHPA and H_2SO_4 [28, 29]. **Figure 6** illustrates the effect of TR on the copper extraction from copper nitrate solutions using LSM. As TR improved, there was an improvement in the efficiency of this ratio as it improved from 1:15 to 1:10. Because of the increased hold-up of the emulsion, this trend may be known from a potential rise in distribution of globule size. Due to increased globule-size distribution at larger emulsion hold-ups, Sengupta et al. (2006) observed a strong decrease in the extraction percentage of silver ions when TR was raised from 1:6 to 1:4.

The formation of LG (larger-globules) decreases the outer surface areas and increases the effective duration of the pathways of diffusion between the globules, resulting in a low removal rate of Cu. Treatment ratios of 1:15, 1:10 and 1:5 indicate a substantial increase in extraction capacity at which time TR increased from

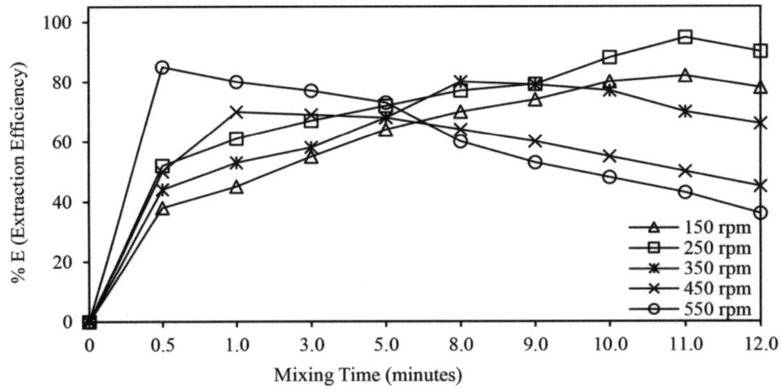

Figure 5.
The effect, of stirring speed on a rate of copper extraction using LSM.

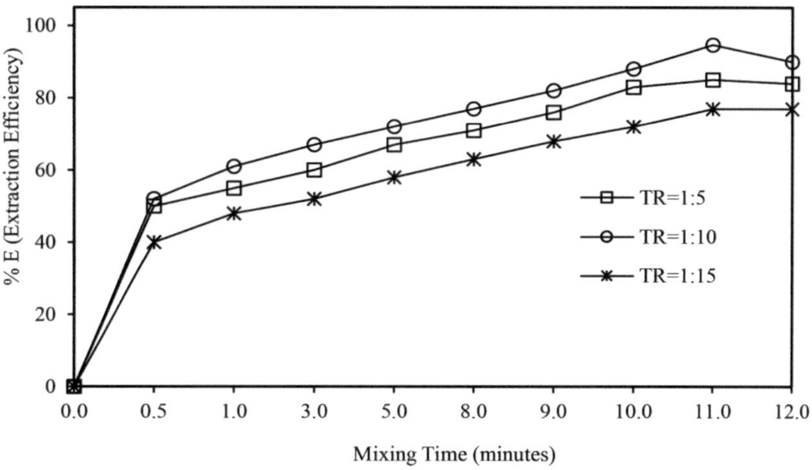

Figure 6.
Effect of (TR) on the Cu-extraction by LSM

1:15 to 1:10, owing to an increase in emulsion retention, the size distribution of the globules tended to shift to LG with a consequent decrease in the pace.

3.4 Effect of changes of initial copper concentration on copper removal efficiency

Using emulsions with O/I = 1/1, span80 = 4 v / v percent of the organic phase and H_2SO_4 = 0.5 M, D2EHPA = 6 percent (v/v), the effect of initial Cu (II) ion concentrations in the feed on the rate of copper extraction was investigated. At 4 and 1:10 respectively, the original (pH) and (TR) were retained. The extraction results are shown in **Figure 7**, which is a plot of the change in copper concentration over time in the feed stage.

Figure 8 demonstrates the pattern of copper loading in LSM along with a quantitative assessment of the quantity of copper stripped in the internal stripping step of the emulsion after a 12-minute contact between the feed and LSM for differences in the initial feed concentration.

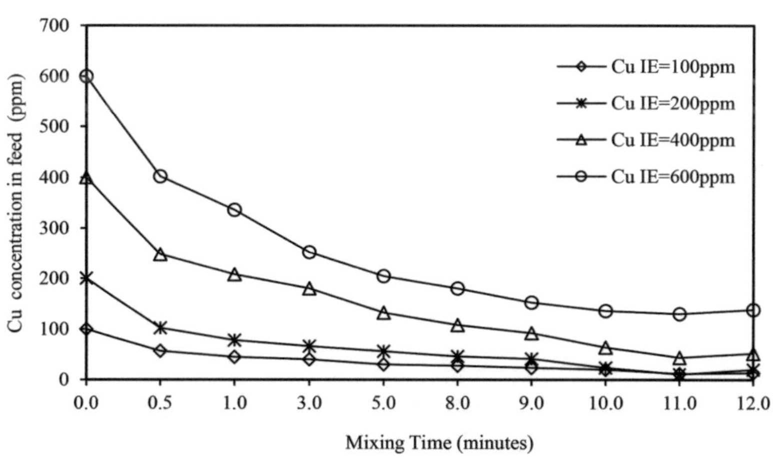

Figure 7.
Effect of initial-feed concentration on rate of copper extraction using LSM (O/I = 1/1, span 80 =,4 v/v%, H_2SO_4=,0.5 M, D2EHPA = 6%, feed concentration≈ 200 mg/L, pH = 4, TR = 1:10, mixing speed = 250 rpm). (Cu IE, initial-concentration of copper, in the external phase).

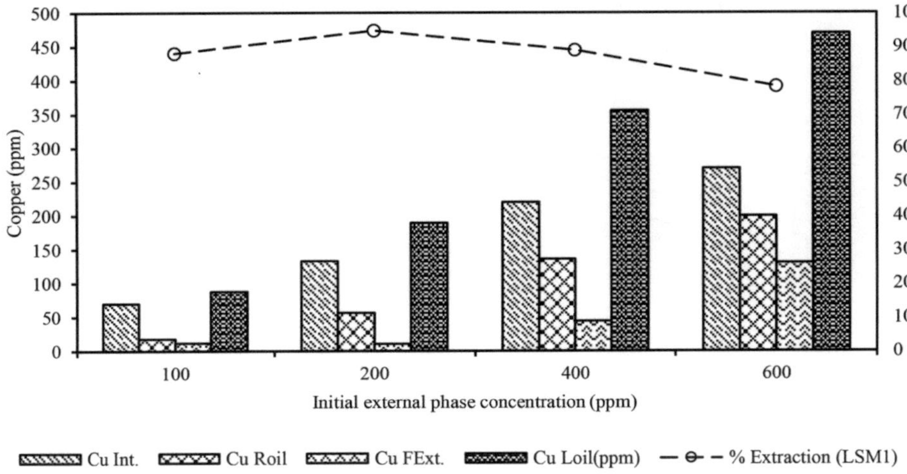

Figure 8.
Copper extraction, stripping patterns in LSM. (Int., internal phase; roil, retained in the oil phase; FExt, final concentration in the external phase.

The extent of copper-extraction-into LSM was also increased as the initial-feed concentration increased.-When Cu loading was low, most of the Cu extracted in the membranes was stripped during the inner process of the membranes. However, the amount of copper stripped during the internal process of the LSMs did not increase significantly at high copper loadings, so most of the copper removed by the LSMs was retained during the membrane phase [4, 21, 22].

From the slow stripping kinetics, as well as the diffusional effects that play an important role in further slowing down the stripping rates, the low percentage of Cu stripping could be recognized. Strong CuIE (Initial copper concentration) values lead to higher copper loads in the LSMs, resulting in rapid saturation of the peripheral internal phase droplets in the emulsion, requiring deeper penetration of the Cu-D2EHPA complex inside the emulsion globules to be stripped.

4. Conclusions

Using a liquid surfactant membrane (LSM), copper Cu (II) extraction from an aqueous process was studied. The membrane consisted of D2EHPA dissolved as a solvent as an emulsifier in kerosene and span80, respectively. The stripping-solution was used for sulfuric acid (H_2SO_4). The optimum conditions for Cu extraction are: (a) 6–8 percent (v / v) concentration of D2EHPA, (b) 4 percent (v/v) concentration of span80, (c) concentration of 0.5 M concentration of H_2SO_4 in the internal phase, (d) 1:1 the internal phase-to-membrane phase ratio; (e) the external phase acidity is 4; (f) the external phase volume is 1/10 of the membrane volume; (g) the extraction time is 11 minutes; and (h) the agitation speed is 250 rpm. The results also showed that many parameters are very important in Cu extraction, stirring speed, D2EHPA concentration, feed concentration and treatment ratio, (2) Cu extraction efficiency (E) is 95 percent at 11 minutes. (3) Small emulsion droplets are produced at the higher agitating velocity of the water /oil/water emulsion, thus increasing the carrier/Cu reaction interface area. However, in order to increase the extraction efficiency, this paper considered a maximum limit (250 rpm); (4) the results showed that the LSM method is a beneficial method for removing Cu from aqueous solution.

Author details

Huda M. Salman* and Ahmed Abed Mohammed
Environmental Engineering Department, College of Engineering,
University of Baghdad, Baghdad, Iraq

*Address all correspondence to: hudamohammad20@gmail.com

References

[1] Hegazi HA. Removal of heavy metals from wastewater using agricultural and industrial wastes as adsorbents. HBRC Journal. 2013;**9**:276-282

[2] Mohammed AA, Selman HM. Liquid surfactant membrane for lead separation from aqueous solution: Studies on emulsion stability and extraction efficiency. Journal of Environmental Chemical Engineering. 2018;**6**:6923-6930

[3] Algureiri AH, Abdulmajeed YR. Removal of heavy metals from industrial wastewater by using RO membrane. Iraqi Journal of Chemical and Petroleum Engineering. 2016;**17**:125-136

[4] Lu D, Chang Y, Wang W, Xie F, Asselin E, Dreisinger D. Copper and cyanide extraction with emulsion liquid membrane with LIX 7950 as the mobile carrier: Part 1, emulsion stability. Metals. 2015;**5**:2034-2047

[5] Hasan M, Selim Y, Mohamed K. Removal of chromium from aqueous waste solution using liquid emulsion membrane. Journal of Hazardous Materials. 2009;**168**:1537-1541

[6] Hochhauser AM, Cussler E. Concentrating Chromium with Liquid Surfactant Membranes. American Institute of Chemical Engineers: AIChE Symposium Series; 1975. pp. 136-142

[7] Rydberg J, Musikas C, Choppin GR. Principles and practices of solvent extraction. M. Dekker New York. 1992

[8] Kongolo K, Mwema M, Banza A, Gock E. Cobalt and zinc recovery from copper sulphate solution by solvent extraction. Minerals Engineering. 2003;**16**:1371-1374

[9] Hu S-YB, Wiencek JM. Copper—LIX 84 extraction equilibrium. Separation Science and Technology. 2000;**35**:469-481

[10] Abdel-Halim S, Shehata A, El-Shahat M. Removal of lead ions from industrial waste water by different types of natural materials. Water Research. 2003;**37**:1678-1683

[11] E.A. Fouad, Zinc and copper separation through an emulsion liquid membrane containing Di-(2-Ethylhexyl) phosphoric acid as a carrier, Chemical Engineering & Technology: Industrial chemistry-plant equipment-process engineering-Biotechnology 31 (2008) 370-376.

[12] P.F. Correia, J.M. de Carvalho, Recovery of 2-chlorophenol from aqueous solutions by emulsion liquid membranes: Batch experimental studies and modelling, Journal of Membrane Science 179 (2000) 175-183.

[13] Mohammed AA, Selman HM. Extraction of Lead ions from aqueous solution by CO-stabilization mechanisms of magnetic Fe2O3 particles and nonionic surfactants in emulsion liquid membrane. Colloids and Surfaces A: Physicochemical and Engineering Aspects. 2019

[14] Fouad E, Ahmad F, Abdelrahman K. Optimization of emulsion liquid membrane for Lead separation from aqueous solutions. Engineering, Technology & Applied Science Research. 2017;7:2068-2072

[15] Zeng L, Wang W, Chen W, Bukirwa C, Yang Y. Experimental and modeling of nickel removal from sulfate solutions by emulsion liquid membrane using PC 88A. Desalination and Water Treatment. 2016;**57**:11184-11194

[16] Ng YS, Jayakumar NS, Hashim MA. Performance evaluation of organic emulsion liquid membrane on phenol

removal. Journal of Hazardous Materials. 2010;**184**:255-260

[17] Correia PF, de Carvalho JM. A comparison of models for 2-chlorophenol recovery from aqueous solutions by emulsion liquid membranes. Chemical Engineering Science. 2001;**56**:5317-5325

[18] Diaconu I, Gîrdea R, Cristea C, Nechifor G, Ruse E, Totu EE. Removal and recovery of some phenolic pollutants using liquid membranes. Romanian Biotechnological Letters. 2010;**15**:5702-5708

[19] Mahmoud HE, Al-Hemiri AA. Minimization of toxic ions in waste water using emulsion liquid membrane technique. Iraqi Journal of Chemical and Petroleum Engineering. 2010;**11**:11-19

[20] Zing-Yi O. N. Oth+mn, M. Mohamad, R. Rashid, Removal performance of lignin compound from simulated pulping wastewater using emulsion liquid membrane process, International Journal of Global Warming. 2014;**6**:270-283

[21] Sengupta B, Bhakhar MS, Sengupta R. Extraction of copper from ammoniacal solutions into emulsion liquid membranes using LIX 84 I®. Hydrometallurgy. 2007;**89**:311-318

[22] Rouhollahi A, Zolfonoun E, Salavati-Niasari M. Effect of anionic surfactant on transport of copper (II) through liquid membrane containing a new synthesis Schiff base. Separation and Purification Technology. 2007;**54**:28-33

[23] Ren Z, Zhang W, Meng H, Liu Y, Dai Y. Extraction equilibria of copper (II) with D2EHPA in kerosene from aqueous solutions in acetate buffer media. Journal of Chemical & Engineering Data. 2007;**52**:438-441

[24] He J, Li Y, Xue X, Ru H, Huang X, Yang H. Extraction of Ce (IV) from sulphuric acid solution by emulsion liquid membrane using D2EHPA as carrier. RSC Advances. 2015;**5**:74961-74972

[25] Irannajad M, Afzali Z, Haghighi HK. Solvent extraction of copper using TBP, D2EHPA and MIBK. Russian Journal of Non-Ferrous Metals. 2018;**59**:605-611

[26] Kumbasar RA. Extraction and concentration of cobalt from acidic leach solutions containing Co–Ni by emulsion liquid membrane using TOA as extractant. Journal of Industrial and Engineering Chemistry. 2010;**16**:448-454

[27] Abdulmajeed YR, Haweel CK, Slaiman QJ. Removal of heavy metals ions from aqueous solutions using biosorption onto amboo. Iraqi Journal Of Chemical And Petroleum Engineering. 2010;**11**:23-32

[28] Kusumastuti A, Syamwil R, Anis S. Emulsion Liquid Membrane for Textile Dye Removal: Stability Study. AIP Conference Proceedings: AIP Publishing; 2017. p. 020026

[29] Perumal M, Soundarajan B, Thazhathuveettil Vengara N. Extraction of Cr (VI) by Pickering emulsion liquid membrane using amphiphilic silica nanowires (ASNWs) as a surfactant. Journal of Dispersion Science and Technology. 2018:1-10

Chapter 9

Hydrocolloids in Dentistry: A Review

Stanley Onwubu and Chibuzor Stellamaris Okonkwo

Abstract

Hydrocolloids are complex polysaccharides that disperse or dissolve in aqueous solution to give thickened or vicious effects. Also hydrocolloids possess high molecular weight. Owing to these characteristics, hydrocolloids have been widely used in various applications. In dentistry, for example, most intricate and precise procedures are made of hydrocolloids and are found in its simplest material to the most complex material such as impression making, fillings, separating media, electro-polishing etc. The two common hydrocolloids widely used in dentistry are reversible (agar) and irreversible (alginate) materials. Hence, this chapter bring to the forefront their preparations, uses and storage for optimal results and application.

Keywords: agar, alginate, dentistry, hydrocolloids, impression materials

1. Introduction

Hydrocolloids are colloidal systems wherein the colloid particles are hydrophilic polymers dispersed in water and depending on the quantity of water available that can take place in different states, e.g., gel or sol (liquid) It is an intermediate between a solution and a suspension which can be distinguished from solutions using the Tyndall effect [1]. Hydrocolloids materials are available in the form of viscous liquids in the "sol" state or the form of semi-solid substances of a gelatinous consistency. Without a filler, the gel would lack stability and would have a slimy surface covered with synerate exudate [2]. They can be either irreversible (single-state) or reversible hydrocolloids (transiting from gel-sol-gel on the application of heat) [3]. Owing to their unique properties, hydrocolloids have found wide and useful applications in various fields including, dentistry, medicine and the food industries. For instance, hydrocolloids such as Xanthan, gum Arabic, Pectin are added to food as additives due to their gelling, viscosity, and stabilizing properties [4, 5]. The aforementioned hydrocolloids could significantly reduce human appetite in acute settings due to the ability to form gelation in the stomach when ingested. An important rheological property of fibers within the intestine is viscosity, which is thought to account for beneficial physiological responses in relation to appetite regulation, glycemic and lipidemic control [4, 5]. In Medicine, studies have been carried out with different strategies and approaches or a combination of both as hydrocolloid gels have found some potentials in bone regeneration in the delivery of osteo-inductive factors, bone-forming cells, or a combination of both [2].

Recently, hydrocolloid (alginate) gels have also been actively investigated for their ability to mediate the regeneration and tissue engineering of different tissues and organs, including skeletal muscle, nerves, the pancreas, and liver. Current strategies for skeletal muscle regeneration include cell transplantation, growth factor delivery, or a combination of both approaches [6, 7]. Also, Studies within the pharmacological field have demonstrated how alginate-antacid formulations can decrease post-prandial symptoms by neutralizing the acidity of gastric contents by forming a gel-like barrier to displace the "acid pocket" from the oesophagogastric junction and protect the oesophageal and gastric mucosa with controlled released drug products used as model system for mammalian cell culture in biomedical studies [6]. In Dentistry, for example, hydrocolloids are widely used in the fabrication of dental and maxillofacial prostheses impression due to their biocompatibility with the tissues, ease of use, physical properties and hydrophilicity with the oral tissues [8]. Other areas of hydrocolloids applications include orthopaedic structures and stone models in surgical cases [8]. This chapter aims to discuss the different hydrocolloids used in dentistry, their preparations, uses and storage for optimal results and application.

1.1 Overview of hydrocolloids application in dentistry

Hydrocolloids were the first elastic materials to be used in the Dentistry [2]. Elastic impression materials commonly used in the dental field include reversible hydrocolloids (agar-agar), irreversible hydrocolloids (alginate), and other synthetic and elastomeric materials such as polysulfide, polyether's, and silicone [2]. The properties and abilities of hydrocolloids materials enables the replication of the oral tissue with little or no deformity on withdrawal while abiding to both manufacturer and mechanical stipulations of its manipulation prior to being loaded on a tray to produce the gel or sol form [3].

Agar discovered by Sears in 1937 was the first hydrocolloids used in dentistry for making impressions to circumvent the cumbersome procedure and oral lacerations of using impression compound [9]. Agar is a vegetable colloid derived from seaweed found on the sea coast of Japan, a jelly-like substance softened when heated and solidifies when cooled [9]. However, the technique of using agar was complicated because of the need for special heaters and tempering Jars for heating and holding prior to use, syringes and water-cooled trays, even though it could be used severally without losing its chemical and physical properties before been discarded [10].

In 1947, alginate was introduced during the second world war as a result of the scarcity of agar from Japan by the extraction of alginic acid from marine seaweed [10]. Unlike agar which reaction is reversible; alginate reaction was chemical which resulted in irreversible hydrocolloids when the alginate gels are mixed with water [10]. Furthermore, the physical, mechanical, biocompatibility and fatigue properties and most importantly the hydrophilic nature that allows hydrocolloids to capture accurate impressions in the presence of some saliva or blood [10].

Equally significant, its low wetting angle makes it easy to capture full or partial arch impressions moderate ability to reproduce the detail and costs relatively little compared with other elastomeric impression materials [10]. Despite this, hydrocolloids materials are not accurate enough for fixed partial dentures but are used for partial framework impressions to the modelling materials, ability to adapt to the oral tissues and the formation of an elastic resilient film [3, 11].

1.2 Types of hydrocolloids use in dentistry

Generally, hydrocolloids used in dentistry can be typified as either reversible (agar) and or irreversible (alginate). This section, therefore, focuses on the properties, composition and application of these two hydrocolloids materials.

1.2.1 Agar

Agar hydrocolloid has remained an excellent, cost-effective impression material since its discovery in 1937 from seaweed found on the coasts of Japan, and thus, has been used widely for the replication/duplication of models [12]. It is a reversible hydrocolloid which can repeatedly pass between highly viscous gel and low viscosity sol through heating and cooling [3]. In terms of its chemical composition and structure, agar is the sulfuric ester of a linear polymer of galactose extracted from seaweed [3].

1.2.1.1 Composition

The components of the agar gels are 12–15% agar, 1% potassium sulphate to ensure a proper set of the gypsum material poured in the impression, 0.2% borax as a strengthener for the gel, 0.1% alkyl benzoate as an antifungal during storage, and 85% water (**Table 1**). Borax and agar retard the set of gypsum products, so potassium sulphate is added to cancel out their effect [13].

The composition described in **Table 1** may differ slightly depending on the dispersing medium for the gel; which could be either loading on an impression tray or a syringe. When fine details of preparation are needed, a less concentrated gel type is used in a syringed. A more concentrated gel is used to in water-cooled tray to form the bulk of the impression. Agar possesses relatively good elastic recovery, reproduction of details, pleasant tasting and easy to clean up. But it cannot be used to produce electroplated pies due to its dissolution inside the electrolytic bath [14]. Agar is a technique sensitive impression material due to its low tear strength of 27.6 KPa. Agar is dimensionally unstable due to the loss of water from the agar gels even when stored at 100% humidity. The consequence of this is an inaccurate model if left for a while before the cast is poured [9]. Agar hydrocolloids are supplied as sticks or gel and require specific equipment for its manipulation before the impression making process. Thereby making the process cumbersome but can be reused once the setup is done. Although agar hydrocolloid is an inexpensive

Material	Composition (Approximate percentage)	Purpose
Agar	12–15%	Colloidal particle as basis of the gel
Potassium Sulphate	1%	Ensures set of gypsum material
Borax	0.2%	Strengthens the gel
Alkyl Benzonate	0.1%	Antifungal agent
Water	85%	Dispensing medium for the colloidal suspension

Table 1.
Composition of agar gels.

impression material with very good accuracy, its use has declined over the years due to the inability to pour impressions immediately, low dimensional stability, ease of manipulation and water-cooled impression trays and the inability to produce electroplated dies [9].

1.2.2 Alginate

Alginate is an irreversible hydrocolloid largely used in dentistry [15]. It is mainly used for diagnostic and planning in the rehabilitation of oral, orthodontics and maxillofacial prostheses [16–18]. The advantage of alginate materials is that it is easy to manipulate, cheap and provides a good level of comfort for patients without the need for specialized instruments and equipment [17, 19].

1.2.2.1 Composition and setting reactions

Discovered in 1945, as a substitute for agar whose importation was hampered by the outbreak of the Second World War. Alginates are salts of alginic acid, a polysaccharide extracted from the cell walls of brown algae (washed, ground and chemically treated, especially the pulp) belonging to the Phaeophyceae family, widespread especially in America [20]. Like agar, alginic acid, chemically known as anhydro-B-D-mannuronic acid has a high molecular weight (30,000 to 200,000) linear polymer [9].

The extracted alginic acid is then converted into a salt (alginate) of sodium, calcium, potassium or magnesium. Although alginate is insoluble in water, its alkaline salts are water-soluble. The production process of sodium alginate from brown algae can be done in two ways; using the calcium alginate method or the alginic acid method [21]. To extract alginic acid, the algae are placed in a sodium carbonate bath, exploiting the solubility of alkaline alginates in water. The alginic acid is recovered from the obtained solution by precipitation with hydrochloric acid or sulfuric acid [21]. The difficulty of the processes lies in the required physical separations; such as in the filtration of muddy residues from viscous solutions or in the separation of gelatinous precipitates that retain a large amount of liquid in their structure, resisting filtration and centrifugation [2].

The alginate impression materials for dental use contain several additives such as sodium alginate, calcium sulphate, trisodium phosphate, diatomaceous earth, zinc oxide, and potassium titanium fluoride, all in the form of a powder [2]. They are irreversible hydrocolloids because the picking reaction is a chemical reaction of irreversible precipitation therefore they cannot return in sol form using physical means, such as temperature, as with reversible hydrocolloids.

The chemical reaction occurs two times: a first phase called 'slowing' and a second phase called 'setting'. Initially, the powder is mixed with water, a sol is formed and the sodium or potassium salts of alginic salts react with the calcium sulphate [2] to allowing crosslinking of the alginic salts [9, 22, 23]. After the sodium phosphate has reacted, the remaining calcium sulphate reacts with sodium alginate to form insoluble calcium alginate that forms a gel with water which acts as a catalyst. There are many commercial variations of alginate that vary in consistency, setting time, elasticity, strength, and dimensional stability; manufacturers also add fillers, which impact on its properties, application, setting time, and pouring time [22]. The standard composition of alginate is as described in **Table 2**.

The alginates available on the market can be of two types: fast setting (hardening time of 1–2 min) or normal setting (setting time between 2 and 5 min). The setting time depends on the composition (water/powder ratio, where increasing the powder accelerates the hardening reaction) and the temperature at which mixing

Material	Percentage (Approximate)	Purpose
Sodium or potassium alginate	15–20%	Colloidal particles as basis of the gel
Calcium sulphate dihydrate	14–20%	Creates irreversible gel with alginate
Potassium sulphate	10%	Ensures set of gypsum materials
Trisodium phosphate	2%	Retarder to control setting time
Diatomaceous earth	55–60%	Filler to increase thickness and strength
Other additives: chemical indicators	Very small quantities	Colour change
• Organic glycols • Flavoring agents • Coloring agents • Disinfectants		Reduce dust when powder is handled Improve taste of material Provide pleasant colors Cause antibacterial action

Table 2.
Composition and properties of alginate use in dentistry.

takes place [2]. The reaction that causes the alginate impression materials to form makes use of the different solubilities of the sodium, potassium, ammonium and calcium H_2O % Na & alginate % $CaSO_4$ ÿ (paste) Ca & alginate % Na_2SO_4 (gel) salts of alginic acid in water.

The setting reaction is a chemical reaction between Sodium Alginate and Calcium Sulphate, where:

$$2Na_3PO_4 + 3CaSO_4 \rightarrow Ca_3(PO_4)2 + 3Na_2SO_4 \quad (1)$$

This reaction (1) can be retarded with Calcium Phosphate, which acts as a retarder, thereby increasing the setting time and obtain a type I (fast set) or type II (Normal set) setting time [20].

The irreversible hydrocolloids, which are the most commonly used, are a mixture of manual or mechanized techniques through the union of powder and water [2]. Alginate impression materials are easy to use and manipulate without specialized equipment but can be mixed manually or mechanically, it is less expensive and has more rapid setting times. The reaction time and the setting time can be controlled with the temperature of the water used. They are slightly flavoured and in recent formulations, have colour indicators according to the phases of the chemical reaction [2].

1.3 Summary

What started as a trial in the 19th century gradually became a benchmark in the history of dentistry and has today found its way into different aspects of medical, pharmaceutical and food industries with more studies on how to improve its effectiveness for optimal use. We are currently in a technologically advanced era which is gradually employing the use of CAD/CAM technologies for the diagnosis and treatment of patients which is still very expensive due to the cost of the equipment and specialized training required for the operators to interface with it. Amid all these, studies are still being carried out on convectional irreversible hydrocolloids to improve their physical, mechanical and biological properties [24].

Recent studies have shown that the dimensional stability of hydrocolloids has been improved upon with the materials which have extended cast pouring times [25]. The incorporation of disinfectant gels into the powder which when mixed with water will dissolve thereby preventing surface inaccuracies when soaked/sprayed with disinfectants [25]. Dust-free particles with the use of glycerine making the powder denser and the two-sol system like elastomeric materials to reduce inaccuracies due to annual mixing [25]. Chromatic products to indicate the different chemical reactions within the sol [25].

An important advancement is the use of agar-alginate laminate for making impressions which give better accuracy, thereby, eliminating the water-cooled trays for agar impressions. Certainly improving the quality and definition of these materials would be possible to expand their use with benefits for patients. Also, the reduced setting time and their single-footprint technique will provide added benefits for dentists in terms of time available for manipulation. The prospect is that these materials will continue to evolve as has happened since the 40s, thus producing high-performing impression materials [2].

1.4 Conclusion and recommendation

Despite the advancement in science and technology, hydrocolloids have remained relevant in dentistry, particularly as an impression material. The resilient properties of hydrocolloids coupled with the simplicity of use and biocompatibility with the oral tissues had endeared it in dental practices. While hydrocolloids had some inherent disadvantage in its properties, the advancement in material science and modification of hydrocolloids with other additivities has improved their properties and usefulness in the oral care practice.

Conflict of interest

The authors declare no conflict of interest.

Author details

Stanley Onwubu[1]* and Chibuzor Stellamaris Okonkwo[2]

1 Department of Dental Sciences, Durban University of Technology, South Africa

2 Shehu Idris College of Health Sciences and Technology, Makarfi Kaduna, Nigeria

*Address all correspondence to: profstan4christ@yahoo.com

References

[1] Hansson O, Eklund J. A historical review of hydrocolloids and an investigation of the dimensional accuracy of the new alginates for crown and bridge impressions when using stock trays. Swedish Dental Journal 1984; 8: 81-95.

[2] Gabriele C, Luca F, et al. Alginate Materials and Dental Impression Technique: A Current State of the Art and Application to Dental Practice Mar Drugs. 2019 Jan; 17(1). doi: 10.3390/md17010018 PMCID: PMC6356954 PMID: 30597945

[3] Madhavan S. A Review on Hydrocolloids-Agar and Alginate. J. Pharm. Sci. Res Cuddalore. 2015; 7: 704-707.

[4] Thornton, A.J.; Alsberg, E.; Albertelli, M.; Mooney, D.J. Shape-defining scaffolds for minimally invasive tissue engineering. Transplantation **2004**; 77: 1798-1803.

[5] Ma, H.L.; Hung, S.C.; Lin, S.Y.; Chen, Y.L.; Lo, W.H. Chondrogenesis of human mesenchymal stem cells encapsulated in alginate beads. Journal of Biomedial Materials Research Part A **2003**, 64, 273-281.

[6] Park, H.; Kang, S.W.; Kim, B.S.; Mooney, D.J.; Lee, K.Y. Shear-reversibly cross-linked alginate hydrogels for tissue engineering. Macromolecular Bioscience **2009**, 9, 895-901.

[7] Saxena, A.K.; Marler, J.; Benvenuto, M.; Willital, G.H.; Vacanti, J.P. Skeletal muscle tissue engineering using isolated myoblasts on synthetic biodegradable polymers: Preliminary studies. Tissue Engineering **1999**, 5, 525-532. [CrossRef] [PubMed]

[8] Vogel, A.B.; Kilic, F.; Schmidt, F.; Rübel, S.; Lapatki, B.G. Dimensional accuracy of jaw scans performed on alginate impressions or stone models: A practice-oriented study. Journal of Orofacial Orthopedics **2015**, 76, 351-365. [CrossRef] [PubMed]

[9] Homer Vernon Reed: Quintessence International Volume 21, Number 3/1990

[10] Walker MP, Burckhard J, Mitts DA, Williams KB. Dimensional change over time of extended-storage alginate impression materials. The Angle Orthodontist 2010; 80(6): 1110-5.

[11] https://www.edinformatics.com/math_science/hydrocolloids.htm (internet) Retrived on 2020/09/13

[12] Rishi D, Patel MTK, Charles J. Goodacre, Myron S. Winer MS. An in vitro investigation into the physical properties of irreversible hydrocolloid alternatives. The Journal of Prosthetic Dentistry 2010 Nov:104(5):325-32.

[13] Rodrigues SB. et al Influence of delayed pouring on irreversible hydrocolloid properties. Braz. oral res. [Internet]. 2012 Oct (cited 2020/09/10) 26(5): 404-409. Available from: http://www.scielo.br/scielo.php?script=sci_arttext&pid=S1806-83242012000500005&lng=en. https://doi.org/10.1590/S1806-83242012000500005

[14] Sweta P., Sharayu N, Anjali B., Yogita D. Recent Advances in Elastomeric Impression Materials. http://www.easpublisher.com/easjdom/ Copyright @ 2019: Received: 05.09.2019 Accepted: 11.09.2019 Published: 26.09.2019

[15] Sedda M, Casarotto A, Raustia A, Borracchini A. Effect of storage time on the accuracy of casts made from different irreversible hydrocolloids. The Journal of Contemporary Dental Practice 2008; 9(4): 59-66.

[16] Torassian G, Kau CH, English JD, Powers J, Bussa HI, Marie Salas- Lopez A, et al. Digital models vs plaster models using alginate and alginate substitute materials. The Angle Orthodontist 2010; 80(4): 474-81

[17] Rudd KD, Morrow RM, Strunk RR. Accurate alginate impressions. The Journal of Prosthetic Dentistry 1969; 22(3): 294-300.

[18] Fokkinga WA, Witter DJ, Bronkhorst EM, Creugers NHJ. Clinical fit of partial removable dental prostheses based on alginate or polyvinyl siloxane impressions. The International Journal of Prosthodontics 2017; 30(1): 33-7.

[19] Imbery TA, Nehring J, Janus C, Moon PC. Accuracy and dimensional stability of extended-pour and conventional alginate impression materials. Journal of the American Dental Association (1939) 2010 Jan; 141(1):32-9.

[20] Spoto, G. Materiali e Tecnologie Odontostomatologiche; AriesDue: Milano, Italy, 2013; pp. 150-153, 154-196, ISBN 978-88-98789-00-9.

[21] Mc Huh D J. 1987 Productgion, properties and uses of Alginates FAO Fish pp 288:58-115. http//www. FAO. org/docrep/x5822E/X5822E00.htm

[22] Nandini VV, Venkatesh KV, Nair KC. Alginate impressions: A practical perspective. Journal of Conservative Dentistry 2008 Jan-Mar; 11(1):37-41.

[23] Igarashi, T.; Iwasaki, N.; Kasahara, Y.; Minami, A. A cellular implantation system using an injectable ultra-purified alginate gel for repair of osteochondral defects in a rabbit model. Journal of Biomedial Materials Research Part A **2010**, 94, 844-855.

[24] Onwubu, S.C., Mdluli, P.S., Singh, S. and Ngombane, Y., 2019. Alginates in Evolution of Restorative Dentistry. *Alginates: Applications in the Biomedical and Food Industries*, p.125.

[25] Ramon Vaz da COSTA; Monique Gonzaga Silva VALENTE; Sicknan Soares da ROCHA Analysis of the dimensional stability of extended-storage irreversible hydrocolloids Rev Odontol Bras Central 2017; 26(76): 7-10 ISSN 1981-3708\

Chapter 10

Application of Colloids and Its Relevance in Mineral Engineering

Abhyarthana Pattanaik and Rayasam Venugopal

Abstract

Mineral engineering is an interdisciplinary branch which includes many branches like physics, chemistry, math and sub branches like instrumentation, chemical engineering, mechanical engineering, geology etc. Amongst the various separation/beneficiation techniques of mineral processing, froth flotation is one of the most important fines beneficiation technique, which depends upon the surface and colloid chemical phenomena as the basis of selectivity. The method of separation relies on the surface state and colloidal chemistry of the ore particles and chemical reagents. Adsorption at the mineral solution interface is of major importance for the behaviour of mineral particles in the solution and for successful flotation performance. Adsorption of simple ions determine the change of the particle surface and electrochemical properties of the pulp/slurry phase and therefore affect the colloidal stability and the adsorption behaviour of reagent on the mineral surface. This chapter describes in detail about the role, importance and application of colloidal chemistry in mineral processing especially froth flotation. Froth flotation will remain a key unit operation for the treatment of low-grade ore fines for the decades to come with the overarching challenge as the need of the hour is to modify and improve existing process conditions so as to maintain an acceptable grade and recovery response for the feed whose liberation is more finer, more complex association of minerals and of lower grade.

Keywords: mineral processing, iron minerals, colloids chemistry, surface chemistry, adsorption, zeta potential, interfacial chemistry

1. Introduction

Mineral Engineering is a multidisciplinary and inter disciplinary branch, which includes many other branches and sub branches. Due to depletion of high-grade ores and liberation size being finer, particles are approaching colloidal size range, mineral processing is becoming more and more applied colloid chemistry. Colloid chemistry is inevitably involved in all aspects of mineral processing, ranging from rheological phenomena in grinding, in froth flotation in which selective adsorption of various chemical additives (flotation collectors, flocculants, dispersants, depressants etc.) onto mineral surfaces, dewatering, tailings management and analysing the forces which control the stability of dispersion as well as the wettability of mineral surfaces. Surface phenomena control both the dewatering of "fines" and area involved in dust suppression. So, to understand the colloidal science and colloidal chemistry in mineral engineering, a basic knowledge on surface science (surface physics, surface chemistry, surface mineralogy etc.) should be there. Colloidal solution is formed by dispersing one phase into another and the phase change can be easily described in terms of attractive and repulsive interactions in the dispersed phase. Different mineral beneficiation processes along with the application of colloid chemistry are depicted in **Figure 1.**

Along with the application in Mineral industry, colloidal chemistry has the application in non-mineral industry also in deinking, waste water treatment etc. We can't just use the word colloid science, in this modern era, interdisciplinary is more accurate word for Mineral Engineers as it has been justified. Fine mineral particles (colloidal particles), when present in slurry exert an affect on the ionic environment surrounding the particle. To understand or to get an extensive knowledge about colloidal systems in mineral processing, a knowledge on basic surface science is required.

So, as flotation, flocculation are (physico-chemical surface based separation processes), involves the application of certain surfactants in the slurry phase, hence valuable understanding of intermolecular forces at the interfaces in mandated.

1.1 Colloidal chemistry in froth flotation

Amongst the various beneficiation/upgradation technologies of Mineral processing, froth flotation is the most innovative and ingenious process development for the treatment of low-grade ore fines, slimes and tailings [2–4]. The basis of separation in froth flotation process rely on the surface state and colloidal chemistry of the particles and chemical reagents [5, 6]. Adsorption of the surfactants/reagents at the mineral solution interface is of the major importance for the behaviour of mineral particles in the solution and for the successful execution of the froth flotation [7].

Grinding of ore for liberation inevitably results distribution of particle sizes. Colloidal chemistry also plays a role between particle sizes and contact angle in the physio chemical processes [8].

Flotation process is not fully interpretable and remains a challenge, as it requires a good understanding of the interactions involved between the major phases (macro processes) and the number of inter related events (micro processes), as represented in **Figure 2.**

Adsorption of simple ions will determine the change of surface characteristics of the particles in the pulp/slurry phase and therefore affect the colloidal stability and the adsorption behaviour of reagents on the mineral surface. A comprehensive understanding of the flotation colloid chemistry is essential to enhance flotation performance and adapt them for treatment of more complex and low-grade ores.

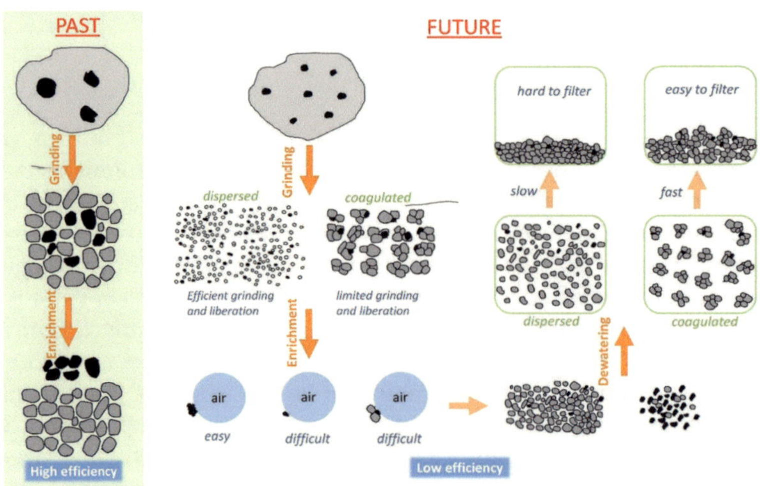

Figure 1.
Important role of Colloid chemistry in mineral processing [1].

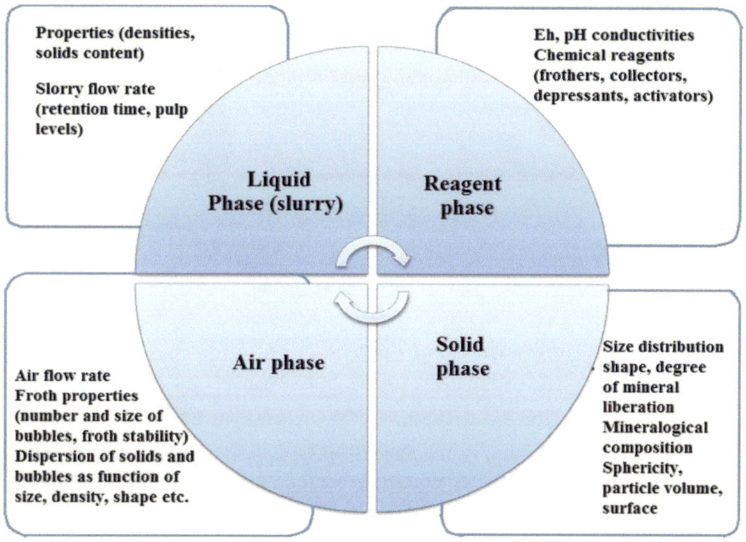

Figure 2.
Different colloidal phases and interfaces of froth flotation [7].

Colloidal chemistry plays an important role in flotation particularly surface and interfacial forces (Vander Waals, electrical double layer, hydrophobic, hydrodynamic, hydration, and adhesion forces) are of great importance in order to understand interparticle, interbubble and bubble-particle interaction mechanisms. However due to complexity of flotation system and difficulty in experimental verification and due to various particle size range, the physicochemical principles of flotation colloid interactions are still not fully understood.

2. Surface energy in Froth flotation

The concept of surface energy provides a satisfactory basis for explaining a wide range of "capillary phenomena", many of which are pertinent to the flotation process.

1. Th e shape s of liquid drops, bubbles, or menisci at equilibrium, either with or without deformation by gravity.

2. The internal pressure difference across curved interfaces.

3. The vapour pressure over curved interfaces of droplets or small particles.

4. The growth of particles (or bubbles) at the expense of smaller ones, either through the vapour phase or via a solvent.

5. The creepage or retraction of a film of liquid over another liquid or solid.

6. The establishment of an equilibrium angle of contact of a liquid on a solid substrate — as governed by Young' s equation

7. The accumulation of one or more components at an interface — adsorption.

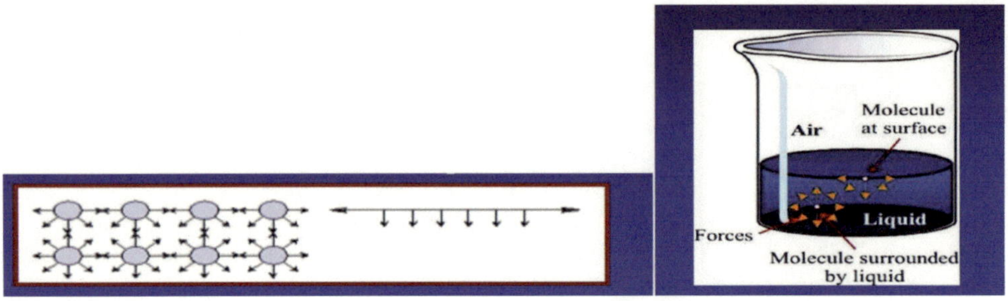

Figure 3.
Surface and Interfacial tensions for the molecules in bulk liquid.

According to molecular theory, different crystal structures have different surface energy [9–11]. Interface is the boundary between two or more phases exist together. The properties of the molecules forming the interface are different from those in the bulk that these molecules are forming an interfacial phase. Several types of interface can exist depending on whether the two adjacent phases are in solid, liquid or gaseous state.

2.1 Interfacial tension

In the liquid state, the cohesive forces between adjacent molecules are well developed. There are various forces present between surface and interfaces molecules presented in **Figure 3.**

Cohesive forces are present with the other molecules which are situated below or adjacent to them. Adhesive forces are being developed with the molecules of other phases in the interface. Surface tension (γ) is the force per unit length that must be applied parallel to the surface so as to counter balance the net inward pull. Interfacial tension is the force per unit length existing at the interface between two different phases.

3. Methods for characterization of colloidal slurry phases

Potential determining ions [H^+ and OH^-] at the interface plays a very important role in determining electrical charge at the surface in the colloidal system [12–15]. There are many diagnostic methods such as particle size analysis and turbidity measurements, surface force measurement with atomic force microscope (AFM) and surface force apparatus (SFA) are direct approaches for studying colloidal interactions, which could provide insights into molecular mechanisms of operating colloidal forces with high resolution [1]. After invention of these advanced techniques, these have become the most important tools in colloid and interface science to directly measure the interactions between particles and liquid substrates, and even the single molecular force involved in the rupture of a single chemical bond and the stretching of polymer chains. These are not only limited to hard or non-deformable surfaces, are also used to measure the forces between one solid particle and an air bubble or an oil drop. Although progress has been made in studying bubble-particle interactions using AFM in recent years, the deformation of the bubble by both the hydrodynamic and surface forces remains a bottleneck in determining the absolute separation and the surface forces.

4. Parameters affecting colloidal stability in mineral processing

Mineral processing is influenced in many ways by colloidal and surface forces (**Figure 4**).

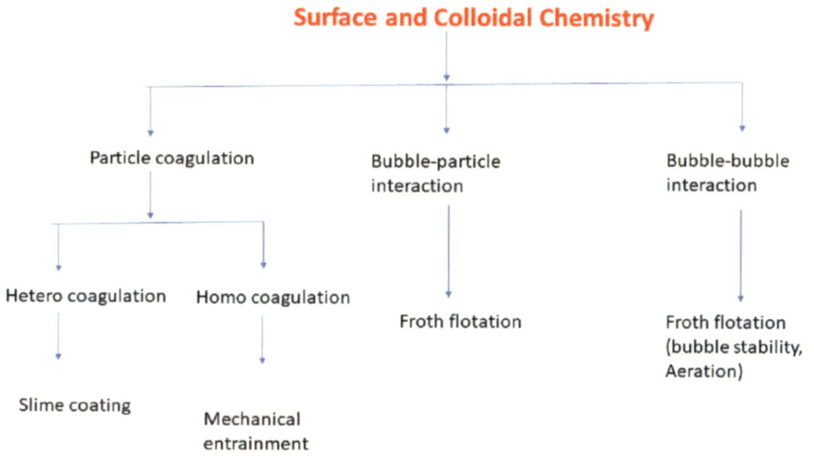

Figure 4.
Colloidal chemistry in Mineral Processing.

Rheology is dominated by particle coagulation and thus affects the effectiveness of grinding. Selective homocoagulation/flocculation is advantageous by decreasing mechanical entrainment or dewatering for fine particle flotation of valuables. Flotation of fine particles also depends on the interaction of bubble-bubble and bubble-particle interaction. However, for thickening/dewatering, hetero-coagulation/flocculation that is detrimental to selective flotation by slime coating is desirable.

4.1 Colloidal stability

The stability of colloidal system is mainly affected by interactions among the particles. The coagulation kinetics of multicomponent colloidal systems with regard to the different electrical double layers for each dispersed species will vary.

The reduction of diffuse layer potential by surfactant adsorption can result in coagulation. Another important factor affecting the coagulation of mineral particles is hydrophobicity and thus affecting the rheology of suspensions. This is particularly true when the particles are sufficiently hydrophobic, being able to counteract electrostatic repulsion through strong hydrophobic attractive forces.

Colloidal interactions of slimes particles (<10 μm) requires careful attention, because it results in slime coating and hence diminishing the separation performance.

4.1.1 Methods to study colloidal stability

Many advancements have been made in the past several decades for accurate and direct measurements of forces acting between particles as a function of the surface separation. Although the findings were consistent with the colloidal forces double-layer hypothesis, deviation was found at very short distances, mainly due to hydration forces or surface roughness.

The presence of cations in the aqueous process, which makes it possible to adsorb hydrated cations on the solid surface, is responsible for the resulting hydration force between the two mineral surfaces and is a significant flotation factor.

On the other hand, for predicting hetero-coagulation such as slime coating, zeta potential calculation can be used.

No attraction, weak attraction, or strong attraction between different mineral fines can be distinguished from the zeta potential measurements for a slurry mixture system containing various mineral fines.

Zeta potential, electric double layer is first prescribed by Helmoltz [16]. The counter ions in an electrical double layer can exchange with ions of the same sign from the solution. In a similar way, the moment of counter ions occurs, with the application of electric field because of the concept of surface conductivity.

4.1.1.1 Atomic Force Microscopic (AFM) analysis

The bubble-particle attachment forces can be calculated by attaching a bubble to a stationary solid surface and a particle to an AFM cantilever. Strong long range attractive forces can be determined for a hydrophobic particle and an air bubble before any double-layer and van der Waals forces can be established, with the gas bubble acting like a hydrophobic surface.

The attractive hydrophobic attraction between the solid-water and the water-gas interfaces is assumed to be the main driving force for film rupture and the attachment of air bubbles to hydrophobic mineral particles, taking into account the repulsive existence of electrostatic repulsion and van der Waals forces between particles and air bubbles encountered in flotation. The development of a dimple due to a hydrodynamic pressure greater than the internal bubble pressure is highly dependent on the velocity of the air bubble's approach. Although the higher bubble approach velocity leads to a more pronounced dimple, in order to evaluate the film thickness of the first dimple and the form of the film for hydrophobic solid surface systems, the surface hydrophobicity needs to be taken into account. By changing the reaction conditions such as electrochemical potential and solution pH, it is possible to maximise the alteration of hydrophobicity with/without collector.

4.1.2 Effect of water chemistry

As water is a polar liquid and moderate conductor of electricity, potential difference will not be observed in the absence of electric current. The chemistry of water has a great influence on the interaction forces in aqueous solution between solid surfaces. The colloidal stability of various slurry phases is influenced by various ions (monovalent, divalent, trivalent, etc.) present in the water. Cations function as a binder to bridge various charged surfaces in certain instances. If present in process water, reagents/surfactants bind with cations to minimise the amount of free cations in the liquid, are able to mitigate the undesirable effect of cations and thus promote the release of mineral particles. The water chemistry of the aqueous system where the attachment occurs is strongly influenced by bubble-particle attachment. Electrolytes in water compress the electrical double layer for the already hydrophobic particles and thus lower the energy barrier created during the collision between hydrophobic particles and air bubbles, which is advantageous for the bubble particle attachment. Knowledge about the surface properties and zeta potential is very important for flotation. The literature may define the ideal situation for different ores, but due to the presence of different interfering ions, the actual situation of different ores with varied mineralogy differs. Therefore, for flotation process effectiveness, the particle size distribution proportions of coarse and fine size groups, surface features of bulk and size fractions of different mineral species and their interactions with the reagents need to be understood and closely monitored.

4.1.3 Effect of reagents

Flotation reagents play the most important role in flotation, as the hetero-coagulation that could contribute to the loss of liberated valuables in the tailings due to slime coating will affect less than the optimum quantity, while the optimum quantity helps to selectively homo-coagulate fine valuable particles, increasing the kinetics of fine particle recovery.

4.1.3.1 Collector

Collectors are reagents which render minerals' surface hydrophobic by adsorption. The selection of an appropriate collector is critical for selective separation of valuable minerals from gangue minerals. Collectors can be classified into non-ionic, anionic and cationic depending on their ionic charge and active ion participation.

4.1.3.2 Frothers

Frothers are organic compounds that dissociate into ions at the air-water interface and decrease the surface tension, thus stabilising the froth consisting of a multitude of mineral-laden air bubbles and inducing the mineralized surface buoyancy effect.

4.1.3.3 Modifiers

These are chemical compounds applied to the pulps of flotation to strengthen the collector-mineral Adsorption, that is, selectivity enhancement. This may be achieved by either (a) creating an environment or revitalizing the floatability of the desired mineral, (b) by suppressing the flotation activity of the undesired mineral (at a particular stage of flotation operation), (c) by removing the deleterious elements which hinder effective flotation of desired minerals or (d) by providing proper hydrophobicity for the selective adsorption between the mineral and collector.

There are different types of modifiers.

I. Activators

II. Depressants

III. Dispersants

IV. pH regulators

Along with flotation, colloidal chemistry plays an important role in flocculation. Inorganic salts and synthetic high-molecular - weight polymers are commonly used in the treatment of mine waste and oil sands tailings to coagulate/flocculate the solid particles. Following its biomedical applications by settling experiments, the feasibility of using highly biocompatible glycopolymers in solid–liquid separation has been explored.

4.1.4 Effect of surface charge

Yield stress measurements are used in combination with electrophoretic and zero charge point measurements as a function of solution pH to detect the surface charge characteristics of particles that play a very important role in froth flotation. Although the coagulation peak, the isoelectric point and the zero charge point all converge at the same value for isotropic minerals, they spread across a wide range of anisotropic minerals, and with the addition of anionic polymer as a dispersant to the mineral suspensions, this disparity between isotropic and anisotropic minerals is even accentuated.

5. Conclusions

In mineral processing, colloidal interactions play an important role.

Developments of different techniques allow us to research the phenomenon of colloidal interactions, including slime coating, homocoagulation, particle and bubble coalescence, and mineral attachment of air bubbles.

Hence, in the analysis of colloidal and surface force in mineral processing, it is important to use complementary techniques.

To control the state of colloidal dispersions by creating favourable conditions, it is of paramount importance to study the colloidal interactions in a relevant system, including such considerations as different electrolyte concentrations, the presence of divalent cations, the co-existence of various cation species and natural surfactants in industrial process water, different pH values, the presence of reagents such as collectors, dispersants, and activators, and the existence of other surfactants. It can be reiterated that lower solution pH, higher salinity, and higher ionic concentration affects the repulsive forces present in a colloidal solution, made the adhesion forces stronger, while both decreasing long-range repulsive forces and increasing adhesion forces are beneficial for mineral froth-flotation and flocculation process [17–20].

The factors affecting the surface colloidal chemistry of fines beneficiation process in Mineral Processing are

I. Equilibrium of mineral particles in the pulp phase at a given pH

II. Equilibrium of mineral surfactants in the solution at a given pH

III. Adsorption of reagents on the particle surfaces

IV. Hydrophobic interaction of hydrophobic particles

V. Oil agglomeration of mineral particles

VI. Attachment of air bubbles to hydrophobic particles.

Author details

Abhyarthana Pattanaik* and Rayasam Venugopal
IIT (ISM), Dhanbad, India

*Address all correspondence to: abhyap10@gmail.com

References

[1] Z. Xu, Z. Li, Q. Liu, Recent Advances in Studying Colloidal interactions in Mineral Processing, 2019, 36, 35-53.

[2] Satyananda Patra, Abhyarthana Pattanaik, A.S. Venkatesh, Rayasam Venugopal, "Mineralogical and Chemical Characterization of Low grade Iron ore fines from Barsua Area, Eastern India with Implications on Beneficiation and Waste utilization", Journal of Geological Society of India, 93(4), 2019, 443-454.

[3] Satyananda Patra, Abhyarthana Pattanaik, Rayasam Venugopal, "Characterization of Low-grade Barsua Iron ore Fines and Identification of Possible Beneficiation Strategies", Canadian Metallurgical Quarterly, 58(1), 2019, 28-45.

[4] Abhyarthana Pattanaik, Satyananda Patra, Rayasam Venugopal, "Selection of a viable upgradation strategy through physico-chemical and mineralogical approach: A case study of low-grade Barsua iron ore fines", Transactions of the Indian Institute of Metals, 73, 2020, 47-63.

[5] Pattanaik A, Venugopal R, "Role of Surfactants in Mineral Engineering – An Overview", Surfactants and Detergents, Intechopen Publisher. ISBN 978-1-78984-660-7. 2019. DOI: 10.5772/intechopen.85947

[6] Abhyarthana Pattanaik and Rayasam Venugopal, "Analysis of Reverse Cationic Iron Ore Fines Flotation using RSM-D-Optimal Design – An Approach towards sustainability", Advanced Powder Technology, 29(12), 2018, 3404-3414.

[7] Abhyarthana Pattanaik and Rayasam Venugopal, "Investigation of Adsorption mechanism of Reagents (Surfactants) System and its applicability in Iron ore flotation - An Overview", Colloids and Interface Science Communications, 25, 2018, 41-65.

[8] J. Ralston, "Chapter 6: The influence of particle size and contact angle in flotation" , Developments in Mineral Processing, Elsevier, 1992, ISBN: 978-0-444-88284-4 ISSN: 0167-4528 , 203-224.

[9] R. Crawford, L.K. Koopal and J. Ralston, Colloids and Surfaces, 1927, 27, 57.

[10] G.D. Parafitt and C.H. Rochesters, 'Adsorption from solution at the solid/liquid interface, Academic press, London, 1983.

[11] P. Somasundaram and B.M. Moudgil, Reagents in Mineral Technology, Surfactant Science series, Vol 127, Marcel Dekker, New York, 1988.

[12] R.J. Pugh and J.A. Kitchener, Journal of Colloid and Interfacial Science, 35, 1971, 656.

[13] R.J. Pugh and J.A. Kitchener, Journal of Colloid and Interfacial Science, 38, 1972, 656.

[14] R.J. Pugh, Journal of Colloid Polymer Science, 252, 1974, 400.

[15] Janusz S. Laskowski, "Chapter 7: An Introduction: Physio chemical methods of separation," Developments in Mineral Processing, Elsevier, 1992, ISBN: 978-0-444-88284-4 ISSN: 0167-4528 , 225-241.

[16] H. Helmholtz, 1879, Studien über electrische Grenzschichten,

[17] N. Arbiter and E.K.C Willams, Conditioning in oleic acid flotation, Fine Particles Processing , E.d. Somasundaram, P., AIME, 1980, Vol 1, 802-831.

[18] Soto, H. and Iwakasi, I. , Selective flotation of phosphates from dolomite using cationic collectors. I. Effect of collectors and nonpolar hydrocarbons, International Journal of Mineral Processing, Vol. 16 , 1986 , 3-16.

[19] Seitz, R. A. , An analysis of the theory and industrial practice of coal flotation, PhD dissertation (unpublished) , Michigan Technological University, 1986.

[20] R.J. Pugh, "Chapter 8: Selective coagulation of colloidal mineral particles," Developments in Mineral Processing, Elsevier, 1992, ISBN: 978-0-444-88284-4 ISSN: 0167-4528, 243-276.

Chapter 11

Magnetic Iron Oxide Colloids for Environmental Applications

Alvaro Gallo-Cordova, Daniela Almeida Streitwieser, María del Puerto Morales and Jesús G. Ovejero

Abstract

This chapter deals with magnetic colloids with catalytic properties for the treatment of polluted waters and the efficient production of fuel alternatives. This kind of materials presents great advantages such as high surface/volume ratio, reproducibility, selectivity, ability to be magnetic harvested, functionalizable surfaces (e.g. with tunable pores and selective chelators deposited on them), high efficiencies and reusability. In particular, this chapter will consider the case of magnetic iron oxide colloids, which can be easily synthesized at low cost, are biocompatible and presents a well-developed surface chemistry. The most common techniques for the synthesis and functionalization of these magnetic nanoparticles will be reviewed and summarized. The iron oxide nanoparticles present outstanding properties that can be exploited in different aspect of the wastewater treatment such as heavy metals and organic pollutants removal by ionic exchange or adsorption, and degradation of the contaminants by advanced oxidation processes, among others. In the field of alternative energies, they have also been used as catalysts for biofuels production from oil crops, in Fischer-Tropsch reactions for liquid hydrocarbons and many other processes with potential environmental impact.

Keywords: magnetic colloids, iron oxide nanoparticles, renewable energies, water remediation, biofuels, pollutant, degradation, adsorption

1. Introduction: Environmental challenges

The incessant deterioration of the environment caused by anthropogenic activities, including industrial ones, has been an issue of great concern over the last few decades. The modern society demands the development of novel technological solutions able to create a more efficient and eco-friendly industry. Nanotechnology has the potential to improve traditional environmental remediation technologies through cleaner processes at a reduced cost. This global "nanorevolution" has located engineered nanomaterials and in particular, magnetic iron oxide colloids, under the spotlight for environmental applications such as water treatments and renewable energies solutions.

1.1 Water management

Water management has emerged as a global issue while the governmental agencies in many countries are stepping up to combat climate change. Pollution of oceans, water sources eutrophication, pollution of effluents by heavy metals and industrials wastes, the spread of desserts and the restricted access to drinking water are all challenges that demand the development of alternative treatment techniques and cleaner industrial processes [1].

Water and wastewater treatment generally includes up to four different stages that encompasses chemical, physical and biological processes [2]. Usually, wastewater needs to go through a preliminary treatment aimed to easily separate large residues, and in some cases a pre-aeration process. Once pretreated, the following stage known as primary treatment consists on processes of sedimentation and smaller sieving. The secondary treatment comprises more complex biological and physicochemical techniques. Finally, in a tertiary treatment, water is disinfected and processed to adequate the biologic oxygen demand and heavy metals concentrations by simple methods like adsorption or filtration [3]. Magnetic iron oxides colloids are a well-known alternative for adsorption processes mentioned in ternary treatments, since they present a high relative surface area and ease of functionalization that increase their adsorption capacity, selectivity and facilitate the separation by magnetic harvesting [4, 5].

Typically, the secondary water treatment stage requires more efforts and presents more challenging inconveniences [6]. Depending on the wastewater effluent composition, this stage may comprise different processes either physicochemical or biological. Biological treatments like aerobic and anaerobic processing are usually implemented to efficiently eliminate and remove organic matter by transforming it into harmless compounds. These treatments may fail when the effluent is not biodegradable with a relation between chemical oxygen demand and biological oxygen demand above 4 (COD/BOD>4) [7]. In those cases, a physicochemical alternative results more convenient, being advanced oxidation processes (AOPs) the most promising and exploitable ones [8]. These AOPs are based on the in-situ generation of highly oxidative hydroxyl radicals or other oxidative species able to purify water by mineralizing the harmful organic matter into CO_2, water, salts or inorganic acids. The oxidative species can be generated with the help of oxidizing agents, irradiation or with a catalyst [9, 10]. In the last years, many efforts have been place to improve these AOPs by combining them with photochemical, electrochemical and catalytic techniques and looking for cheaper, eco-friendly and more efficient agents that enable the transference of these technologies to industrial processing of wastewater [11]. In this sense, iron oxide colloids have been proposed as a very interesting catalyst for the degradation of contaminants by AOPs, through the Fe-mediated Fenton and Fenton-like reactions [12].

1.2 Fuels and alternatives

Fossil fuels are a non-renewable energy resource produced by organic matter from different living beings accumulated hundreds of millions of years at the bottom of lakes or sees with very little oxygen and covered with several layers of sediment. Humanity has known about the existence of fossil fuels since ancient times, becoming the main energy source during the industrial revolution [13]. At present, fossil fuel along with natural gas, are still fundamental for modern's society economy and alternative sources have not yet been found to replace them. The current energy model, based on these fuels, presents serious concerns of sustainability, either due to the emissions of polluting greenhouse gases or the economic and political tensions and therefore, there is a need to search for new alternative energies [14, 15]. In the last decade, different resources have been used to overcome the problematic carried out by fossil fuels, among these are solar, wind, water, natural gas, coal and biomass. **Figure 1** shows how some alternative fuels can be processed

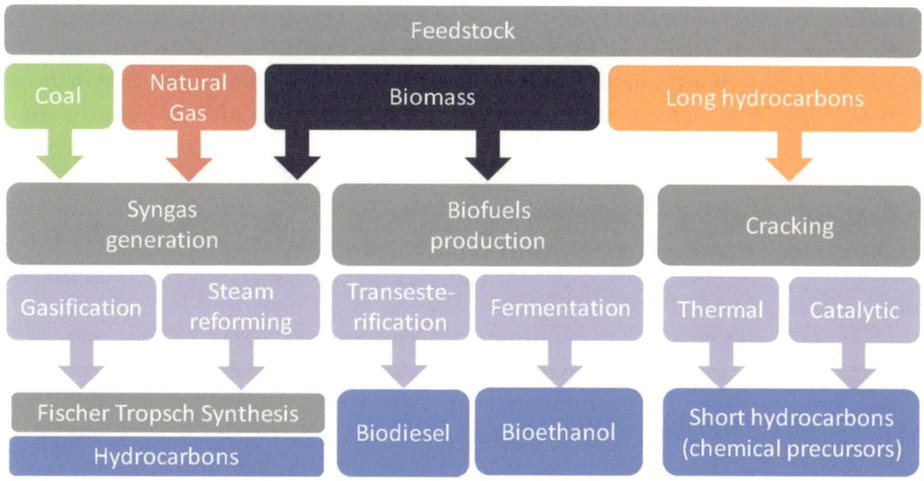

Figure 1.
Different feedstock for alternative fuels.

from different feedstock. The idea of waste valorization comes as a strong alternative for fuel and bio-fuels production. It is possible to reuse, recycle and compost waste materials and convert them into fuels. This is a way to overcome the environmental impact of common fossil fuels while taking advantage of useless wastes like food and wood residues, agricultural and municipal waste and used engine oil, among others [16, 17].

Biodiesel is a synthetic fuel obtained from natural components such as vegetable oils or animal fats through a transesterification reaction. It is mainly used for the preparation of diesel substitutes and can be mixed with it for commercial purposes. The use of biodiesel offers many advantages against traditional fossil fuels. As this combustible is synthesized from vegetable sources, such as rape, soy or sunflower seeds, it can be considered as an environmentally friendly fuel. It is even possible to manufacture it from recycled oils produced by different food industries [18–20]. Moreover, biodiesel generates less emissions of polluting gases and harmful substances like soot or benzenes. During the last decade the efficient production of biodiesel has been of great importance, and novel catalytic pathways, using for example nanocatalysts, are currently explored to increase the production yields [21].

Another alternative for fuel production comes from the waste valorization for hydrocarbons synthesis. Fischer-Tropsch process obtains gasoline from synthesis gases (syngas). This process was developed in the early 20's to obtain liquid fuels from coal as raw material as an alternative for fossil fuels [22]. But the syngas can be obtained from more sustainable sources, being biomass the one with the less environmental impact. The Fischer-Tropsch alternative process consists in transforming dry biomass into hydrocarbons via gasification with oxygen and using the syngas generated in the process as organic source [23]. As any chemical reaction, an efficient catalyst could lead this process to better yields of the desired product.

Finally, a third strategy to obtain alternative fuels consist on taking advantage of highly pollutant wastes such as used engine oils to produce liquid hydrocarbons. By thermal cracking, it is possible to break down the long, branched and cyclic chains, to obtain less heavy hydrocarbons that are in the order of the 15 to 20 carbons, which are usually the compounds of diesel [24].

Two important parameters play a fundamental role in the enhanced production of alternative fuels: the efficient heating and the catalyst separation. Both of them are usually expensive and time consuming and there is still a lack of efficient processes to provide a profitable alternative. The use of nanocatalysts in all these processes has been extensively studied since they make possible to work at moderated temperatures and increase mass transfer. Recent works in this area have shown that magnetic nanoparticles based on iron oxide could be a powerful tool to address many of the limitations exposed for the efficient waste valorization for fuels production [25, 26].

1.3 Magnetic colloids as alternative materials

In both applications, wastewater treatment processes and catalytic biodiesel production at industrial scale, it is crucial the use of efficient and inexpensive materials. This is the case of magnetic iron oxide nanoparticles (IONPs) that, as mentioned before, can be used in the former case as adsorbents for heavy metals or organic compounds and for the advanced oxidation of organic matter, and for the latter case, they have demonstrated a high-performance as reaction catalysts [5, 27].

It is essential to design and produce an efficient colloid for the mentioned environmental processes taking into account parameters such as the particle size and the colloidal stability. It should be emphasized that in the case of these colloids, the magnetic properties of the nanoparticles provide important advantages over other commonly used materials as it is the possibility of easy separation by using a magnetic gradient and heating them under an alternating magnetic field [28, 29]. However, magnetic properties are also responsible for the formation of aggregates and agglomerates due to the magnetic interactions inter-particles that reduce the specific surface area and the colloidal stability, limiting the efficiency and possible re-use of the particles [30]. Therefore, it results crucial to design colloids with nanoparticle sizes in the nanometer range (<100 nm) and coatings that provide them electrostatic and, if possible, steric repulsion, to keep magnetic interactions at the minimum and assuring in this way long term colloidal stability.

2. Environmental applications of magnetic iron oxide colloids

Figure 2 summarizes in a schematic way some of the applications of IONPs colloids that can be beneficial for environmental conservation. Compared to conventional macroscale materials such as activated carbons or zeolites, IONPs can achieve similar performance with minimum masses reduction on materials and energy costs. To understand how important IONPs are turning out to be, in this section the most recent works on this field will be reviewed.

2.1 Wastewater treatment agents

IONPs have been extensively studied for water purification and remediation of wastewaters. The three most important properties for these applications are the surface area/volume ratios, the possibility of magnetic harvesting and their surface reactivity. Among the possible uses of IONPs in this field, some of the most highlighted are the adsorption for the removal of contaminants and, less-known is the advanced oxidation processes (AOPs) for its degradation [31–33].

Adsorption is a surface phenomenon where the molecules of a sorbate are bound to the solid surface of the sorbent. In this phenomenon, mass transfer has a remarkable significance in the three steps involved: external diffusion, pore diffusion and surface reaction. In the first step, the adsorbate is transported from the bulk phase to the external surface of the sorbent. The pore diffusion refers to the transport through the sorbent pores so it can get to the final step were the sorbate molecules are attached to the internal surface, for the surface reaction [34]. Once

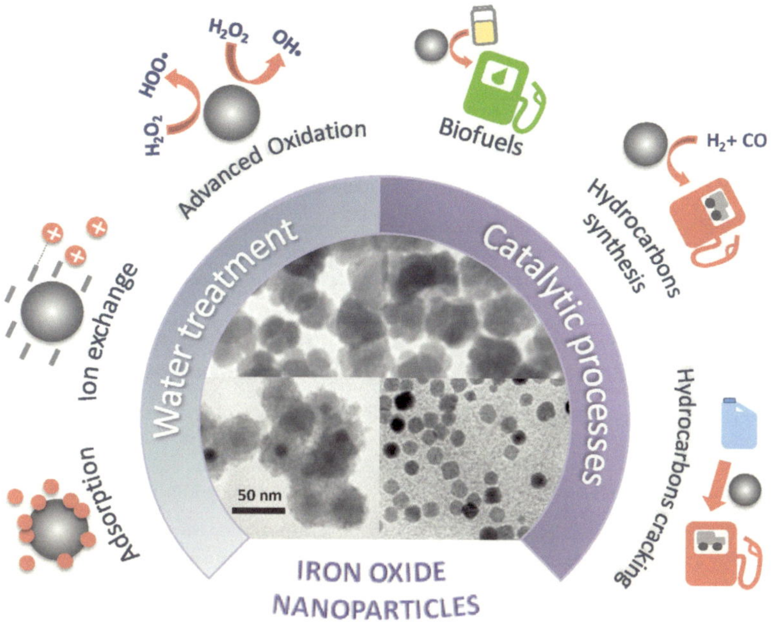

Figure 2.
Examples of applications of magnetic iron oxide colloids with positive impact on the environment.

adsorbed, the pollutants can be removed by magnetic collection using permanent magnets or alternatively degraded by the characteristic surface chemistry of iron oxide. As mentioned before, AOPs are based on physicochemical processes capable of producing profound changes in the chemical structure of pollutants. The concept was initially established by Glaze and collaborators, who defined AOPs that involve the generation and use of powerful transient species, mainly the hydroxyl radical (HO•) [35]. This radical, and others like HO•, OOH• can be generated by photochemical means (including sunlight) or by other forms of energy and are highly effective for the oxidation of organic matter. Some toxic pollutants that are not very susceptible to oxidation, such as metal ions or halogenated compounds, requires the use of chemical catalyst such as iron oxide [36]. An additional advantage of AOPs, is that they do not only change the pollutant phase but transform it chemically by its complete mineralization with no sludge generation [37].

Table 1 summarizes some recent works in the field of adsorption and AOPs using different iron oxide nanoparticles formulations. As it can be observed, IONPs are widely used in this matter due to their unique properties. Their applications for water remediation range from the elimination of inorganic and organic pollutants to the elimination of bacteria, proving that IONPs are materials with great versatility. Between the inorganic contaminants that can be removed by magnetic colloids, the most common are arsenic (As), cadmium (Cd), chromium (Cr), copper (Cu), lead (Pb) and zinc (Zn), which can cause, above their maximum levels in water, adverse health effects [38]. On the other hand, organic compounds like azo dyes, which present ecological hazardous effects due to their capacity of obstruct light within streams causing distress in aquatic environments, can also be removed using magnetic colloids [39]. Another class of organic pollutants that have been intensively studied for many researchers in the last decade are the emerging contaminants [40]. IONPs have demonstrated their ability to successfully remove many of these novel contaminants like for example pharmaceutical active chemicals, pesticides, personal cleaning compounds and others proving to be an efficient tool for environmental remediation [41, 42].

Material	Process	Highlights	Ref.
Fe_3O_4 immobilized on sand	Chromium adsorption	Supportive matrix for nanoparticles to avoid agglomeration and enhance adsorption.	[43]
Fe/La oxide microspheres	Arsenic adsorption	SiO_2 template used for the microsphere synthesis and a double shell material composed of Iron oxide and Lanthanum oxide	[44]
Resin-based hydrated iron oxide	Organic and inorganic compounds	Simultaneous removal of ρ-arsanilic acid and adsorption of arsenic.	[45]
Pristine γ-Fe_2O_3 nanoparticles	Arsenite adsorption	Material presented binding affinity to arsenite when coexisting with arsenate.	[46]
Fe_3O_4/Douglas fir biochar	Pharmaceuticals adsorption	Byproduct waste of syn-gas production was used for developing the adsorbent. Caffeine, ibuprofen and acetylsalicylic acid were removed in a fast equilibrium process.	[47]
Zr metal organic framework immobilized onto Fe_3O_4@SiO_2	Fungicides adsorption	Adsorbent with maximum adsorption capacities for triclosan and triclocarban (476 and 602 mg/g). Material recycled up to 11 times.	[48]
Chitosan-polyglycidol coated iron oxide	Methylene blue adsorption	Interactions between the coating molecules where analyzed. It was proved that this material can be used for dye removal	[49]
α-Fe_2O_3/lignosulfonate composite (no-magnetic)	Cadmium adsorption	The incorporation of organics onto magnetic sorbent can improve the adsorption process of heavy metals	[50]
Fe_3O_4 grafted with β-cyclodextrin	Bacteria adsorption	Pathogenic bacteria in water can be adsorbed and removed by magnetic harvesting processes.	[51]
Fe_3O_4/activated carbon	Combined adsorption and Fenton oxidation of chlorophenol	Iron oxide was impregnated over porous activated carbon. Adsorption followed the intraparticle diffusion model and 90% degradation was achieved. 5 times recycled.	[52]
Fe_3O_4 immobilized on graphene oxide	Fenton-like degradation of emerging pollutants and arsenic	Simultaneous degradation of ρ-arsanilic acid and adsorption of arsenic.	[53]
Fe_3O_4/SiO_2 coated with Polyethylene and polyacrilyc acid	Photo-Fenton for bacteria inactivation	*E. coli* bacteria photo-Fenton inactivation was achieved at natural pH	[54]
(Ag_3PO_4)-(Fe_3O_4)@ activated biochar	Sonocatalytic degradation of Rhodamine B and Bisphenol A	H_2O_2 production was achieved by pyrolysis of water molecules on catalyst surface. Degradation of synthetic dyes, endocrine-disrupting compounds/pharmaceutical active chemicals, and chlorinated compounds was tested.	[55]

Table 1.
Different iron oxide nanoparticles as adsorbents or catalysts for water remediation (references taken from recent works).

To maximize the adsorption capacity of the magnetic colloids, several parameters can be adjusted to the compound to be removed, that include the surface chelated molecules, the surface charge and the surface area available for the adsorption, being the later one determined by the nanoparticle size and the porosity of the coating.

IONPs can be efficiently chelated with certain molecules like (3-Aminopropyl) triethoxysilane (APTES) to modify the surface charge and promote the heavy metals adsorption process. Gallo et al. showed how the efficiency of IONPs increased with the increasing positive surface charge [56]. Moreover, the way IONPs are chelated can influence their selectivity for certain compounds. Removal of uranium from nuclear power polluted wastes has become a major issue in water processing since it can cause severe health and ecological problems. Helal et al. developed and efficient IONPs nanosorbent coated with APTES and succinyl-β-cyclodextrin molecules to increase its selectivity for uranium [57]. For other heavy metal ions such as chromium usually found in its anionic form ($HCrO^{4-}$, $Cr2O_7^{2-}$), the ionic attraction with its sorbent is an important parameter to be considered.

Silica coating of magnetic nanoparticles is an innovative way to modify the porosity of the sorbents surface and has been widely tested for the adsorption and degradation of different pollutants. Gallo *et al.* designed an interesting mesoporous silica coated IONPs for the adsorption of heavy metals and organic compounds [58]. They observed an interesting remark in which mesopores, growth with porogenic agents (e.g. octadecyltrimethoxysilane), present more affinity for the adsorption of organic compounds than heavy metals, in spite of having larger molecules. In this sense, it is not only possible to optimize the IONPs surface area but also their selectivity for specific compounds. On the other hand, as proved by Wu, *et al.* it is possible to grow the IONPs by spontaneous infiltration over a mesoporous SiO_2 template, which results in a much faster and easy way of producing these hybrid materials for the degradation of harmful azo dyes [59]. Even though the active sites available where occupied by IONPs and the adsorption decreased, they observed that the removal was enhanced by the Fenton-like degradation, proving that the hybrids are efficient agents for dyes remediation. Likewise, an interesting approach for wastewater treatment using IONPs@SiO_2 is to chelate the shell to increase selectivity for certain compounds. Uranium selectivity of cobalt ferrite nanoparticles coated with SiO_2 was studied by Huang *et al.* where they decorated the shell with 2-Phosphonobutane-1,2,4-tricarboxylic acid to increase the affinity sorbent/sorbate [60]. The SiO_2 coating was performed by the Stöber method and by coexisting ions tests they proved that the silica matrix can be efficiently chelated to improve selectivity. These examples show the importance of the physico-chemical nature of the coating on the stability, adsorption capacity and catalytic activity of the IONPs [61].

2.2 Catalysts for alternative energies

The use of magnetic iron oxide nanoparticles as catalyst supports dates back to the 70's when Robinson *et al.* reported the synthesis of enzymatic biocatalysts supported over magnetic iron oxides. The initial interest for the use of iron oxide was to facilitate the catalyst recovery and the immobilization of the catalyst in the reactor with magnetic fields [62]. Since then, the interest for IONPs as catalyst has emerged for an extensive list of chemical reactions including the ones that contribute to diminishing environmental harmful effects triggered by anthropogenic activities. Some of these reactions are based on developing alternatives to the highly pollutant use of fossil fuels, *e.g.* biodiesel production, Fischer-Tropsch synthesis and catalytic cracking of used engine oil, among others.

The design of IONPs for this application, should prevent mechanical breakdown of the catalyst and increase its lifetime by avoiding the possible particle growth or sintering during the process. One way is by introducing IONPs in mesoporous materials, assuring better catalytic process due the relatively large pores with high surface area that facilitate mass transfer and increase the concentration of active sites per mass of material. A recent study of Wei, *et al.* consisted in the Fischer-Tropsch synthesis catalyzed by a mesoporous iron oxide nanoparticle-decorated graphene oxide [63]. Here, they showed that the designed hierarchically mesoporous material can hinder the contact of the syngas with the active sites, highlighting that the improved mesoporous structure of the IONPs is extremely beneficial for reactants access and products release. Zhang *et al.* postulated that maybe metals atoms with multiple valences, including Fe and their oxides, result partially reduced during pyrolysis processes and generate oxygen vacancies that might transform some volatile biomass compounds into bio-oil [64]. The high specificity in the catalytic cracking of spent engine oils, reducing the undesired aromatics and high molecular weight constituents in a produced diesel-fuel, has been demonstrated also for simple natural magnetite particles in the micrometer range [65].

More complex nanostructures, like $CaO@Fe_3O_4$ composites have been shown to increase yields in the transesterification reaction of vegetable oils with no need of additional base compounds. It is also possible to improve the biodiesel production by immobilizing enzymes like lipase over IONPs. In this way there will be no need of purification after the reaction as these catalysts decrease yields of toxic byproducts [66]. Furthermore, Teo *et al.* prepared a highly recyclable $CaSO_4/Fe_2O_3$-SiO_2 catalyst for biodiesel production, showing its efficiency in the reaction [67].

In spite of the promising results in the use of magnetic colloids in this area, the literature is quite limited, and it is needed to strengthen the efforts in the years ahead to evaluate the potential of these catalysts in the efficient production of alternative fuels. It should be mentioned that besides the benefits of high surface areas, high selectivity and specificity and the ability to be functionalized, IONPs can offer a selective heating at the nanoparticle surface under an alternating magnetic field that may enhance the reaction rate and yield as will be described in Section 4.

3. Preparation of iron oxide magnetic colloids

The past two decades have seen tremendous advances in the synthesis and application of IONPs that take advantage of their distinct properties and functionalities. As seen in **Figure 3**, the economical perspectives for the market of magnetite nanoparticles is in continuous growth mainly boosted by their use in biomedical applications but also in fields like energy and wastewater treatment. As the interest for IONPs in different applications rises, the demand of new ways and technologies to produce them also increases.

Iron is one of the most abundant elements in nature presenting multiple crystalline phases with different structural and magnetic properties. Specifically, magnetite

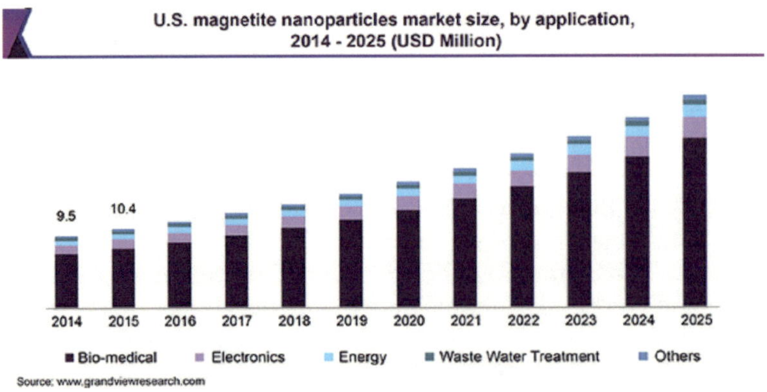

Figure 3.
U.S. market perspectives on the application of magnetite nanoparticles [68].

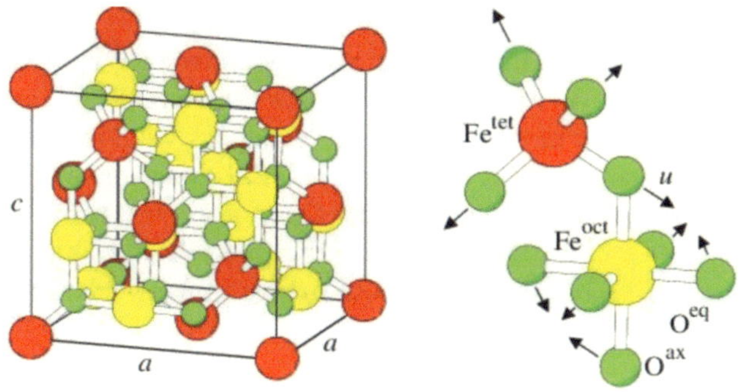

Figure 4.
Typical inverse spinel crystalline structure of magnetite [70]. © IOP Publishing and Deutsche Physikalische Gesellschaft. Reproduced by permission of IOP Publishing. CC BY-NC-SA.

(Fe_3O_4) and maghemite (γ-Fe_2O_3, the oxidized form of magnetite) have an interesting magnetic crystalline structure that can be described as a cubic inverse spinel structure with O_2 shaping an fcc structure and Fe cations in the tetrahedral and octahedral sites, as presented in **Figure 4**. These are the two most common phases used in environmental applications due to their high magnetic susceptibility and good chemical stability. Other iron oxides like hematite (α-Fe_2O_3) or goethite (α-FeOOH) frequently found in nature, present poor magnetic properties although they present unique and different advantages for other specific applications [69].

Different methods of synthesis have been studied and optimized so far in order to improve the physicochemical features of the magnetic iron oxide colloids. By selecting the proper method and controlling their key parameters (solvent, temperature, reaction time...) it is possible to generate IONPs of specific morphologies, size distributions and to control their colloidal stability. Each synthesis method presents specific advantages and drawbacks, therefore none of them can be declared as the universal method for producing IONPs. Between top-down or bottom-up approaches, the last ones are the most commonly used for large scale production because they offer a better control on the production of uniform nanoparticles with less defects, more homogeneous in shape, and better short and long range ordering (better crystallinity). This bottom-up category can be subclassified as a function of the reaction media, as aqueous synthesis (coprecipitation, hydrothermal and electrochemical, among others) or organic synthesis (thermal decomposition, polyol process, etc.) [71, 72]. In this section we will focus our attention in three different methods for comparison purposes: coprecipitation, thermal decomposition and polyol-based hydrothermal method.

3.1 Synthesis techniques

3.1.1 Coprecipitation

This is probably the most common and simplest method of synthesis of IONPs. Magnetic nanoparticles of magnetite or maghemite can be produced by coprecipitation of a stoichiometric mixture of Fe (II) and Fe (III) salts in an alkaline medium such as sodium hydroxide or ammonium hydroxide, for example. It is possible to obtain particles with diameters between 5 and 15 nm by controlling the synthesis key parameters like pH, temperature, addition rate of iron precursors and concentration of precursors [73]. For magnetite formation the overall reaction can be described as Eq. (1):

$$Fe^{2+} + 2Fe^{3+} + 8OH^{-} \rightarrow Fe_3O_4 + 4H_2O \quad (1)$$

The complete precipitation of magnetite is obtained in pH ranging from 8 to 14 with Fe^{3+}/Fe^{2+} ratios of 2:1, usually in an oxygen free environment to avoid premature Fe^{2+} oxidation. As magnetite usually is sensitive to oxidation by air, for environmental applications results much better to work with maghemite that will preserve its properties throughout the processes [72]. Magnetite is transformed to maghemite by heating up to 250°C or by acid treatment with nitrate/nitric acid. The main problem using this synthesis method is that the IONPs obtained, usually present a wide particle size distribution and poor crystallinity since they are prepared at temperatures below 100°C.

3.1.2 Thermal decomposition

This synthesis technique is based on the decomposition of organometallic precursors of Fe in high boiling point organic solvents in the absence of oxygen and the presence of massive amount of surfactants. It is possible to finely tune the size and shape of the IONPs just by controlling the boiling temperature of the solvent, the reactivity and concentration of the iron precursor, and the surfactants (typically fatty acids) chain lengths. The high reaction temperature used in thermal decomposition (>300°C), creates IONPs with narrow size distributions and excellent crystallinity in a range size between 5 and 100 nm [74, 75]. Due to the presence of organic surfactants (oleic acid, oleylamine…), the raw product has hydrophobic character and forms stable dispersions in many organic solvents like hexane, cyclohexane and toluene. However, these particles cannot be dispersed in water, being necessary a second step, like a ligand exchange reaction or the coating with an amphiphilic polymer, to transfer the synthesized IONPs to aqueous medium [76].

3.1.3 Polyol method

The polyol synthesis method was specifically designed for the development of nanostructured materials. Polyols are a family of solvents whose characteristics and properties (boiling temperature, viscosity, polarity) depends mainly on the length and alcohol substitution of methylene chains. They take advantage of the boiling temperatures of multivalent alcohols in their liquid phase to fix the temperature of reaction. It is interesting to highlight that the boiling temperature increases with the number of –OH moieties, the same with the molecular weight, viscosity and polarity [77, 78]. The main advantage of the use of polyols is that they provide reaction temperatures like the ones obtained in organic media, but the obtained IONPs are hydrophilic and can be dispersed in water like the ones produced in aqueous media.

Due to the diversity of polyols, it is possible to control the reaction temperature, just by selecting one with the interested boiling point. These temperatures can range between 200 and 320°C. Besides, glycol chains of the polyols can be used to control the particle sizes. **Figure 5** shows the TEM images of different IONPs with their size distribution obtained by Hachani *et al.* [79] where they proved that the length of the polyol chain is strictly related to the size of the obtained particles. They confirmed by thermal gravimetric analyses that each polyol was attached to the surface supporting the crucial role of the solvent on the growth of the IONPs. Moreover, thanks to the high polarity of the polyols, many common metallic salts are soluble and can be used as precursors for the synthesis of magnetite or other cobalt or zinc ferrites [80].

This well-known synthesis method was firstly described in 1989 by Fievet *et al.* when they synthesized metal powders of gold, copper, cobalt and lead in the micrometer range [81]. In that seminal work, they carried out reactions in a polyol-based media from the ionic form of each oxide, hydroxide or salt. In general, they

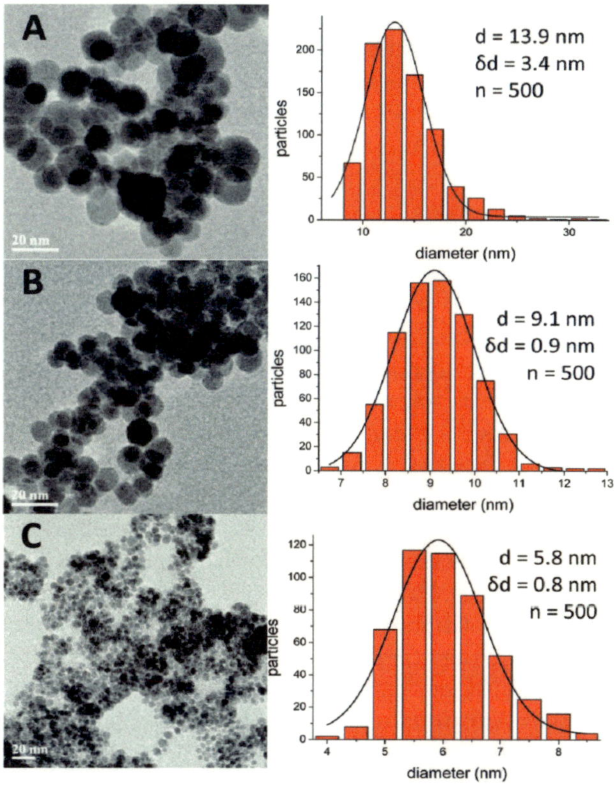

Figure 5.
Iron oxide nanoparticles synthesized with different polyols: (A) Tetraethylene glycol, (B) triethylene glycol and (C) diethylene glycol [79] - Published by The Royal Society of Chemistry.

demonstrated that polyols can act simultaneously as solvents, reducing agents and in certain cases as protective agents. Moreover, surfactant abilities of polyols are usually considered weak due to their relatively low molecular weighted molecules, thus they can be easily removed and exchanged by specific functional groups depending on the IONPs application [81].

An interesting approach to synthesize single (5–15 nm) and multicore (20–300 nm) IONPs and other metal ferrites is to combine the polyol method with a microwave assisted heating or with high pressure autoclaves. The microwave assisted polyol method is a versatile technique with improved yields, shorter residence times and highly reproducibility. The polyol molecules are able to adsorb microwave radiation due to their high polarity, with dielectric constants ranging from 20 to 45 [80, 82]. On the other hand, the polyol synthesis performed in a high pressure autoclave can decrease reaction times in a well-sealed environment and control of the size and the aggregation of the particles by controlling the synthesis parameters [83].

3.2 Colloidal stabilization

Magnetic attraction between nanoparticles compromise the colloidal stability of the suspension inducing agglomeration or even large precipitates [72, 84]. Depending on the application, the lack of colloidal stability may reduce the efficiency of the material. Therefore, it is necessary to develop compatible coatings that increase either the electrostatic or the steric repulsion between IONPs. Usually, IONPs are coated with polymers, surfactant agents, ligands, or inorganic materials like noble metals, oxides or silica that can prevent the dissolution of the particle, stabilize them at the working pH or introduce functional groups for the attachment of specific molecules. For environmental applications, tuning the nanoparticle surface, either in aqueous media by using small molecules or inorganic coating, or in organic media using surfactants, has been shown to improve its adsorption and catalytic capacities.

3.2.1 Small molecules

The modification of the nanoparticle surface with small molecules having a carboxylic or phosphate group with high coordination capacity to the iron atoms assures long term colloidal stability. The variety of molecules used is immense, some of the most commonly used are phosphonates, dimercaptosuccinic acid (DMSA), 11-mercaptoundecanoic acid or citric acid, small molecules with charged functional groups that provide excellent electrostatic stability.

If the IONPs were synthesized in aqueous media, the coating molecules can be introduced directly in the reaction media or preferably in a second step after the synthesis of the nanoparticles. In the case of IONPs obtained by the thermal decomposition method, it is possible to introduce functional groups by ligand exchange during the water transference. For example, in the coating with DMSA, one of the carboxylic groups of this molecule would replace the one at the surface of the IONPs with oleic acid, and the other carboxylic groups will remain facing outwards providing high negative surface charge in a wide pH range and a carboxylic functional group for further functionalization depending on the application [85].

3.2.2 Engineered silica coating

Coatings with silica (SiO_2) have now become a promising and important pathway for the development of coated magnetic colloids for different applications due to its biocompatibility, stability, easy conjugation with different functional groups that offer high selectivity and specificity [72]. Most of the SiO_2 coating strategies for magnetic colloids result in core-shell structures with an ionic positive charge that activate the surface and avoids aggregation (isoelectric point = 2–3). This diamagnetic coating reduces the magnetization per gram of material, but also increase the colloidal stability avoiding aggregation issues that results inconvenient for many applications [86].

The coating routes to obtain IONPs@SiO_2 can be divided into three categories: pre-synthesized silica matrices, in-situ fabrication of core-shell structures, and silica coating in already synthesized nanoparticles. This last one been the most common technique, where Stöber method is the easiest pathway to obtain homogeneous particles by hydrolysis and polycondensation of tetraethyl orthosilicate under alkaline conditions with temperatures above 60 °C [86]. SiO_2 layer over the IONPs can also be growth by a sol–gel process, where the silica shell is limited by a water-in-oil reverse microemulsion [87]. With these processes it is possible to control the shell thickness and to design a matrix with enhanced properties for specific applications.

It is also possible to design a high surface area Fe_3O_4@SiO_2 nanostructures where mesopores can be potentiated by porogenic agents that allow its in-situ formation through the SiO_2 shell. In this approach, nanoparticles are first coated by reverse microemulsion to add a first protective silica layer and the porogenic-doped shell is added by the Stöber method in a secondary step [58]. A schematic pathway for these kind of approach is showed in **Figure 6** were it is also pointed out how the porosity of silica engineered structures can be incremented by creating a hollow structure with hydrothermal or etching methods [88]. This example shows how the multitude of designing parameters of this kind of grafting molecules convert the Fe_3O_4@SiO_2 nanocomposites in a versatile material for environmental processes.

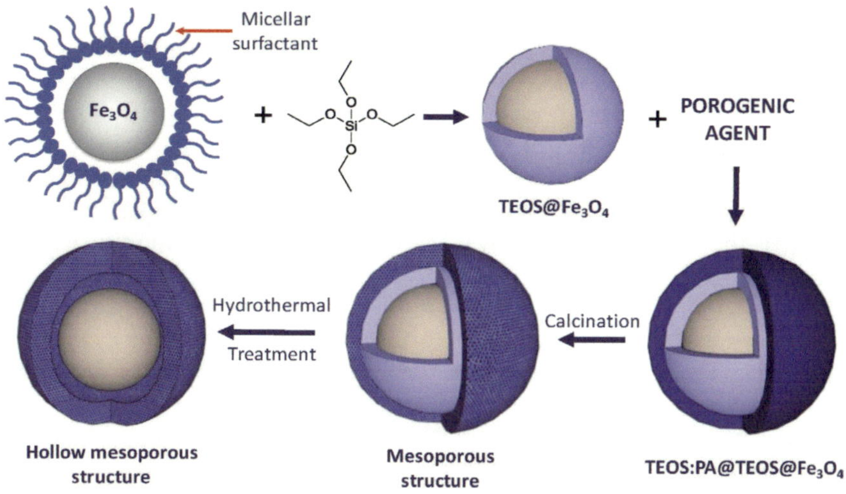

Figure 6.
Grafting iron oxide nanoparticles with engineered silica structures.

3.2.3 Organic coatings

Since many alternative energies processes are performed with oils or organic solvents, it is also important to develop IONPs soluble in organic media. Just as hydrophobic particles can undergo a ligand exchange to be redispersed in water, hydrophilic ones can be coated with molecules that allow their dispersion in organic media. For example, an interesting way to increase the stability in organic solvents is to add, in an additional step after the synthesis of the nanoparticles, a surfactant with a hydrophobic end. This step consists in the mixture of the surfactant on the IONPs aqueous suspension at ~80 °C and is frequently used for ferrofluids preparation. Oleic acid is one the most common compounds used for these purposes as it is a fatty acid formed by a terminal carboxylic acid group (–COOH) and a long hydrophobic alkyl chain. The resulting particles are quite stable in many organic solvents like hexane, toluene, cyclohexane, etc. [84].

For biodiesel production some studies have shown that iron oxides catalysts modified with polymers presented higher efficiencies [89]. Calcining organic compounds over nanoparticles surfaces can lead to a high surface area material with carbonaceous residues for enhanced adsorption and catalytic properties.

4. Promising magnetic features

The unique magnetic features of the iron oxide colloids represent one of the most exploitable characteristics of these materials for developing novel applications in the environmental area. In this section, we will review the physical principles of the superparamagnetic behavior observed in magnetic particles at the nanoscale and how it can be used in water remediation and biofuel generation.

At the macroscale, the magnetic materials minimize their magnetic energy breaking the alignment of their atomic moments into regions of coherent magnetization known as magnetic domains. The size of these domains depends on the anisotropy and saturation magnetization of the material. When the size of the solid is reduced to the size of a single magnetic domain, all the atomic magnetic moments of the material are oriented in the same direction and rotate coherently with the applied field (macrospin approximation). If we continue decreasing their size, we observe the characteristic magnetic response of small magnetic nanoparticles known as superparamagnetism. For magnetite nanoparticles smaller than 50 nm, the thermal fluctuations observed at room temperature are able to disorder the moments, cancelling the global magnetization of the sample. Consequently, in the absence of a magnetic field, the superparamagnetic nanoparticles present no remanent magnetization, avoiding the instability problems related to magnetic aggregation. However, when a magnetic field is applied, for example by approaching a magnet, the nanoparticles recover their magnetism with a high susceptibility and will be dragged towards the magnet proximity [90, 91].

Interestingly, when the superparamagnetic nanoparticles are subjected to an alternating magnetic field, they are able to absorb the magnetic energy and dissipate it as heat. The applied AMF forces the inversion of the spins of the atoms in a hysteretic process. During this process of magnetization reversal, the AMF energy will be transformed into heat increasing the temperature of the close environment of the nanoparticles. The way in which IONPs dissipate energy depends on the relaxation process and it is a function of the particle size, magnetic anisotropy and the viscosity of the media. The two principal relaxation mechanisms reported are Brown and Néel [92].

In the first mechanism, Brownian relaxation, the magnetic moment rotates with the particle within the medium, thus it is only observed when the particles are dispersed in a liquid medium. In this case, the time required to reverse moments by this mechanism (τ_B) depends on the hydrodynamic volume (V_h), the viscosity of the solvent where the particles are located (η) and the absolute temperature (T), as shown by the following expression, Eq. (2) where k_B is Boltzmann's constant [93]:

$$\tau_B = \frac{3V_h\,\eta}{k_B T} \tag{2}$$

On the other hand, the Néel mechanism describes the relaxation of the magnetic moment within the particle crystal axis. This mechanism is always present, and it is the only one that intervenes in the relaxation of magnetic moments when the particles are in compacted powder or in a frozen liquid where they cannot physically rotate. The expression for the relaxation time (τ_N) of the magnetic moments by Néel mechanism is as follows in Eq. (3):

$$\tau_N = \tau_0 e^{K_{eff}V_{MAG}/k_B T} \tag{3}$$

where, T is the temperature, V_{MAG} the magnetic volume of the particle, K_{eff} the energy by unit of volume needed to reverse the magnetic moment orientation (effective anisotropy) and τ_0 the inverse value of the Larmor frequency [94].

When the particles are small (<~20 nm), we can consider that $\tau_N << \tau_B$ and the relaxation of the magnetic moment takes place by Néel relaxation. On the other hand, for larger nanoparticles in which the magnetic moment is blocked in the direction of the easy axis of magnetization within the particle, it is satisfied that $\tau_B << \tau_N$, and the main relaxation mechanism is the Brownian rotation. The superparamagnetic IONPs that are usually employed in environmental applications are in an intermediate range in which both relaxation mechanisms might be present [94].

This superparamagnetic behavior is beneficial for wastewater treatment and catalysis in two aspects. On one hand, IONPs can be dispersed in the absence of a magnetic field without problems of magnetic aggregation and later be separated with a magnet once they have achieved their purpose like adsorbed a specific pollutant on their surface. On the other hand, they can be selectively heated under an AMF using moderate field conditions, which can rise reaction yields and shorten residence times.

4.1 Boosting environmental processes

Apart of the fact that IONPs can be efficiently recovered with an external magnetic field, facilitating its regeneration and reuse, environmental processes like the presented in previous sections can also be enhanced by taking advantage of the magnetic heating power of IONPs in the presence of AMF. In the case of adsorption with IONPs, Rivera *et al.* presented the improvement of the adsorption capacity of chromium under an AMF [95]. Here, they showed much higher adsorption yields for IONP systems heated up with AMF than with common thermal heating even though both systems were set at the same global temperature. The heat generated by the IONPs is dissipated in their surface what generates local temperature much greater than those measured in the reaction media, giving rise to better adsorption yields. Furthermore, in a more recent work they used the same principle to improve the reaction yields of the methylene blue degradation in the presence of IONPs and AMF [96].

The advanced oxidation processes have also been benefited by the use of IONPs, although the influence of the AMF has not been deeply studied yet. **Figure 7** shows the reaction mechanism that these particles can undertake in combination with hydrogen peroxide using the potential of IONPs as in situ nanoheaters. There are just a few references on this matter, where typical studies only focus in the increasing degradation yields with the increasing temperatures in common thermal reactors but they do not take advantage of the IONPs selective heating. Among the limited references on the subject, Munoz *et al.* were also able to prove that the catalytic wet peroxide oxidation of antibiotic sulfamethoxazole presented rates significantly faster with an AMF than in a typical CWPO system [97].

Recently, magnetic colloids have been used as catalysts to enhance biomass hydrodeoxygenation reaction with magnetic induction heating, proving that this heating can provide a better environment for the reaction to take place in [98]. Furthermore, the potential heating of these catalysts have only been analyzed in a few environmental reaction mechanisms. One of them is the CO_2 methanation presented by Rivas-Murias *et al.* where they achieved conversions >95% using a cobalt ferrite catalyst under a 93 kHz and 53 mT AMF with a SAR value of 270 W/g [99].

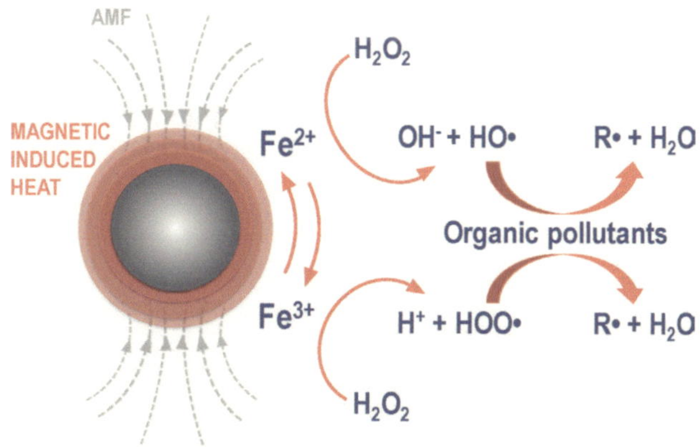

Figure 7.
An environmental catalytic process: Advanced oxidation of organic pollutants using iron oxide nanoparticles under an alternating magnetic field (AMF). (R: degradation products).

Our research group has recently tested the potential of IONPs as AOPs catalyst and local heating source on the decontamination of landfill leachate and complex textiles wastewater. The mineralization efficiency of the AOP of industrial wastewater was increased by magnetic induction heating of the nanocatalyst resulting in much quicker degradations in the presence of AMF compared to conventional heating. In general, magnetic induction-driven processes are a promising alternative for the improvement of well-known chemical reactions and real wastewaters remediation techniques.

5. Conclusion

This chapter shows the wide range of possible applications of magnetic iron oxide colloids in the field of environmental applications. It probes how iron oxide nanoparticles are excellent platforms to overcome many of the current technological challenges of this area. The chapter reviews the most recent references on water treatment and alternative fuels production in which iron oxide colloids has been used as treatment agents or catalysts. We have covered the areas of adsorption, advanced oxidation processes, hydrocarbon synthesis, biofuels and catalytic cracking of long chain hydrocarbons. Using iron oxide colloids has been proved to be a promising alternative for these processes and recent works in the field have shown an excellent performance of this material.

However, in order to obtain a competitive material, it is important to control parameters such as the particle size, coating and stability of the magnetic colloids. The analysis made here provides details of the most common synthesis and colloidal stabilization methods to create magnetic iron oxide colloids for this application.

The prospection of using iron oxide colloids tends to take advantage of the presence of Fe atoms at the surface working as catalysts for the degradation of contaminants through Fenton reactions, and their magnetic properties, not only for magnetic separation but also for the possibility of heating under an alternating magnetic field, enhancing catalytic and remediation processes.

Acknowledgements

This research was funded by the Spanish Ministry of Economy and Competitiveness under grant - MAT2017-88148R (AEI/FEDER, UE), PIE-201960E062 project and the EU project H2020-FETOPEN- RIA 829162, HOTZYMES. This study was also supported by USFQ with Collaboration Grant 2018 N° 11197 and PoliGrant 2018-2019 N° 12501. Finally, we acknowledge support of the publication fee by the CSIC Open Access Publication Support Initiative through its Unit of Information Resources for Research (URICI).

Conflict of interest

"The authors declare no conflict of interest."

Author details

Alvaro Gallo-Cordova[1,2*], Daniela Almeida Streitwieser[2], María del Puerto Morales[1] and Jesús G. Ovejero[1*]

1 Institute of Materials Science of Madrid, ICMM/CSIC, Sor Juana Inés de la Cruz 3, 28049 Madrid, Spain

2 Institute for the Development of Alternative Energies and Materials, IDEMA, Chemical Engineering Department, Universidad San Francisco de Quito USFQ, Av. Diego de Robles and Vía Interoceánica, EC 170901 Quito, Ecuador

*Address all correspondence to: alvaro.gallo@csic.es and jesus.g.ovejero@csic.es

References

[1] N. Ferronato, V. Torretta, Waste mismanagement in developing countries: A review of global issues, Int. J. Environ. Res. Public Health. 16 (2019) 1060. https://doi.org/10.3390/ijerph16061060.

[2] P.R. Rout, T.C. Zhang, P. Bhunia, R.Y. Surampalli, Treatment technologies for emerging contaminants in wastewater treatment plants: A review, Sci. Total Environ. 753 (2021) 141990. https://doi.org/10.1016/j.scitotenv.2020.141990.

[3] S. Bhojwani, K. Topolski, R. Mukherjee, D. Sengupta, M.M. El-Halwagi, Technology review and data analysis for cost assessment of water treatment systems, Sci. Total Environ. 651 (2019) 2749-2761. https://doi.org/10.1016/j.scitotenv.2018.09.363.

[4] F. Lu, D. Astruc, Nanocatalysts and other nanomaterials for water remediation from organic pollutants, Coord. Chem. Rev. 408 (2020) 213180. https://doi.org/10.1016/j.ccr.2020.213180.

[5] P. Xu, G.M. Zeng, D.L. Huang, C.L. Feng, S. Hu, M.H. Zhao, C. Lai, Z. Wei, C. Huang, G.X. Xie, Z.F. Liu, Use of iron oxide nanomaterials in wastewater treatment: A review, Sci. Total Environ. 424 (2012) 1-10. https://doi.org/10.1016/j.scitotenv.2012.02.023.

[6] P. Gautam, S. Kumar, S. Lokhandwala, Advanced oxidation processes for treatment of leachate from hazardous waste landfill: A critical review, J. Clean. Prod. 237 (2019) 117639. https://doi.org/10.1016/j.jclepro.2019.117639.

[7] A. Popat, P. V. Nidheesh, T.S. Anantha Singh, M. Suresh Kumar, Mixed industrial wastewater treatment by combined electrochemical advanced oxidation and biological processes, Chemosphere. 237 (2019) 124419. https://doi.org/10.1016/j.chemosphere.2019.124419.

[8] K. Paździor, L. Bilińska, S. Ledakowicz, A review of the existing and emerging technologies in the combination of AOPs and biological processes in industrial textile wastewater treatment, Chem. Eng. J. 376 (2019) 120597. https://doi.org/10.1016/j.cej.2018.12.057.

[9] M. Usman, S.A. Cheema, M. Farooq, Heterogeneous Fenton and persulfate oxidation for treatment of landfill leachate: A review supplement, J. Clean. Prod. 256 (2020) 120448. https://doi.org/10.1016/j.jclepro.2020.120448.

[10] V. Augugliaro, M. Litter, L. Palmisano, J. Soria, The combination of heterogeneous photocatalysis with chemical and physical operations: A tool for improving the photoprocess performance, J. Photochem. Photobiol. C Photochem. Rev. 7 (2006) 127-144. https://doi.org/10.1016/j.jphotochemrev.2006.12.001.

[11] P. Khurana, S. Thatai, Sapna, D. Kumar, Destruction of recalcitrant nanomaterials contaminants in industrial wastewater, Elsevier Inc., 2019. https://doi.org/10.1016/B978-0-12-814673-6.00006-1.

[12] G. Boczkaj, A. Fernandes, Wastewater treatment by means of advanced oxidation processes at basic pH conditions: A review, Chem. Eng. J. 320 (2017) 608-633. https://doi.org/10.1016/j.cej.2017.03.084.

[13] S.N. Jorgenson, J.C. Stephens, B. White, Environmental education in transition: A critical review of recent research on climate change and energy education, J. Environ. Educ. 50 (2019) 160-171. https://doi.org/10.1080/00958964.2019.1604478.

[14] V. Ediger, An integrated review and analysis of multi-energy transition from fossil fuels to renewables, Energy Procedia. 156 (2019) 2-6. https://doi.org/10.1016/j.egypro.2018.11.073.

[15] A. Qazi, F. Hussain, N.A.B.D. Rahim, G. Hardaker, D. Alghazzawi, K. Shaban, K. Haruna, Towards Sustainable Energy: A Systematic Review of Renewable Energy Sources, Technologies, and Public Opinions, IEEE Access. 7 (2019) 63837-63851. https://doi.org/10.1109/ACCESS.2019.2906402.

[16] R. Sindhu, P. Binod, R.B. Nair, S. Varjani, A. Pandey, E. Gnansounou, Waste to wealth, Elsevier, 2020. https://doi.org/10.1016/b978-0-444-64321-6.00009-4.

[17] M. Ouadi, M.A. Bashir, L.G. Speranza, H. Jahangiri, A. Hornung, Food and Market Waste-A Pathway to Sustainable Fuels and Waste Valorization, Energy and Fuels. 33 (2019) 9843-9850. https://doi.org/10.1021/acs.energyfuels.9b01650.

[18] A. Callegari, S. Bolognesi, D. Cecconet, A.G. Capodaglio, Production technologies, current role, and future prospects of biofuels feedstocks: A state-of-the-art review, Crit. Rev. Environ. Sci. Technol. 50 (2020) 384-436. https://doi.org/10.1080/10643389.2019.1629801.

[19] T.M.I. Mahlia, Z.A.H.S. Syazmi, M. Mofijur, A.E.P. Abas, M.R. Bilad, H.C. Ong, A.S. Silitonga, Patent landscape review on biodiesel production: Technology updates, Renew. Sustain. Energy Rev. 118 (2020) 109526. https://doi.org/10.1016/j.rser.2019.109526.

[20] D. Singh, D. Sharma, S.L. Soni, S. Sharma, P. Kumar Sharma, A. Jhalani, A review on feedstocks, production processes, and yield for different generations of biodiesel, Fuel. 262 (2020) 116553. https://doi.org/10.1016/j.fuel.2019.116553.

[21] V.C. Akubude, K.N. Nwaigwe, E. Dintwa, Production of biodiesel from microalgae via nanocatalyzed transesterification process: A review, Mater. Sci. Energy Technol. 2 (2019) 216-225. https://doi.org/10.1016/j.mset.2018.12.006.

[22] A.P. Steynberg, Introduction to Fischer-Tropsch technology, Elsevier B.V., 2004. https://doi.org/10.1016/s0167-2991(04)80458-0.

[23] R.G. dos Santos, A.C. Alencar, Biomass-derived syngas production via gasification process and its catalytic conversion into fuels by Fischer Tropsch synthesis: A review, Int. J. Hydrogen Energy. 45 (2020)18114-18132. https://doi.org/10.1016/j.ijhydene.2019.07.133.

[24] R. Khan, I. Ahmad, H. Khan, M. Ismail, K. Gul, A. Yasin, W. Ahmad, Production of diesel-like fuel from spent engine oil by catalytic pyrolysis over natural magnetite, J. Anal. Appl. Pyrolysis. 120 (2016) 493-500. https://doi.org/10.1016/j.jaap.2016.06.022.

[25] H.N. Pandya, S.P. Parikh, M. Shah, Comprehensive review on application of various nanoparticles for the production of biodiesel, Energy Sources, Part A Recover. Util. Environ. Eff. 0 (2019) 1-14. https://doi.org/10.1080/15567036.2019.1648599.

[26] S.E. Alavi, M.A. Abdoli, F. Khorasheh, F. Nezhadbahadori, A. Bayandori Moghaddam, Nanomaterial-assisted pyrolysis of used lubricating oil and fuel recovery, Energy Sources, Part A Recover. Util. Environ. Eff. 00 (2020) 1-15. https://doi.org/10.1080/15567036.2020.1807655.

[27] D. Astruc, Introduction: Nanoparticles in Catalysis, Chem. Rev. 120 (2020) 461-463. https://doi.org/10.1021/acs.chemrev.8b00696.

[28] M.I.A. Abdel Maksoud, A.M. Elgarahy, C. Farrell, A.H. Al-Muhtaseb,

D.W. Rooney, A.I. Osman, Insight on water remediation application using magnetic nanomaterials and biosorbents, Coord. Chem. Rev. 403 (2020) 213096. https://doi.org/10.1016/j.ccr.2019.213096.

[29] Q. Zhang, X. Yang, J. Guan, Applications of Magnetic Nanomaterials in Heterogeneous Catalysis, ACS Appl. Nano Mater. 2 (2019) 4681-4697. https://doi.org/10.1021/acsanm.9b00976.

[30] L. Gutiérrez, L. De La Cueva, M. Moros, E. Mazarío, S. De Bernardo, J.M. De La Fuente, M.P. Morales, G. Salas, Aggregation effects on the magnetic properties of iron oxide colloids, Nanotechnology. 30 (2019) 112001. https://doi.org/10.1088/1361-6528/aafbff.

[31] K. Qiao, W. Tian, J. Bai, L. Wang, J. Zhao, Z. Du, X. Gong, Application of magnetic adsorbents based on iron oxide nanoparticles for oil spill remediation: A review, J. Taiwan Inst. Chem. Eng. 97 (2019) 227-236. https://doi.org/10.1016/j.jtice.2019.01.029.

[32] R. Bhateria, R. Singh, A review on nanotechnological application of magnetic iron oxides for heavy metal removal, J. Water Process Eng. 31 (2019) 100845. https://doi.org/10.1016/j.jwpe.2019.100845.

[33] D. Jaspal, A. Malviya, Composites for wastewater purification: A review, Chemosphere. 246 (2020) 125788. https://doi.org/10.1016/j.chemosphere.2019.125788.

[34] X. Guo, J. Wang, A general kinetic model for adsorption: Theoretical analysis and modeling, J. Mol. Liq. 288 (2019) 111100. https://doi.org/10.1016/j.molliq.2019.111100.

[35] W.H. Glaze, J.W. Kang, D.H. Chapin, The chemistry of water treatment processes involving ozone, hydrogen peroxide and ultraviolet radiation, Ozone Sci. Eng. 9 (1987) 335-352. https://doi.org/10.1080/01919518708552148.

[36] J. Wang, S. Wang, Reactive species in advanced oxidation processes: Formation, identification and reaction mechanism, Chem. Eng. J. 401 (2020) 126158. https://doi.org/10.1016/j.cej.2020.126158.

[37] B.C. Hodges, E.L. Cates, J.-H. Kim, Challenges and prospects of advanced oxidation water treatment processes using catalytic nanomaterials, Nat. Nanotechnol. 13 (2018) 642-650.

[38] R. Bhateria, R. Singh, A review on nanotechnological application of magnetic iron oxides for heavy metal removal, J. Water Process Eng. 31 (2019) 100845. https://doi.org/10.1016/j.jwpe.2019.100845.

[39] A. Tkaczyk, K. Mitrowska, A. Posyniak, Synthetic organic dyes as contaminants of the aquatic environment and their implications for ecosystems: A review, Sci. Total Environ. 717 (2020) 137222. https://doi.org/10.1016/j.scitotenv.2020.137222.

[40] T. Deblonde, C. Cossu-Leguille, P. Hartemann, Emerging pollutants in wastewater: A review of the literature, Int. J. Hyg. Environ. Health. 214 (2011) 442-448. https://doi.org/10.1016/j.ijheh.2011.08.002.

[41] R. Gusain, K. Gupta, P. Joshi, O.P. Khatri, Adsorptive removal and photocatalytic degradation of organic pollutants using metal oxides and their composites: A comprehensive review, Adv. Colloid Interface Sci. 272 (2019) 102009. https://doi.org/10.1016/j.cis.2019.102009.

[42] A.A. Yaqoob, T. Parveen, K. Umar, M.N.M. Ibrahim, Role of Nanomaterials in the Treatment of Wastewater: A Review, Water 12 (2020) 495. https://doi.org/10.3390/w12020495

[43] J. Sorwat, A. Mellage, A. Kappler, J.M. Byrne, Immobilizing magnetite onto quartz sand for chromium remediation, J. Hazard. Mater. 400 (2020) 123139. https://doi.org/10.1016/j.jhazmat.2020.123139.

[44] S. Yan, K. Zhou, Y. Li, Q. He, L. Xia, S. Liu, H. Li, D. Liang, S. Song, Efficient removal of As(V) from diluted aqueous solutions by Fe/La oxide magnetic microspheres, J. Clean. Prod. 273 (2020) 123134. https://doi.org/10.1016/j.jclepro.2020.123134.

[45] B. Liu, Z. Liu, H. Wu, S. Pan, X. Cheng, Y. Sun, Y. Xu, Effective and simultaneous removal of organic/inorganic arsenic using polymer-based hydrated iron oxide adsorbent: Capacity evaluation and mechanism, Sci. Total Environ. 742 (2020) 140508. https://doi.org/10.1016/j.scitotenv.2020.140508.

[46] S. Molinari, M. Magro, D. Baratella, G. Salviulo, J. Ugolotti, J. Filip, M. Petr, J. Tucek, G. Zoppellaro, R. Zboril, F. Vianello, Smart synthetic maghemite nanoparticles with unique surface properties encode binding specificity toward AsIII, Sci. Total Environ. 741 (2020) 140175. https://doi.org/10.1016/j.scitotenv.2020.140175.

[47] A.S. Liyanage, S. Canaday, C.U. Pittman, T. Mlsna, Rapid remediation of pharmaceuticals from wastewater using magnetic Fe_3O_4/Douglas fir biochar adsorbents, Chemosphere. 258 (2020) 127336. https://doi.org/10.1016/j.chemosphere.2020.127336.

[48] J. Ma, S. Li, G. Wu, M. Arabi, F. Tan, Y. Guan, J. Li, L. Chen, Preparation of magnetic metal-organic frameworks with high binding capacity for removal of two fungicides from aqueous environments, J. Ind. Eng. Chem. 90 (2020) 178-189. https://doi.org/10.1016/j.jiec.2020.07.010.

[49] A. Iovescu, G. Stîngă, M.E. Maxim, M. Gosecka, T. Basinska, S. Slomkowski, D. Angelescu, S. Petrescu, N. Stănică, A. Băran, D.F. Anghel, Chitosan-polyglycidol complexes to coating iron oxide particles for dye adsorption, Carbohydr. Polym. 246 (2020) 1-13. https://doi.org/10.1016/j.carbpol.2020.116571.

[50] Q. Liu, J. Tang, X. Li, Q. Lin, R. Xiao, M. Zhang, G. Yin, Y. Zhou, Effect of lignosulfonate on the adsorption performance of hematite for Cd(II), Sci. Total Environ. 738 (2020) 139952. https://doi.org/10.1016/j.scitotenv.2020.139952.

[51] H. Eibagi, K. Faghihi, M. Komijani, Synthesis of new environmentally friendly poly(urethane-imide)s as an adsorbent including β-cyclodextrin cavities and attached to iron nanoparticles for removal of gram-positive and gram-negative bacteria from water samples, Polym. Test. 90 (2020) 106734. https://doi.org/10.1016/j.polymertesting.2020.106734.

[52] Z. Duan, W. Zhang, M. Lu, Z. Shao, W. Huang, J. Li, Y. Li, J. Mo, Y. Li, C. Chen, Magnetic Fe_3O_4/activated carbon for combined adsorption and Fenton oxidation of 4-chlorophenol, Carbon N. Y. 167 (2020) 351-363. https://doi.org/10.1016/j.carbon.2020.05.106.

[53] S. Wu, D. Yang, Y. Zhou, H. Zhou, S. Ai, Y. Yang, Z. Wan, L. Luo, L. Tang, D.C.W. Tsang, Simultaneous degradation of p-arsanilic acid and inorganic arsenic removal using M-rGO/PS Fenton-like system under neutral conditions, J. Hazard. Mater. 399 (2020) 123032. https://doi.org/10.1016/j.jhazmat.2020.123032.

[54] L. Fernández, J. González-Rodríguez, M. Gamallo, Z. Vargas-Osorio, C. Vázquez-Vázquez, Y. Piñeiro, J. Rivas, G. Feijoo, M.T. Moreira, Iron oxide-mediated photo-Fenton catalysis in the inactivation of enteric bacteria present in wastewater effluents at neutral pH, Environ. Pollut. 266 (2020)

115181. https://doi.org/10.1016/j.envpol.2020.115181.

[55] B.M. Jun, Y. Kim, Y. Yoon, Y. Yea, C.M. Park, Enhanced sonocatalytic degradation of recalcitrant organic contaminants using a magnetically recoverable Ag/Fe-loaded activated biochar composite, Ceram. Int. 46 (2020) 22521-22531. https://doi.org/10.1016/j.ceramint.2020.06.012.

[56] A. Gallo-Cordova, M. del P. Morales, E. Mazarío, Effect of the surface charge on the adsorption capacity of chromium(VI) of iron oxide magnetic nanoparticles prepared by microwave-assisted synthesis, Water (Switzerland). 11 (2019) 1-12. https://doi.org/10.3390/w11112372.

[57] A.S. Helal, E. Mazario, A. Mayoral, P. Decorse, R. Losno, C. Lion, S. Ammar, M. Hémadi, Highly efficient and selective extraction of uranium from aqueous solution using a magnetic device: Succinyl-β-cyclodextrin-APTES@maghemite nanoparticles, Environ. Sci. Nano. 5 (2018) 158-168. https://doi.org/10.1039/c7en00902j.

[58] A. Gallo-Cordova, J. Lemus, F.J. Palomares, M.P. Morales, E. Mazarío, Superparamagnetic nanosorbent for water purification: Assessment of the adsorptive removal of lead and methyl orange from aqueous solutions, Sci. Total Environ. 711 (2020) 134644. https://doi.org/10.1016/j.scitotenv.2019.134644.

[59] Z. Wu, W. Zhu, M. Zhang, Y. Lin, N. Xu, F. Chen, D. Wang, Z. Chen, Adsorption and Synergetic Fenton-like Degradation of Methylene Blue by a Novel Mesoporous α-Fe_2O_3/SiO_2 at Neutral pH, Ind. Eng. Chem. Res. 57 (2018) 5539-5549. https://doi.org/10.1021/acs.iecr.8b00077.

[60] Y. Huang, H. Zheng, H. Li, C. Zhao, R. Zhao, S. Li, Highly selective uranium adsorption on 2-phosphonobutane-1,2,4-tricarboxylic acid-decorated chitosan-coated magnetic silica nanoparticles, Chem. Eng. J. 388 (2020) 124349. https://doi.org/10.1016/j.cej.2020.124349.

[61] A. Jawed, V. Saxena, L.M. Pandey, Engineered nanomaterials and their surface functionalization for the removal of heavy metals: A review, J. Water Process Eng. 33 (2020) 101009. https://doi.org/10.1016/j.jwpe.2019.101009.

[62] P.J. Robinson, P. Dunnill, M.D. Lilly, The properties of magnetic supports in relation to immobilized enzyme reactors, Biotechnol. Bioeng. 15 (1973) 603-606. https://doi.org/10.1002/bit.260150318.

[63] Y. Wei, L. Yan, C. Ma, C. Zhang, S. Sun, X. Wen, Y. Yang, Y. Li, Mesoporous Iron Oxide Nanoparticle-Decorated Graphene Oxide Catalysts for Fischer-Tropsch Synthesis, ACS Appl. Nano Mater. 3 (2020) 7182-7191. https://doi.org/10.1021/acsanm.0c01522.

[64] C.T. Zhang, L. Zhang, Q. Li, Y. Wang, Q. Liu, T. Wei, D. Dong, S. Salavati, M. Gholizadeh, X. Hu, Catalytic pyrolysis of poplar wood over transition metal oxides: Correlation of catalytic behaviors with physiochemical properties of the oxides, Biomass and Bioenergy. 124 (2019) 125-141. https://doi.org/10.1016/j.biombioe.2019.03.017.

[65] R. Khan, I. Ahmad, H. Khan, M. Ismail, K. Gul, A. Yasin, W. Ahmad, Production of diesel-like fuel from spent engine oil by catalytic pyrolysis over natural magnetite, J. Anal. Appl. Pyrolysis. 120 (2016) 493-500. https://doi.org/10.1016/j.jaap.2016.06.022.

[66] H.R. Mardani, M. Forouzani, N. Moradi, Z. Kheibarian, A comparative study on physicochemical properties of two nanomagnetic compounds CaO@Fe_3O_4 and Fe_3O_4@CaO and their

catalytic role on biodiesel preparation, Environ. Prog. Sustain. Energy. 39 (2020) 1-11. https://doi.org/10.1002/ep.13395.

[67] S.H. Teo, A. Islam, E.S. Chan, S.Y. Thomas Choong, N.H. Alharthi, Y.H. Taufiq-Yap, M.R. Awual, Efficient biodiesel production from Jatropha curcus using $CaSO_4/Fe_2O_3$-SiO_2 core-shell magnetic nanoparticles, J. Clean. Prod. 208 (2019) 816-826. https://doi.org/10.1016/j.jclepro.2018.10.107.

[68] Grand View Research, No Title, Magn. Nanoparticles Mark. Size, Share Trends Anal. Rep. By Appl. Bio-Medical, Electron. Energy, Wastewater Treat. By Reg. Segm. Forecast. 2019 - 2025. (2019). https://www.grandviewresearch.com/industry-analysis/magnetite-nanoparticles-market (accessed October 27, 2020).

[69] J. Lian, X. Duan, J. Ma, P. Peng, T. Kim, W. Zheng, Hematite (α-Fe_2O_3) with Various Morphologies: Ionic Liquid-Assisted Synthesis, Formation Mechanism, and Properties, 3 (2009) 3749-3761.

[70] M. Friák, A. Schindlmayr, M. Scheffler, Ab initio study of the half-metal to metal transition in strained magnetite Ab initio study of the half-metal to metal transition in strained magnetite, New J. Phys., 9 (2007) 5. https://doi.org/10.1088/1367-2630/9/1/005.

[71] X.M. Lin, A.C.S. Samia, Synthesis, assembly and physical properties of magnetic nanoparticles, J. Magn. Magn. Mater. 305 (2006) 100-109. https://doi.org/10.1016/j.jmmm.2005.11.042.

[72] A.G. Roca, L. Gutiérrez, H. Gavilán, M.E. Fortes Brollo, S. Veintemillas-Verdaguer, M. del P. Morales, Design strategies for shape-controlled magnetic iron oxide nanoparticles, Adv. Drug Deliv. Rev. 138 (2019) 68-104. https://doi.org/10.1016/j.addr.2018.12.008.

[73] N.D. Kandpal, N. Sah, R. Loshali, R. Joshi, J. Prasad, Co-precipitation method of synthesis and characterization of iron oxide nanoparticles, J. Sci. Ind. Res. (India). 73 (2014) 87-90.

[74] G. Singh, V.S. Myasnichenko, W.R. Glomm, New insights into size-controlled reproducible synthesis of anisotropic Fe_3O_4 nanoparticles: the importance of the reaction environment , Mater. Adv. 1 (2020) 1077-1082. https://doi.org/10.1039/d0ma00275e.

[75] T.D. Clemons, R.H. Kerr, A. Joos, Multifunctional magnetic nanoparticles: Design, synthesis, and biomedical applications, Elsevier Ltd., 2019. https://doi.org/10.1016/B978-0-12-803581-8.10462-X.

[76] S.I.C.J. Palma, M. Marciello, A. Carvalho, S. Veintemillas-Verdaguer, M. del P. Morales, A.C.A. Roque, Effects of phase transfer ligands on monodisperse iron oxide magnetic nanoparticles, J. Colloid Interface Sci. 437 (2015) 147-155. https://doi.org/10.1016/j.jcis.2014.09.019.

[77] R.J. Joseyphus, K. Shinoda, D. Kodama, B. Jeyadevan, Size controlled Fe nanoparticles through polyol process and their magnetic properties, Mater. Chem. Phys. 123 (2010) 487-493. https://doi.org/10.1016/j.matchemphys.2010.05.001.

[78] G. Hemery, A.C. Keyes, E. Garaio, I. Rodrigo, J.A. Garcia, F. Plazaola, E. Garanger, O. Sandre, Tuning Sizes, Morphologies, and Magnetic Properties of Monocore Versus Multicore Iron Oxide Nanoparticles through the Controlled Addition of Water in the Polyol Synthesis, Inorg. Chem. 56 (2017) 8232-8243. https://doi.org/10.1021/acs.inorgchem.7b00956.

[79] R. Hachani, M. Lowdell, M. Birchall, A. Hervault, D. Mertz, S. Begin-Colin, N.T.K. Thanh, Polyol synthesis,

functionalisation, and biocompatibility studies of superparamagnetic iron oxide nanoparticles as potential MRI contrast agents, Nanoscale. 8 (2016) 3278-3287. https://doi.org/10.1039/c5nr03867g.

[80] Á. Gallo-Cordova, A. Espinosa, A. Serrano, L. Gutiérrez, N. Menéndez, M. Del Puerto Morales, E. Mazarió, New insights into the structural analysis of maghemite and (MFe_2O_4, M = Co, Zn) ferrite nanoparticles synthesized by a microwave-assisted polyol process, Mater. Chem. Front. 4 (2020) 3063-3073. https://doi.org/10.1039/d0qm00460j.

[81] F. Fievet, J.P. Lagier, M. Figlarz, Preparing Monodisperse Metal Powders in Micrometer and Submicrometer Sizes by the Polyol Process, MRS Bull. 14 (1989) 29-34. https://doi.org/10.1557/S0883769400060930.

[82] M.E.F. Brollo, S. Veintemillas-Verdaguer, C.M. Salván, M.D.P. Morales, Key parameters on the microwave assisted synthesis of magnetic nanoparticles for MRI contrast agents, Contrast Media Mol. Imaging. 2017 (2017) 8902424. https://doi.org/10.1155/2017/8902424.

[83] H. Gavilán, E.H. Sánchez, M.E.F. Brollo, L. Asín, K.K. Moerner, C. Frandsen, F.J. Lázaro, C.J. Serna, S. Veintemillas-Verdaguer, M.P. Morales, L. Gutiérrez, Formation Mechanism of Maghemite Nanoflowers Synthesized by a Polyol-Mediated Process, ACS Omega. 2 (2017) 7172-7184. https://doi.org/10.1021/acsomega.7b00975.

[84] B.I. Kharisov, H.V.R. Dias, O. V. Kharissova, A. Vázquez, Y. Peña, I. Gómez, Solubilization, dispersion and stabilization of magnetic nanoparticles in water and non-Aqueous solvents: Recent trends, RSC Adv. 4 (2014) 45354-45381. https://doi.org/10.1039/c4ra06902a.

[85] S.I.C.J. Palma, M. Marciello, A. Carvalho, S. Veintemillas-Verdaguer, M. del P. Morales, A.C.A. Roque, Effects of phase transfer ligands on monodisperse iron oxide magnetic nanoparticles, J. Colloid Interface Sci. 437 (2015) 147-155. https://doi.org/10.1016/j.jcis.2014.09.019.

[86] C. Li, C. Ma, F. Wang, Z. Xi, Z. Wang, Y. Deng, N. He, Preparation and biomedical applications of core-shell silica/magnetic nanoparticle composites, J. Nanosci. Nanotechnol. 12 (2012) 2964-2972. https://doi.org/10.1166/jnn.2012.6428.

[87] G. Ennas, M.F. Casula, G. Piccaluga, S. Solinas, M.P. Morales, C.J. Serna, Iron and iron-oxide on silica nanocomposites prepared by the sol-gel method, J. Mater. Res. 17 (2002) 590-596. https://doi.org/10.1557/JMR.2002.0083.

[88] C. Adhikari, A. Mishra, D. Nayak, A. Chakraborty, Drug delivery system composed of mesoporous silica and hollow mesoporous silica nanospheres for chemotherapeutic drug delivery, J. Drug Deliv. Sci. Technol. 45 (2018) 303-314. https://doi.org/10.1016/j.jddst.2018.03.020.

[89] X. Liu, L. Lei, Y. Li, H. Zhu, Y. Cui, H. Hu, Preparation of carriers based on magnetic nanoparticles grafted polymer and immobilization for lipase, Biochem. Eng. J. 56 (2011) 142-149. https://doi.org/10.1016/j.bej.2011.05.013.

[90] S.M. Devi, A. Nivetha, I. Prabha, Superparamagnetic Properties and Significant Applications of Iron Oxide Nanoparticles for Astonishing Efficacy—a Review, J. Supercond. Nov. Magn. 32 (2019) 127-144. https://doi.org/10.1007/s10948-018-4929-8.

[91] T.D. Clemons, R.H. Kerr, A. Joos, Multifunctional magnetic nanoparticles: Design, synthesis, and biomedical applications, Elsevier Ltd., 2019. https://doi.org/10.1016/B978-0-12-803581-8.10462-X.

[92] V. Marghussian, Magnetic Properties of Nano-Glass Ceramics, 2015. https://doi.org/10.1016/b978-0-323-35386-1.00004-9.

[93] M. Jeun, Y.J. Kim, K.H. Park, S.H. Paek, S. Bae, Physical contribution of néel and brown relaxation to interpreting intracellular hyperthermia characteristics using superparamagnetic nanofluids, J. Nanosci. Nanotechnol. 13 (2013) 5719-5725. https://doi.org/10.1166/jnn.2013.7524.

[94] J. Dieckhoff, D. Eberbeck, M. Schilling, F. Ludwig, Magnetic-field dependence of Brownian and Néel relaxation times, J. Appl. Phys. 119 (2016) 043903. https://doi.org/10.1063/1.4940724.

[95] F.L. Rivera, F.J. Palomares, P. Herrasti, E. Mazario, Improvement in Heavy Metal Removal from Wastewater Using an External Magnetic Inductor, (2019).

[96] F.L. Rivera, F.J. Recio, F.J. Palomares, J. Sánchez-Marcos, N. Menéndez, E. Mazarío, P. Herrasti, Fenton-like degradation enhancement of methylene blue dye with magnetic heating induction, J. Electroanal. Chem. (2020) 114773. https://doi.org/10.1016/j.jelechem.2020.114773.

[97] M. Munoz, J. Nieto-Sandoval, E. Serrano, Z.M. De Pedro, J.A. Casas, CWPO intensification by induction heating using magnetite as catalyst, J. Environ. Chem. Eng. 8 (2020) 104085. https://doi.org/10.1016/j.jece.2020.104085.

[98] J.M. Asensio, A.B. Miguel, P. Fazzini, P.W.N.M. Van Leeuwen, B. Chaudret, Hydrodeoxygenation Using Magnetic Induction : High-Temperature Heterogeneous Catalysis in Solution, Angew. Chem. Int. Ed. Engl., 58 (2019) 11306-11310. https://doi.org/10.1002/anie.201904366.

[99] B. Rivas-murias, J.M. Asensio, N. Mille, B. Rodríguez-gonzμlez, P. Fazzini, J. Carrey, B. Chaudret, V. Salgueiriço, Magnetically Induced CO_2 Methanation Using Exchange-Coupled Spinel Ferrites in Cuboctahedron-Shaped Nanocrystals, Angew. Chem. Int. Ed. Engl., 59 (2020) 15537-15542. https://doi.org/10.1002/anie.202004908.

Chapter 12

Colloidal Nanocrystal-Based Electrocatalysts for Combating Environmental Problems and Energy Crisis

Roshan Nazir, Abhay Prasad, Ashish Parihar, Mohammed S. Alqahtani and Rabbani Syed

Abstract

The serious threat that human beings face in near future will be shortage of fossil fuel reserves and abrupt changes in global climate. To prepare for these serious concerns, raised due to climate change and shortage of fuels, conversion of excessive atmospheric CO_2 into valuable chemicals and fuels and production of hydrogen from water splitting is seen most promising solutions to combat the rising CO_2 levels and energy crises. Amoung the various techniques that have been employed electrocatalytic conversion of CO_2 into fuels and hydrogen production from water has gained tremendous interest. Hydrogen is a zero carbon-emitting fuel, can be an alternative to traditional fossil fuels. Therefore, researchers working in these areas are constantly trying to find new electrocatalysts that can be applied on a real scale to deal with environmental issues. Recently, colloidal nanocrystals (C-NCs)-based electrocatalysts have gained tremendous attention due to their superior catalytic selectivity/activity and durability compared to existing bulk electrodes. In this chapter, the authors discuss the colloidal synthesis of NCs and the effect of their physiochemical properties such as shape, size and chemical composition on the electrocatalytic performance and durability towards electrocatalytic H_2 evolution reaction (EH_2ER) and electrocatalytic CO_2 reduction reactions (ECO_2RR). The last portion of this chapter presents a brief perspective of the challenges ahead.

Keywords: colloidal nanocrystals, electrocatalysis, size control, shape control, CO_2 reduction reactions, H_2 evolution reaction

1. Introduction

A clean environment and sustained energy resources are essential for future generations. With growing concerns for both dwindling traditional fossil fuels and global warming, there is an urgent need to develop renewable and environmentally benign alternatives to address these issues of mankind [1–3]. Currently, humans are mainly dependent on fossil fuels and thus extract carbon from the geosphere and put it into the atmosphere where it causes global warming. There are two methods that can be helpful in preventing and reducing carbon emissions. The first and convenient way to stop carbon emissions is to move towards zero carbon-emitting resources. In view of this, hydrogen (H_2) produced by photo/electrocatalytic water splitting has shown great potential to become the fuel of the future. The merits have been attributed to its high energy density and it produces only one by-product of water upon combustion [4]. Thus H_2, which is a zero carbon-emitting fuel, can be a promising solution to the mitigation of climate change. The second method is to capture carbon from the atmosphere and then store it back into the geosphere. However, the geosphere sequestration of CO_2 has no financial benefits. In contrast, chemical transformation of excess accumulated CO_2 from air into valuable industrial products, such as fuels (methanol, ethanol), hydrocarbon (methane, ethylene) and chemicals (formic acid, acetic acid), is an effective way to solve both global warming and energy crises [5, 6]. Furthermore, it has economic significance from an industrial point of view. However, carbon-di-oxide reduction (CO_2R) is a highly cumbersome and non-specific process. So far, several approaches, such as biochemical, chemical, thermal, photochemical and electrochemical catalysts have been explored to achieve aspirated activity and selectivity in this region [7–9]. Nonetheless, unlike other catalytic system electrocatalysts have gained tremendous attention due to its easy operation at ambient temperature and pressure. In addition, the selectivity of the product can be obtained by just adjusting reaction conditions, such as redox potential, electrode, and electrolyte, pH temperature and so on. The main advantage of using electrocatalysts is that they can be powered by renewable energy sources that emit zero carbon. For all these reasons, many research activities have shifted to the areas of EH_2ER and ECO_2RR (**Figure 1**) [10].

Over the years of time metal, metal oxide, metal sulphide have shown great promise in ECO_2RR and EH_2ER using electrocatalysis phenomenon. These electrocatalysts are being considered as a promising system that would be able to operate on a real scale without polluting environment. Whereas, there are still many limitations that are associated with electrocatalysts, such as high cost, poor product selectivity, high overpotential and low stability [11]. Colloidal Nanocrystal (C-NC) based electrocatalysts have become indispensable to overcome these limitations to certain extent owing to their larger surface to volume ratio, precise shape, long-term durability, and the plethora of configurations [12, 13]. These factors are important in influencing their efficiency, selectivity, and durability for EH_2ER and ECO_2RR. For example, variation in the size and/or shape causes alteration in reactivity at

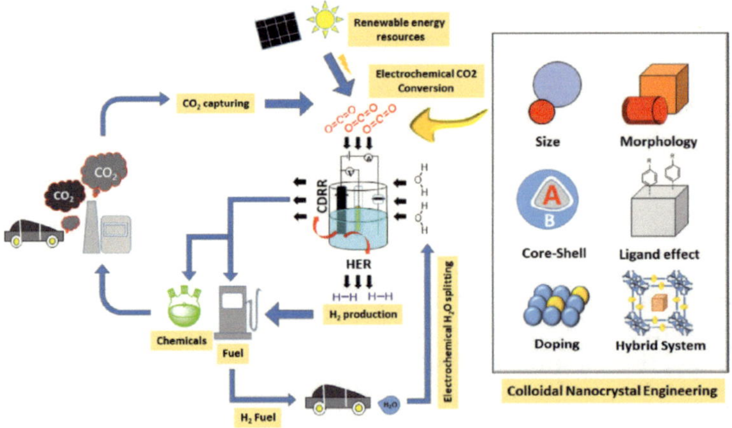

Figure 1.
Electrochemical CO_2 conversion into fuel and H_2 production by using colloidal nanocrystal-based electrocatalysts.

different locations (edges/corners/faces) of the C-NC due to changes in a specific atomic-arrangement of active centers and crystal surface energy. The impact of these features will be discussed in detail as this chapter unfolds.

The purpose of this chapter is to elaborate on recent research developments and challenges in the field of heterogeneous C-NCs-based electrocatalysts for ECO_2RR and EH_2ER. In the first part of this chapter, colloidal synthesis of nanocrystals will be discussed. The second part of this chapter will address the structural aspects, such as size, shape, and composition, are important in tuning the catalytic efficiency, selectivity, and durability of NC-based catalysts for ECO_2RR. Here a brief introduction of effect of ligand functionalization and effect of MOF/NCs hybrid system on ECO_2RR activity will be also discussed. In the third part of this chapter, the role of C-NCs-based catalysts on EH_2ER, its activity and stability is given. Moreover, the detailed mechanism of EH_2ER is also discussed in this part. Finally, Authors have given extractive commentary that sheds light on the future perspective of ECO_2RR and EH_2ER in conclusion.

2. Colloidal synthesis of NCs

The C-NC is an inorganic material with a size of 1–100 nm and surface covering of protecting capping agents like polymer and surfactants molecules. Generally, the inorganic part exhibits characteristic features, such as optical, electrical, magnetic, and catalytic, that can be tuned by changing their physicochemical parameters, while surface capping guarantees the stabilization of these structures and paves the way for synthesizing more complex structures [14, 15]. The physical parameters like morphology and chemical composition of C-NCs can be easily adapted by varying their reaction parameters like monomer concentration forming inorganic core of NCs and judicious choice of capping substances for surface covering. Over the last two-three decades, researchers have gained good control over synthesis of high-quality and cost-effective NCs with uniform morphology and chemical constituents using colloidal synthesis [16–19]. The C-NC_S approach has not only enhanced efficiency, selectivity of NCs, but also improved their service life. So far, researchers have found many commercial applications of C-NCs in various fields ranging from life sciences to the material world. One of the striking applications of C-NCs is in the field of biological imaging of cells, where quantum dots are used owing to their excellent fluorescent properties and also they do not photo-chemically bleach out like organic dyes [20]. Recently, quantum dots are being used commercially in LED displays also known as *QLED*-displays [21]. In addition, C-NCs-based photo/electro-catalysts for ECO_2RR and EH_2ER are being developed to solve the energy crisis and global warming. However, their uses at economical scale in this area is still facing challenges. Deep insights of C-NCs synthesis and the effect of C-NCs physiochemical parameters on their electrocalytic properties need to be investigated for their successful applications at the economical level.

In general, C-NCs can be synthesized in both water and organic solvents. However, synthesizing a broad spectrum of NCs requires different reaction conditions that are much more feasible to achieve in organic solvents compared to water that is mainly used in the synthesis of noble metal C-NCs [14]. Therefore, in this section, authors will focus on organic phase C-NCs synthesis. Generally, C-NCs synthesis requires three major elements: 1) precursor molecules or building blocks forming inorganic core of NCs, 2) capping agents, and 3) organic solvents. Capping agents sometime act as solvent. The process of nanocrystal formation starts with transformation of precursor molecules into unstable and reactive species or monomers that usually occurs at quite high temperature. Thereafter, these monomers

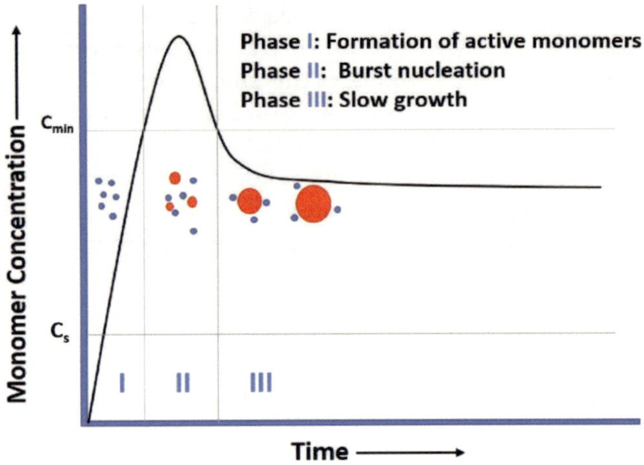

Figure 2.
The LaMer mechanism based formation of active monomers, burst nucleation, and subsequent slow growth of colloidal nanocrystals. Adapted from [23].

lead to formation of C-NCs whose growth is mainly influenced by capping agents. The crystallization of C-NCs can be best understood using widely accepted LaMer mechanism as depicted in **Figure 2** [22–25]. Based on this mechanism, NCs development considered to have three major steps. In the first phase, the precursor molecules are converted to the reactive species, or monomer, and then eventually reach to a supersaturated phase (I), where no particle or second phase is still visible. In the next step, reactive species concentrate to the critical limit of supersaturation (phase II), at which a thermodynamically feasible state (**Cmin**) is developed for nucleation, followed by monomers to form initial seed for nucleation. In the second phase, the supersaturation again drops at some point due to instant nucleation, reducing the monomer concentration below **Cmin** that triggers the third phase (III) of the mechanism, that is, the growth of NCs. In this phase, because of monomer concentrations remain below C_{min}, therefore, NCs grow without forming further nuclei until they attain an equilibrium state. During the growth phase of C-NCs, capping agents also play an important role in determining the final morphology of NCs. First and foremost, the capping agent should have a tendency to adsorb on the surface of the growing NCs.

Secondly, capping agents are required to bind in such a way that it can desorb and adsorb on the surface of growing NCs during growth process, making growing NCs surface accessible for reactive monomers, yet surface covering of capping agents, overall, stabilize NCs [14, 26].

2.1 Size control in C-NCs synthesis

During NC synthesis, control over size uniformity is a prominent feature of contemporary synthetic methods. The LaMer mechanism discussed earlier not only explains the formation of NCs, but is also an approach to synthesize C-NCs with narrow size distributions. The essential element of LaMer approach to synthesize uniform-sized C-NCs is to divide crystallization into two disparate events; nucleation and growth. As discussed previously, the nucleation occurs for short periods of time, also known as burst nucleation, triggering a different growth phase where all nuclei then grow at the same rate without generating extra nuclei. The formation of extra nuclei during the growth phase can cause differences in the size of C-NCs because the newly formed nuclei will lag behind the previously growing NCs in growth kinetics. The several methods utilize LaMer mechanism to grow uniform size C-NCs [24, 27]. The first example is seed-mediated growth method where reaction medium is introduced pre-developed nuclei at low monomer concentration to inhibit secondary nucleation. Therefore, these pre-developed nuclei further grow up to the desired size of C-NCs without creating extra nuclei. However, in order to obtain narrow size distribution of NCs, it is still required to control growth phase as well. The second example is hot injection method where the reaction medium at high temperature is rapidly supplemented with the precursor or reducing reagents to create a state of supersaturation that triggers subsequent burst nucleation. The next example is the heating up method where reaction medium pre-treated with precursor, and capping agents is heated at high temperature to induce the LaMer crystallization. Due to the simplicity of this method it is often used to synthesize C-NCs at large scale [28–30].

2.2 Shape control in C-NCs synthesis

Unlike the bulk materials, the physiochemical properties of C-NCs are also strongly dependent on their shape. In this chapter, authors will further discuss how the shape of C-NCs can be adapted to improve the service life, selectivity and efficiency of electrocatalysts. Generally, C-NCs shapes can be tuned by means of both thermodynamic and kinetic controls. In C-NCs synthesis, capping agents are used, primarily, to obtain the desired shape using their specific binding nature on the surface of the growing NCs [31]. If the surface adsorption of capping agents causes the decrement in surface energy of any specific facet, then the obtain shape will be favored by thermodynamically. Whereas, if capping agents serves an obstruction between growing NCs and diffusing monomers then the resulting shape of C-NCs will be governed by kinetic factors [32, 33]. Thermodynamically, the growing NCs attain it most likely shape by reducing its total surface free energy. For example, during the formation of fcc NC, the capping agents first selectively adheres to the (100) plans, which in turn, decreases their surface free energy [32]. These selective adsorptions onto the (100) plans causes transformation of cuboctahedron into cubic structure due the subsequent growth of higher energy facets. Whereas, in the kinetic regime, capping agents selectively adhere to some specific facets to lower their growth rate compare to others, resulting in various NCs shapes. Simultaneously, capping agents inhibit the diffusion of pre-deposited atoms over the NCs surface. In the real situation, however, the final shape of C-NCs is governed by the comparable kinetics of diffusion and deposition of monomers to growing NCs [32].

3. C-NCs-based catalysts for ECO_2RR

3.1 Mechanistic insight of ECO_2RR

Thermodynamically, CO_2 is a quite stable molecule (bond dissociation enthalpy of C=O is ~750 KJ mol^{-1}), so high energy is required for its activation. Moreover, the highest oxidation state of CO_2, causes problems for its selective reduction [34]. A catalyst in this regard is an alternative that offers reactive sites for its selective and rapid transformations. In this chapter, the authors are focusing specially on NC-based electrocatalyst for ECO_2RR. Generally, ECO_2RR involves several proton/electron transfers processes that take place at the cathode (catalyst). This process is considered to have three major stages. First, CO_2 is absorbed on the catalytic surface and its binding strength depends on the structure and composition of the NC as well as the nature of the electrolyte. Once it is absorbed, an electron is transferred

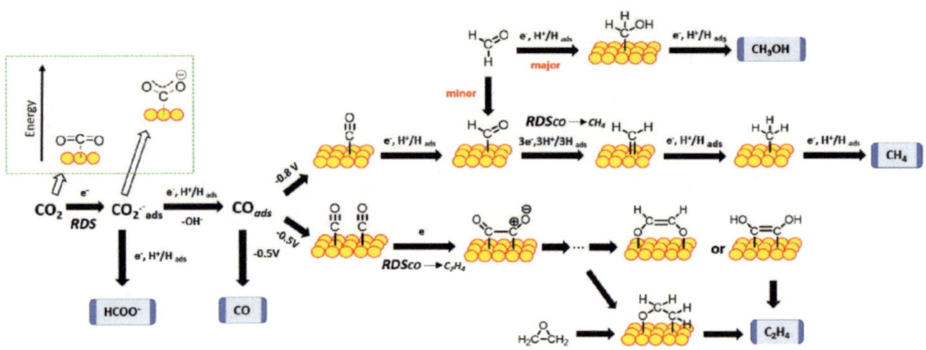

Figure 3.
Reaction pathways leading to the formation of formate, CO, and C–H products are highlighted. Adapted from [36]. Abbreviation: RDS, rate-determining step.

to the CO_2 molecule that produces surface bound $CO_2^{-\cdot}$ Intermediate as depicted in **Figure 3.** This step is known as a rate-limiting step because it requires large reconstitution energy to convert linear CO_2 into twisted form, that is, $CO_2^{\cdot-}$. For this reason, it requires extra potential (overpotential of −1.91 V) for electrochemical CO_2 conversion, even if it is thermodynamically feasible. After the formation of $CO_2^{\cdot}$– its reactivity on the catalytic surface determines resulting product in ECO_2RR. In principle, an optimal binding of the intermediate on the surface of the electrode is required for the rapid electron transfer process, thereby, increasing the selectivity and kinetics of the conversion. The reason for this is, a very strong affinity with intermediates will poison the surface of the electrode, while weak interaction will disrupt the electron transfer process. Sn, In and Pb, for example, represent week interaction with CO_2^-. intermediate, therefore, further reduction leads to production of $HCOO^-$ as resulting product [35]. In comparison, CO_2^-. on the surface of Ag, Au and Zn is reduced to $COOH^*$ which can be further reduced into CO^* [35]. However, CO^* has week affinity towards these metal ions, and thus, gaseous CO is generated. Interestingly, Cu, which has been extensively studied by Hori et al., shows optimal binding with CO^* and uniquely reduces CO_2 in many products including alcohol and hydrocarbons [35]. Whereas, metals, such as Pt and Ni, have a strong affinity for CO*, which prevents further reduction, and for this reason, these metals favor EH_2ER over ECO_2RR [35]. Therefore, optimal interaction between metal-based electrodes and surface bound intermediates has a significant importance in product selection and/or reaction rate.

Furthermore, it was realized that surface properties of electrocatalyst, such as, surface area, roughness, composition, and morphological design have profound influence on efficiency, selectivity and durability of electrodes in electrochemical reactions [37–38]. It is a general understanding that the higher surface area provides good economy of the active center on the surface relative to the bulk, and thus, accelerates CO_2 reduction. Similarly, Cu shows optimal coordination with CO, however, changes in surface structure, such as roughness, may deviate from their normal behavior. For example, Jiang et al. showed that the high population of under-coordinated sites on the rough surface of the Cu leads to the formation of oxygen-containing compounds and hydrocarbons compared to CO due to enhanced interaction with CO* intermediates [38]. However, it is still challenging to adjust the optimal binding energy for intermediates to increase selectivity/reactivity towards ECO_2RR, because of the large number of intermediates and many possible intricate pathways involved. Until now, many bulk metal-based electrodes have been investigated from both a material and structural point of view; however, they are still facing many hurdles, such as: 1) high overpotential of the existing electrodes, 2) obtaining a mixture of products due to poor product selection of the catalyst, 3) deactivation of metal-based electrodes in short periods, 4) high cost of metal-based electrodes such as Ag, Au, Pt etc. discourages their use economically, 5) slow kinetics, or low activity. Beside there is always a competitive reaction EH_2ER to the ECO_2RR based on the thermodynamics. Rapid electron transfer in first step of electrochemical CO_2 reduction may be an effective strategy to suppress EH_2ER, which, in turn, accelerates ECO_2RR. In **Table 1**, the authors have summarized electrochemical Eqs. (1–16) of CO_2R in some valuable products such as CH_4, CH_3OH, HCOOH, etc., with their respective equilibrium potentials in aqueous medium (pH 6.8) at 1 atm. and 25°C with respect to standard hydrogen electrode (SHE) [39, 40]. The equilibrium potential of the ECO_2RR, as provided, corresponds to the equilibrium potential of the EH_2ER.

3.2 C-NCs-based heterogonous catalyst for ECO_2RR

Recent studies have shown that nanometer-sized (1–100 nm) electrocatalyst is not only capable of reducing overpotential, but also shows an improvement in current density for CO_2 conversion. Regardless of the metal, the electronic structure of the catalysts at nanoscale is a key player in determining their efficiency, selectivity and durability for ECO_2RR. Several electronic factors have been determined, such as finite size effects, and the location of the d band center that can tune the binding strength of intermediates, such as CO^*, CHO^*, etc., on the nanoparticle

Products	Thermodynamic half-cell equations	E(V)
Hydrogen	$2H^+ + 2e^- \rightarrow H_2$	**0.000**
C1		
Methane	$CO_2 + 8H^+ + 8e^- \rightarrow CH_4 + H_2O$	0.17
Carbon-mono oxide	$CO_2 + 2H^+ + 2e^- \rightarrow CO + H_2O$	−0.10
Methanol	$CO_2 + 6H^+ + 6e^- \rightarrow CH_3OH + H_2O$	0.03
Formic acid	$CO_2 + 2H^+ + 2e^- \rightarrow HCOOH + H_2O$	−0.02
C2		
Acetaldehyde	$2CO_2 + 10H^+ + 10e^- \rightarrow CH_3CHO + 3H_2O$	0.05
Acetate	$2CO_2 + 8H^+ + 8e^- \rightarrow CH_3COOH + 2H_2O$	−0.26
Ethanol	$2CO_2 + 12H^+ + 12e^- \rightarrow C_2H_5OH + 3H_2O$	0.09
Ethylene Ethylene glycol	$2CO_2 + 12H^+ + 12e^- \rightarrow C_2H_4 + 4H_2O$ $2CO_2 + 10H^+ + 10e^- \rightarrow C_2H_6O_2 + 2H_2O$	0.08 0.20
Glyoxal	$2CO_2 + 6H^+ + 6e^- \rightarrow C_2H_2O_2 + 2H_2O$	−0.16
Glycoaldehyde	$2CO_2 + 8H^+ + 8e^- \rightarrow C_2H_4O_2 + 2H_2O$	−0.03
C3		
Acetone	$3CO_2 + 16H^+ + 16e^- \rightarrow CH_3COCH_3 + 5H_2O$	−0.14
Allyl alcohol	$3CO_2 + 16H^+ + 16e^- \rightarrow C_3H_6O + 5H_2O$	0.11
Propionaldehyde	$3CO_2 + 16H^+ + 16e^- \rightarrow C_3H_6O + 5H_2O$	0.14
1-Propanol	$3CO_2 + 18H^+ + 18e^- \rightarrow C_3H_7OH + 5H_2O$	0.21
Hydroxyacetone	$3CO_2 + 14H^+ + 14e^- \rightarrow C_3H_6O_2 + 4H_2O$	0.46
Oxygen	$O_2 + 4H^+ + 4e^- \rightarrow 2H_2O$	**1.23**

Table 1.
Thermodynamic electrochemical half-cell equations of CO2R products, along with their relative standard redox potential (vs SHE in volt), or E(V) at pH 6.8 [39, 40].

(NP) surface and thus, alter the reactivity of NP-based electrocatalysts [41–43]. In addition, the geometric effect is an important factor that is crucial in stabilizing the specific intermediate during the reduction process, which leads to an increase in selectivity [44, 45]. When using C-NC-based electrocatalysts, electronic and geometric factors can be adapted to increase selectivity, stability and activity using several approaches, such as size adjustment, shape modification, compositional control, surface functionalization and reaction conditions. As discussed earlier, the nucleation and growth process can be optimized by changing the reaction conditions (thermodynamic and kinetic of the reaction) to achieve the desired C-NCs with well-defined composition, and morphology. Therefore, C-NCs-based catalytic systems have been found to be highly promising to lead the catalytic field that can achieve higher selectivity and efficiency than existing systems. Here, our focus will be mainly on to understand how properties like size, shape, morphology, composition and surface functionalization of nanocrystal affect efficiency and durability of C-NCs-based electrocatalysts for ECO_2RR.

3.3 Effect of C-NCs size on ECO_2RR

The size is an important factor for C-NC-based electrocatalysis because different size of C-NCs show different activity/selectivity towards ECO_2RR. Several studies have shown that alteration in size at nanoscale can affect both atomic distributions at various reaction sites (plans, edges, corners) and electronic structure of NCs, changing their catalytic characteristics. Therefore, the search for the optimal size C-NC showing the best efficiency, selectivity and durability requires contemporary research in this area. So far, to explore the effect of size towards ECO_2RR, different sizes of electrocatalysts have been evaluated, however, our focus here is towards C-NC-based electrocatalysts. Colloidal synthesis of NCs is an excellent approach to study different sizes of NCs while keeping other factors such as shape and composition stable. Loiudice et al. synthesized spherical and cubical Cu C-NCs in the size range of 7.5–27 nm and 24–63 nm respectively for ECO_2RR [46]. Their study has revealed unprecedented correlation between size of NCs and their electrochemical activity as well as selectivity for ECO_2RR. The activity of Cu C-NCs increases as size of NCs within same morphology decreases, however, this does not hold while comparing cubical NCs with spherical. For example, the 44 nm cube has higher current density than 27 nm sphere. To understand this phenomenon, the propensity of Cu (100) facet towards ethylene production can provide a better understanding. Based on previous findings, it is widely accepted that the Cu (100) plane is selective for the electrochemical reduction of CO_2 in ethylene. Additionally, edges in ethylene production are thought to be responsible for the absorption and stabilization of an important intermediate (i.e., COOH *) as well as inhibiting EH_2ER. In this study, the X-ray diffraction pattern has shown a higher contribution of the Cu (100) plane in the nanocube than in the nanosphere and Cu foil. Interestingly, among all Cu C-NCs, a size of 44 nm exhibited highest selectivity with 50% FE for ethylene and overall 80% for ECO_2RR over EH_2ER. This can be ascribed to changes in atomic configuration at different sites of C-NC due to differences in size. As nanocube move from smaller to larger sizes, the number of atoms at the edges and corners decreases, however, the number of atoms at the plane site increases. As a result, it came close to the morphology of a single crystal, where all the atoms of the surface are populated on the (100) plane. And, as previously stated edges also play an important role in electrochemical reduction of CO_2 into C_2 products. Therefore, an optimal balance between ratio of edge to plane site in 44 nm Cu C-NC, suggesting not only its unique selectivity for C_2H_4, but also an overall activity towards ECO_2RR.

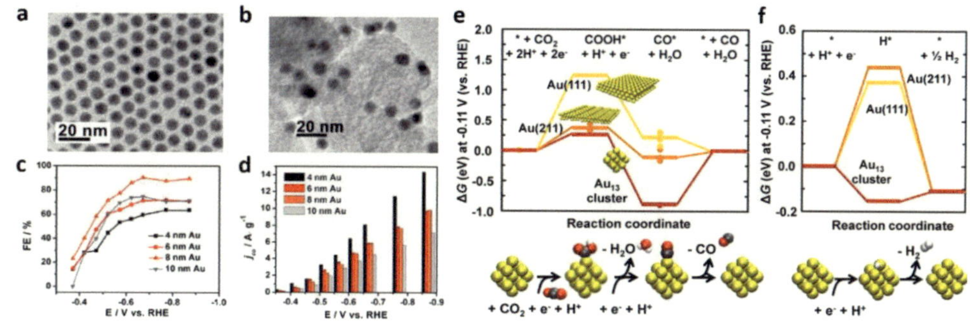

Figure 4.
TEM images of (a) the 8 nm Au NPs and (b) the C-Au NCs. (c) Potential-dependent FEs of the C-Au on electrocatalytic reduction of CO_2 to CO. (d) Current densities for CO formation (mass activities) on the C-Au at various potentials. Free energy diagrams for electrochemical reduction of (e) CO_2 to CO and (f) protons to hydrogen on Au(111) (yellow symbols), Au(211) (orange symbols), or a 13-atom Au cluster (red symbols) at −0.11 V. reprinted with the permission from [47]. Copyright © 2013, American Chemical Society.

Similarly, Zhu et al. investigated the non-monotonic size-dependent selectivity of Au C-NCs for CO_2 reduction into CO [47]. Among Au C-NCs of size 4, 6, 8, 10 nm synthesized, the 8 nm exhibited a highest selectivity for CO production with FE of ~90% at −0.67 V vs. RHE as shown in **Figure 4(c)**. Based on the DFT calculations, the group concludes that dominance of edge sites at 8 nm C-NC facilitates selective CO_2 reduction in hydrocarbons, whereas depletion of corner sites inhibits EH_2ER. They have further reported the unique selectivity of Au_{13} C-NC towards EH_2ER. Although Au_{13} facilitates the formation of COOH* intermediate as compared to Au (211) and Au (111) facet, a stronger binding with CO* intermediate would lead to its lower tendency to produce CO_2 reduction products.

Additionally, as shown in **Figure 4(f)**, the free energy of formation ($\Delta Gfor_{mation}$) of H* intermediates on Au_{13} is lower than $\Delta Gfor_{mation}$ of COOH*, suggesting the generation of H_2 at low overpotentials. Furthermore, this study is found to correspond to a size dependence study on the small size (2–15 nm) Cu C-NCs for ECO_2RR. For particle sizes at 5–15 nm, significantly higher activity and selectivity were found for H_2 and CO than for hydrocarbons. Moreover, particles smaller than 5 nm in size showed an exponential increase in the formation of H_2 and CO relative to hydrocarbons. This unique property of small C-NCs is due to an increase in under coordinated catalytic sites, which, in turn, strongly stabilize and bind with H^* and CO^* intermediates. Strong adsorption of H* and CO* intermediates is suggested to prevent subsequent hydrogenation of CO into hydrocarbons, so the increase in H_2 and CO production occurs at small (<5 nm) NCs.

3.4 Effect of C-NCs shape on ECO_2RR

The shape of the C-NC plays an important role in determining the selectivity/activity of the electrocatalysts. Several findings have shown that the shape-dependent C-NC activity/selectivity towards ECO_2RR is typically associated with the presence of specific crystal plane. For example, Suen et al. have revealed that cubic Cu C-NC (C-Cu) with mainly (100) facet shows enhanced selectivity towards C_2 products while octahedron Cu C-NC (O-Cu) with predominantly (111) facets shows selectivity for C_1 products in ECO_2RR [48].

Besides, Zhou et al. have illustrated the correlation between different shapes of C-NCs and their corresponding crystal planes using a stereographic triangle as depicted in **Figure 5(a)** [49]. The lower index surfaces, namely, the (100) (110) and (111) facets that lie on the three vertices of triangle, are present on the nanocrystal

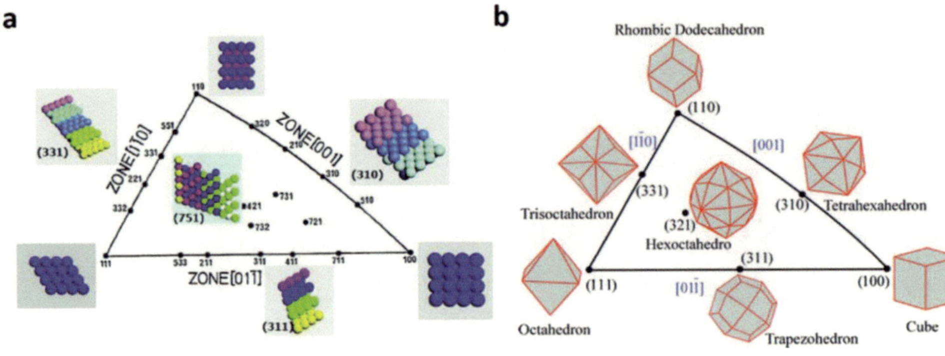

Figure 5.
(a) Unit stereographic triangle of fcc single-crystal and models of surface atomic arrangement. (b) Unit stereographic triangle of polyhedral nanocrystals bounded by different crystal planes. Reproduced form [49] with permission from the Royal Society of Chemistry.

cube, octahedral, and rhombic dodecahedral, respectively. These low index surfaces are said to be less energetic, that is, catalytically less active. However, the plains lying on three sidelines of triangle and one which is residing inside the triangle are known as high index surfaces, that is, catalytically more active and stable. The high index planes (310), (311), (331), and (321) are correlated with tetrahedron, trapezohedron, trisoctahedron, and hexoctahedron, respectively, as illustrated in **Figure 5(b).** The high energy of these polyhedron is attributed to predominance of atomic steps, islands and kinks on their surface. For example, a concave rhombic dodecahedron Au C-NC enclosed with multiple high index planes showed excellent activity and selectivity towards CO formation [50]. In addition, this particular structure of Au C-NC was found durable for longer period of time. Summing up, low coordinating high index planes with CN less than 7 exhibits high catalytic activity, selectivity, stability whereas low index planes with higher CN greater than 6 exhibits lower catalytic activity and stability.

Lattice strain has evolved as another factor that have a significant influence on electrocatalytic properties of C-NCs towards ECO_2RR. It was studied that strain generates a distortion in lattice, which, in turn, alters d-band center of metallic C-NCs. The altered d-band center either facilitates or lowers the adsorption of intermediates, as a result of that changing catalytic properties. Typically, upshifting of the d-band center facilitates adsorption of reaction intermediates on the surface of C-NCs. This is because, as the d band center approaches the Fermi level, which is the highest occupied state, the antibonding orbitals move over it, causing them to empty. This, as a result, strengthens the binding strength of the reaction intermediate on the catalytic surface and, therefore, enhances the reaction kinetics [51]

In general, lattice strain in C-NCs can be induced by shaping C-NC in various morphologies, which leads to the inward displacement of atoms at high-energy locations, such as corners and edges, while the outward displacement of atoms on planes to gain overall crystal stability. These compression and expansion in the atomic arrangement in the NC_S induce aeolotropic strain gradients that may improve the catalytic efficiency/selectivity of C-NCs [45]. Octahedral and Icosahedron C-NC of Pd with similar sizes were examined by huang et al. to investigate the effect of strain on ECO_2RR. They observed that icosahedral/C C-NC shows higher FE (91% with −0.81 V vs. RHE) for CO production than octahedral/C C-NC in ECO_2RR. The molecular simulations and DFT calculations showed that surface strain in icosahedral C-NC enhanced catalytic selectivity due to shifting in d-band center, which, in turn, facilitates absorption of a key intermediate ($COOH^*$)

in ECO_2RR. Thus, surface strain in icosahedral C-NC boots catalytic efficiency and selectivity for CO_2R [52].

3.5 Effect of C-NCs composition on ECO_2RR

The variation in composition of metal C-NCs is another intriguing factor that plays an important role in tuning electrocatalytic efficiency/selectivity towards ECO_2RR. Among other effects of composition, the synergistic effect where the mutual synergy between both electronic and geometric effects determines the activity/selectivity of C-NCs towards ECO_2RR is well known. The electronic effect can be understood by the concept of shifting in d-band center due to alteration in composition. In addition to the electronic effect, the geometrical effect also makes a significant contribution, where the particular atomic arrangement in the active center can modify the binding strength of the reaction intermediate, thereby improving the electrocatalytic efficiency/selectivity towards ECO_2RR [44]. Therefore, to understand the synergistic chemistry between geometric and electronic effects in this section, the electrocatalytic efficiency, selectivity and durability of bimetallic C-NCs and doped C-NCs electrocatalysts towards ECO_2RR will be discussed.

3.5.1 Bimetallic C-NCs

Until now, several bimetallic NCs have been investigated for ECO_2RR. For example, Kortlever et al. have found an optimal composition of a novel $Pd_{70}Pt_{30}/C$ electrocatalyst highly active and selective towards HCOOH production. Moreover, this has a remarkably lower onset potential close to 0 V vs. RHE, making it best catalyst till the date [53]. However, usage of noble high-cost metals discourages its application on economical scale. Incorporation of non-noble metals such as Cu, Ni, Fe, etc., in bimetallic system could serve a better alternative to address high cost and stability of electrocatalyst. Kim et al. synthesized different composition of NCs including Au, AuCu, $AuCu_3$, Au_3Cu, and Cu, which were assembled in a monolayer on glassy carbon while keeping precise control over morphology (Size, Shape etc.) [44]. When considering electronic effect solely, pristine Au NC should have shown higher activity due to its optimal binding with COOH and CO intermediates. Interestingly, they have observed higher activity for Au_3Cu bimetallic C-NC than expected one. Therefore, electronic effect mere does not explain volcanic acti vity correlation for C-NCs. The geometric effect that works synergistically along with electronic effect ensure further stabilization of intermediates, thus, explaining optimal activity of Au_3Cu bimetallic C-NC towards electrochemical CO_2 reduction, among others. Previous studies have shown excellent properties of In based metal catalysts for selective CO_2 conversion into HCOOH. However, these metal catalysts suffer from limited current density and poor stability.

Kown et al. synthesized In_2O_3-ZnO C-NCs that showed excellent selectivity towards formation of HCOOH [54]. The XRD pattern showed that pre reduction of these C-NCs during electrolysis leads to the formation of In-Zn bimetallic C-NCs. Among all $Zn_{1-x}In_x$ NCs, $In_{0.05}Zn_{0.95}$ offered remarkable selectivity for HCOOH production at FE of 95% (−1.2 V vs. RHE) as well as with higher current density. The higher catalytic activity was seen in both $Zn_{1-x}In_xO$ and $Zn_{1-x}In_x$ with decreasing value of x. The XPS data has revealed predominance of O vacancies in bimetallic systems at lower x, which, in turn, decreases thickness of oxide layers in $Zn_{1-x}In_x$. Consequently, conductivity of $Zn_{1-x}In_x$ C-NCs increases at lower x, therefore, facilitates rapid electron transfer processes in ECO_2RR. Thus, highest catalytic activity of $Zn_{0.95}In_{0.05}$ C-NC is attributed to its remarkable conductivity. DFT calculations revealed that tight binding of OCHO* on pristine In, which impedes HCOOH production, has weakened by introduction of Zn in bimetallic C-NC. Therefore, mutual synergies of In with Zn in bimetallic system enhanced its catalytic selectivity in CDRR as compared to In [54]. Guo et al. have shown compositional effect in Cu_3Pt C-NCs is responsible for improved activity and selectivity for CH_4 [55]. It was observed that increasing Cu contents in bimetallic NC leads to desorption of more CO^* intermediates that subsequently gets protonated into CH_4. Moreover, Pt which shows higher affinity for H^+ has significantly accelerated protonation of CO^*. However, higher Cu content beyond ratio of Cu and Pt (3:1) raised CO^* poisoning, and thus, declining in CH_4 production [55].

3.5.2 Dopped C-NCs

In doped C-NCs, extrinsic or intrinsic introduction of impurities can cause a change in the electronic structure in a way that can enhance catalytic efficiency/ selectivity and durability. For example, the incorporation of impurities can provide additional electrons (n-type) or additional vacancies (p-type), thereby, increasing the conductivity of the catalyst that would otherwise be poor in conductivity. Therefore, the increased conductivity may accelerate the electron transfer process in the ECO_2RR. Although many advances have been made in the field of doped C-NCs, some studies conducted in ECO_2RR. Recently, a group led by Kim et al. reported an unprecedented selectivity of vanadium ($_{23}V$) doped In_2O_3 C-NCs towards CH_3OH formation in addition to HCOOH and CO, previously not known with pristine In and In_2O_3 [56]. This can be understood using the commonly suggested scheme for producing CH_3OH shown in the **Figure 3.** The CO^* intermediate needs to be stabilized on the surface to proceed towards CH_3OH formation otherwise it may release as CO gas and terminate the reaction. The introduction of V^{3+} into In_2O_3 strengthens the binding of CO^* intermediate with NCs possibly due to the π-back donation from dopant to intermediate CO^*, and thus, reaction proceeds towards CH_3OH (FE of 15.8% at −0.83 V vs. RHS) [56].

3.6 C-NCs/metal organic framework hybrid

In the past years, molecular organic frameworks (MOFs) have received significant attention in the field of catalysis, including ECO_2RR, where they have been used primarily to provide solid support for molecular catalysts [57, 58]. The combination of MOF with C-NCs creates a class of hybrid materials that have demonstrated great potential to become future class of electrocatalysts with improved efficiency/selectivity and durability. Concerning this, a deep understanding of the material design and working principle of these novel hybrid systems is indispensable prior to implementing them practically. Recently, Guntern et al. investigated catalytic selectivity and durability of Ag C-NCs/Al PMOF hybrid system for electrochemical CO_2 reduction [59]. Based on UV visible and XPS data, it was concluded that electron transfer from MOF to Ag C-NC in this hybrid system increases electron density at Ag C-NC, thereby, facilitates electron transfer to $CO_2^{\cdot-}$ intermediate. As in result, selectivity of Ag C-NCs/Al PMOF towards CO increases than pristine Ag C-NCs while decreases for H_2ER. Additionally, a small contribution of transport events due to diffusion of reactants and products within pores of MOF adds in the selectivity of NC/MOF hybrid towards CO. Besides, NC/MOF hybrid system showed better stability compared to bare Ag C-NCs at lower potential [59].

3.7 Ligand functionalized C-NCs

Capping agents such as, organic ligands, surfactants or polymers are engendering factors of surface anchoring molecules that are crucial in synthesis of C-NCs based catalysts with well-defined shape, uniform size distribution and different composition. After synthesis, these capping agents can significantly alter C-NCs catalytic efficiency in both positive and negative way by staying absorbed on their surface. For example, Wang et al. reported that catalytic activity of Ag NCs modified with capping agents increased by 53-fold compared to pristine Ag NCs towards ECO_2RR [60]. In Several studies it has been shown that presence of surface molecule can affect active site of NC in numerous ways: 1) by perturbing electronic structure of active sites 2) by introducing steric hindrance that impedes diffusion of absorbates reaching at active center 3) by blocking selective facets of C-NCs, which, in turn, could enhance selectivity and activity. Moreover, chiral ligands can be used to produce stereoselective products. Therefore, surface functionalized NCs has opened a new window in the field of electrocatalytic reactions where unprecedented control over efficiency/selectivity can be achieved by ligand design [61, 62]. However, researchers still have a limited understanding of how these anchoring ligands affect the local electronic environment of NC and how the backbone of anchoring ligands regulates reactivity between NC and surrounding reactants, or reaction intermediates.

Pankhurst et al. have tuned ECO_2RR selectivity of Ag C-NCs using different imidazolium ligands [63]. Here they were able to introduce different organic component by varying tail and anchoring groups on imidazole motif. When performed ECO_2RR, the ligand bearing NO_2 anchoring group with octyl tail-found highly selective towards CO with FE of 92%. It was concluded that interaction between cationic imidazolium group with CO_2 increases population of this reactant over the catalytic surface. Furthermore, an optimal chain length of ligand tail-group increased hydrophobicity of surface, and thus, increases selectivity and efficiency for ECO_2RR by inhibiting EH_2ER. However, electronic changes induced by ligand anchoring-group did not improve significantly properties of electrocatalyst. The next example of ligand surface functionalization for ECO_2RR discusses N-heterocyclic-carbine functionalized Au C-NCs (Au C-NC-Cb) from group led by Cho et al. The significant downfield shifting in 13C NMR peaks of NHC reveals strong electron donation from ligand to metal, making Au C-NCs surface electron-rich. As in result, it facilitates the electron transfer process to CO_2, and thus, enhances efficiency of Au C NC Cb relative to bare Au C-NCs [64].

4. C-NCs-based heterogeneous catalyst for EH_2ER

The concept of using electric current to control various chemical reactions achieved much attention, since the time when humankind invented first power resources. Splitting of water to produce hydrogen and oxygen gas started much earlier, but now it has started at large scale in industrial process and seems to play a crucial role in combating the future energy crisis. The increasing demand for energy day by day and the shortage of fossil fuels have encouraged scientists to develop a renewable and clean source of energy.

Hydrogen is considered a clean source of energy because by-product of H_2 combustion is H_2O and the starting material to obtain H_2 is water, therefore, an efficient and clean source to supplant the depleting fossil fuels [65–67].

Moreover, H_2 produces highest energy on combustion of per unit mass relative to any other fuels, thus, leading to become a fuel of future. The H_2 can be produced by electrochemical water splitting reaction which is an endothermic process with a potential of $\Delta E° = 1.23$ V and $\Delta G° = 237.2$ kJ mol^{-1}. Water splitting generates both hydrogen and oxygen. For the process of electrochemical water splitting the reactions that occur at anode are called as oxygen evolution reaction (O_2ER) and the reactions which occurs at cathode are called as H_2 evolution reaction (H_2ER) [68]. Electrocatalysis is actually an attempt to elucidate and predict observable phenomena like overall activity of the reactions that occur on the surface of electrode by the interactions of electrode/electrolyte interface. Development of an efficient electrocatalyst is important to minimize the energy losses during the electrocatalytic splitting of water to produce hydrogen and oxygen gas. **Figure 6** shows a diagrammatic representation of evolution of hydrogen (Depicted to the left-side of **Figure 6**) and oxygen (depicted to the right side of **Figure 6**) gas on the surface of glassy carbon electrode (GCE) after deposition of catalyst on its surface.

$$4H^+ + 4e^- \rightarrow 2H_2$$

$$2H_2O \rightarrow O_2 + 4H^+ + 4e^-$$

$$\text{Net reaction}: 2H_2O \rightarrow 2H_2 + O_2$$

The reactions which are central to hydrogen energy are two types. These are hydrogen evolution ($2H^+ + 2e^- \rightarrow H_2$) and hydrogen oxidation ($H_2 \rightarrow 2H^+ + 2e^-$) reactions. The research of oxidizing and evolving hydrogen was started in 1960 but it gained importance in 1970 and 1990 when the shortage of oil was realized [69]. The most success in this regard was achieved when precious metals like platinum (Pt) were used. Metal NPs on the surface of carbon also showed great success in H_2ER and hydrogen oxidation reaction (H_2OR).

In the world of EH_2ER electrochemistry, recent merge of computational quantum chemistry and nanotechnology have shown great progress in explaining

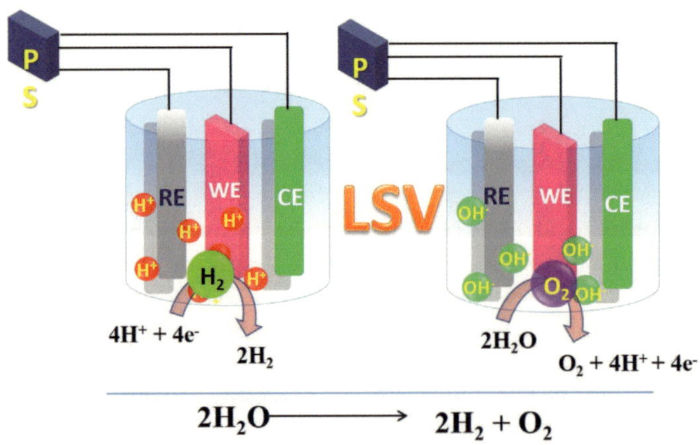

Figure 6.
Diagrammatic representation of formation of hydrogen during hydrogen evolution reaction and formation of oxygen during oxygen evolution reaction on the surface of glassy carbon electrode. Abbreviation: LSV - linear sweep voltammetry, WE - working electrode, RE - reference electrode, CE - counter electrode.

fundamentals and basics of EH_2ER with much emphasize on its utility and storage [51, 70, 71]. Metallic Pt is considered as 'state of the art catalyst' and exhibits small Tafel slope values and extremely high exchange current density (j_0) [72–74]. However, because of high cost and less availability of Pt, a sustainable, cost effective, and stable catalyst needs to be developed. So, there have been efforts to synthesize the EH_2ER active catalysts from the transition metals that are abundant in nature.

4.1 Mechanistic overview of EH_2ER

EH_2ER kinetics has a long history and have been explained in detail [69]. EH_2ER ($2H^+ + 2e^- \rightarrow H_2$) is a process involving a series of electrochemical steps which takes place on the electrode surface and results in the evolution of hydrogen. There are two mechanisms in acidic and basic conditions accepted universally as shown in **Figure 7** [75]. These steps are:

1. Electrochemical hydrogen adsorption (Volmer reaction) (Eq. (1), (2))

$$H^+ + M + e^- \leftrightarrow M - H^* \text{ (acidic solution)} \tag{1}$$

$$H_2O + M + e^- \leftrightarrow M - H^* + OH^- \text{ (alkaline solution)} \tag{2}$$

This step is followed by.

2. Electrochemical desorption (Heyrovsky reaction) (Eq. (3), (4))

$$M - H^* + H^+ \leftrightarrow H_2 \text{ (acidic solution)} \tag{3}$$

$$M - H^* + H_2O \leftrightarrow M + OH^- + H_2 \text{ (alkaline solution)} \tag{4}$$

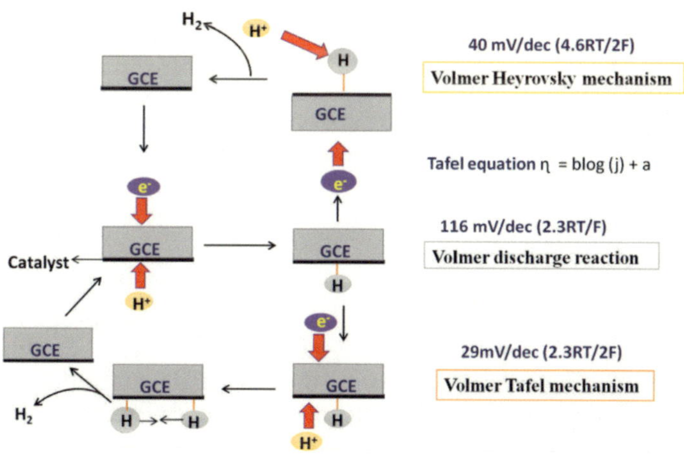

Figure 7.
Mechanism of hydrogen evolution reaction on surface of glassy carbon electrode (GCE). Abbreviation: GCE - glassy carbon electrode.

Or,

3. Chemical desorption or combination reaction (Tafel reaction) (Eq. (5))

$$2\,M - H^* \leftrightarrow 2M + H_2\ (\text{both acidic \& alkaline}) \tag{5}$$

In the above reactions, H* indicates an adsorbed hydrogen atom that has been adsorbed chemically on the surface of electrode (M) at the active site. These reaction pathways are highly dependent on electronic and chemical properties of the electrode surface [76]. Tilak et al., explained that rate controlling steps (1, 2 and 3) is predicted by deducing Tafel slope values from EH_2ER polarization curves [72]. The mechanism and rate determining step is studied by the Tafel slope. Tafel slope is an inherent and interesting property because it gives information about the potential difference required to increase or decrease the current density by 10-fold. Tafel slope is also useful to determine the effectiveness of a catalyst. In order to calculate the Tafel slope the linear portion of the Tafel plots is to be fitted in the Tafel equation (η = b log (j) + a, where η = overpotential, b = Tafel slope, and j = current density) [77, 78]. Theoretical facts about Tafel slope have been derived from Butler-Volmer equation and it is proved for three limited cases. First, if the discharge reaction proceeds very quickly and H_2 is evolved by the rate determining combination reaction (Tafel step). The slope value is 29 mV dec-1 at 25°C (2.3RT/2F). Second, if the discharge reaction proceeds very quickly and H_2 is evolved by the rate determining desorption reaction (Heyrovsky step). The slope value for this step is 40 mV dec-1 at 25°C (4.6RT/3F). Third, if the discharge reaction proceeds very slowly and then the rate determining step will be Volmer step irrespective of the fact whether H_2 is evolved by the combination reaction or the desorption reaction. The Tafel slope is 116 mV dec^{-1} at 25°C (4.6RT/F). The detailed mechanism is shown in **Figure 7**. It is evident that reaction (1) represents chemical adsorption, whereas, reaction (2) and (3) exhibits H atoms desorption from the electrode surface, which are competing with each other. Sabatier and co-workers came with an idea (Sabatier principle) that a better catalyst should not only form a strong bond with absorbed H* and facilitates the proton electron transfer process, but also it should be weak enough in facial bond breaking to assure quick release of H_2 gas [79]. It is difficult to establish a quantitative relationship between energies of H* intermediate and rate of electrochemical reaction owing to absence of directly measured surface-intermediate bonding energy values [80]. However from the perspective of physical chemistry, both for H* adsorption and H_2 evolution on the catalyst surface can be determined from the change in free energy of H* adsorption (ΔGH*) using EH_2ER free energy diagram [81]. According to Sabatier principle, under the condition ΔGH* = 0 will have maximum overall reaction rate (expressed in terms of EH_2ER exchange current density, j0).

4.2 Metal-based C-NCs for EH_2ER

An important correlation between ΔGH* and j0 have been proposed in the form of "volcano curve" for a wide variety of electrode surfaces as illustrated in **Figure 8** [81, 82]. Pt group of metals are the most efficient in the process and that's why are found at the top of the volcano curve, because they have small Tafel slope and quasi zero onset potential.

However, due to high cost of Pt, various research groups have been working on modifying Pt group metals such as engineering the NCs. Crystal plane (110) of Pt NCs has been proved good surface for EH_2ER. Like Pt, Palladium (Pd) NCs have

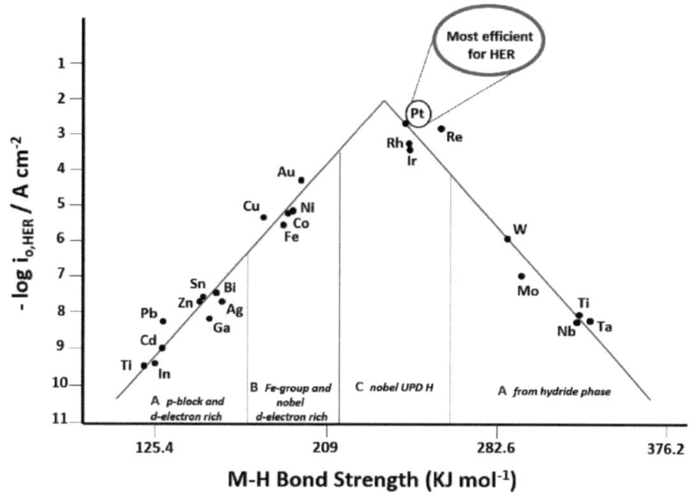

Figure 8.
Volcano plot for log I_o values for HER as a function of M-H bond energy. Adapted from [83].

been showing great promise as it is in the same group and has almost same size and its lattice matches about 0.77% to Pt, good thing about Pd is that it is comparatively cheaper than Pt. Pd has one advantage that it can adsorb hydrogen from both electrolytes and the gas phase. Pd can be loaded on various supports to increase its activity as it alters surface area and electronic conductivity of the nanocrystals [84]. Huang et al. observed that Pd C-NCs deposited on carbon paper substrate has activity higher than Pt black electrode and it required very less catalyst loading that is 0.0106 mg cm^{-2} [85]. One thing to be noted about Pd is that it has higher activity in acidic medium than alkaline medium because of lower Pd-H binding energy and lower activation energy, acid (32.3 ± 0.7 kJ/mol) and base (38.9 ± 3.0 kJ/mol). Ruthenium (Ru) NCs are also being explored for EH_2ER. As the Sabatier principle suggest that catalyst should not have much stronger binding to the hydrogen and should possess moderate binding capacity so desorption is easy, Ru-H follows this trend as it has ~65 kcal/mol energy for Ru-H bond and thus less activation barrier for desorption process [86]. Ru C-NCs are usually used with some support as they have durability problems because of aggregation. One such example where Joshi et al. used Ru C-NCs supported with Tungsten (W). DFT calculations suggest that Ru (0001) has high H_2 binding to surface energy but using W support, it could reduce the H_2 adsorption energy and changes the electronic environment thus making it similar to Pt (111) and increases its activity for EH_2ER process [86]. Baek et al. revealed that Ru when deposited on graphene nanoplatelets (GnP) to form Ru@GnP, its activity usually surpasses that of Pt/C in both acidic and alkaline medium. This happens because it is more stable, possess low Tafel slope (30 mV dec^{-1} in 0.5 M aq. H2SO4; and 28 mV dec^{-1} in 1.0 M aq. KOH) and also has comparatively low overpotential at 10 mA cm^{-2} (13 mV in 0.5 M aq. H_2SO_4; 22 mV in 1.0 M aq. KOH) [87]. Not only Pt, Pd, Ru, metals like Iridium (Ir) are also explored for the EH_2ER process and also earth abundant metals are used but they are prone to corrosion in the presence of alkaline and acidic medium. Thus, the other way is using non-noble metals for the process.

Non-Noble metals follow this trend for the catalytic activity Nickel (Ni) > Molybdenum (Mo) > Cobalt (Co) > W > Iron (Fe) > Copper (Cu) which is calculated using the voltammetric techniques [88]. Ni shows a very good catalytic activity when using in the hybrid form of Ni/NiO/$CoSe_2$ because this composite helps in less resistance to charge transfer. But this hybrid has poor stability as Ni does not work well in acidic medium [89]. Recent studies by Qiu et al. found that when Ni is used with graphene it forms Ni-C bonds which increases the stability as well as the activity of the catalyst and is the best one proved for the process using Ni [90]. Co when embedded with Nitrogen rich CNTs forms a very good catalyst that catalyzed at all pH ranges. The reason for this good activity at all pH ranges is the N-doped content and the structural defect caused by the caused by pyrolysis of the Co-NRCNTs at higher temperatures [91]. These Nitrogen rich Co based catalyst can also be dispersed over the nanofibers for improving the catalytic activity. These particular catalysts also showed good stability for various potential cycles of process [92]. Along with Co and Ni Other Non-Noble metals are also used for the EH_2ER process including Fe, W, Mo but these all face the problem of their stability and there is still a lot to discover in this field.

4.3 Non-nobel metal-based C-NCs for EH_2ER

To avoid using precious noble metals there have been a plethora of reports using non-noble metal-based C-NCs for EH_2ER process. Some of them are transition Metal Oxides (TMOs), transition metal nitrides (TMNs), transition metal carbides (TMCs), transition metal borides (TMBs), transition metal phosphides (TMPs), transition metal dichalcogenides (TMDs). In this section the authors will discuss about the advancement in these types of NCs their advantage and disadvantages all. Although there are huge number of reports on transition metal-based compounds, but Mo and W display very good catalytic activity out of all of them. TMOs are easily available, stable, not harmful to the environment and obviously not precious like noble metals and thus is a good class of EH_2ER catalysts. Out of all the TMOs, Mo (MoO_2) and W (WO_2) based EH_2ER catalysts are best as they have high electrical conductivity as compared to other TMOs the credit goes to their monoclinic and distorted rutile crystal structure [93]. Compact MoO_2 faces have a problem of aggregation and thus various research groups tried to bring changes in the catalyst by decreasing the size of structure, forming hybrid composites, doping the MoO_2 or even introducing surface defects. One such example where Yu et al. encapsulated MoO_2 with phosphorous-doped porous carbon embedded on rGO to form MoO_2@PC-rGO which has a small tafel slope (41 mV dec-^{1}) and less over potential (η_{10} = 64 mV). Here electronic coupling between MoO_2 and rGO along with the porous carbon layer helps in getting rid of the problem of aggregation and improves its efficiency as a EH_2ER catalyst [94]. The other important metal oxide for EH_2ER is WO_2. Reports have revealed when WO_2 have vacancies of oxygen, it provides large number of active sites and thus augmented the EH_2ER process as compared to the compact WO_2 counterpart. One such example where Shen et al. encapsulated WO_x with Carbon on a Carbon support to form WO_x@C/C which has EH_2ER activity comparable to Pt metal with ultra-low overpotential (η_{60} = 36 mV) and very small tafel slope (19 mV dec^{-1}) which is because of the thick carbon shell helping in increase of charge transfer and also changes the Gibbs free energy values of H* for different adsorption sites [95].

TMNs are also other class of noble metal-compound based NCs for EH_2ER. The importance lies in the fact that they possess more contraction of the d bands and density of state near Fermi level because of interaction of negatively charged N atom which expands the lattice and increases it efficiency as a catalyst to make it comparable to noble metals like Pt, Pd [96]. As discussed earlier, whenever NCs are encapsulated with carbon material they tend to avoid aggregation and improve the efficiency as a catalyst. One such example where Shao et al. developed MoN

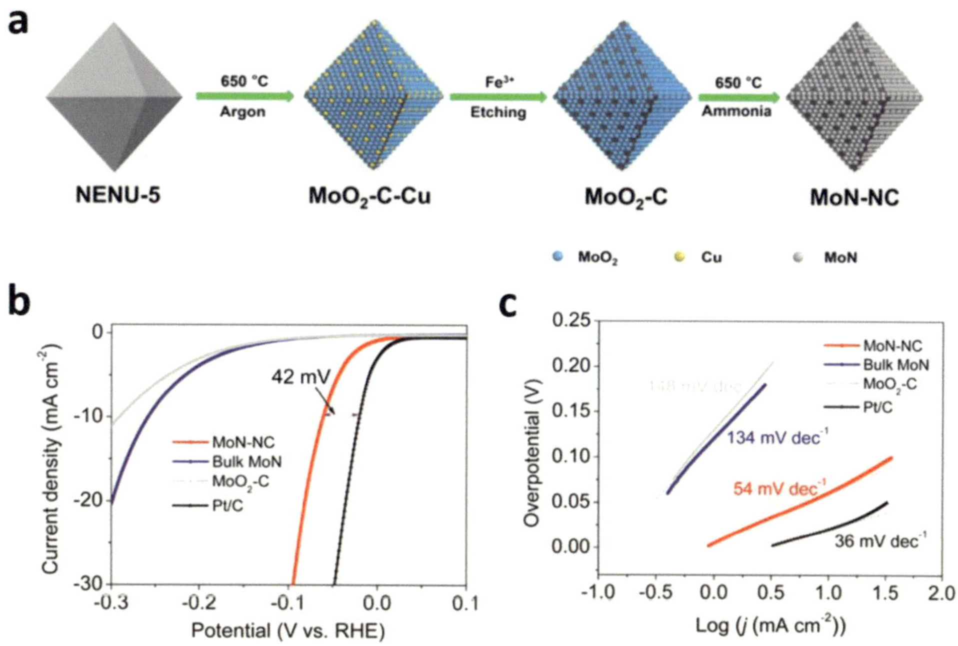

Figure 9.
(a) Procedure for the synthesis of MoN-NC nano-octahedrons derived from Mo-based MOFs (b) polarization curves (iR compensated) in 0.5 M H_2SO_4 at a scan rate of 5 mV s^{-1} and (c) the corresponding Tafel plots of the MoN-NC nano-octahedrons, intermediate MoO_2-C, bulk MoN, and 20% Pt/C catalysts. Reprinted with the permission from [97]. Copyright © 2017, American Chemical Society.

encapsulated with N-doped carbon material to form MoN@NC which exhibited very low overpotential (η_{10} = 62 mV dec^{-1}) and small tafel slope (54 mV) as depicted in **Figure 9 (c)** [97]. WN also possess same features as they exhibit less activity when used as it is but when encapsulated with carbon material its activity increases, one such example is using WN encapsulated with N doped graphene material (WN_xNRPGC) which has high electrocatalytic activity because of the formed hetero-architecture [98]. Ni based TMNs have also been explored a lot as they possess high electrocatalytic activity.

4.4 Metal-alloy-based C-NCs for EH_2ER

Whenever there is introduction of another element in the lattice of the metal synergistic effect comes into effect (intercalation of crystal planes, change in the metal–metal length to have strain and also formation of heteroatom bond to give ligand effect). This result in the change of electronic properties and morphology and hence in the electrocatalytic activity of the catalyst. To reduce loading of noble metals they are doped either with other comparatively cheap noble metal or sometimes with transition metal. The authors will discuss examples of both these types. Pt can be doped with Pd to form 1-D single crystalline material (thickness 3 nm) which increases the Pt utilization efficiency this was done by Liu et al. using solution phase method directed by surfactant [99]. Pt can also be doped with non-noble metals one such example is the use of Pt-Co alloy encapsulated on carbon material, thus possessing high activity as displayed by the small tafel slope of 20 mV dec^{-1}. This catalyst requires very less loading of Pt (ca. 5 wt %) and has activity comparable to Pt/C catalyst [100]. Using the same strategy Pd and Ru can also be doped with either noble metal or transition metals, some examples are Pd-Au catalyst, Pd-Co catalyst, Ru-Co and Ru-Ni [101–104].

Transition metal can be alloyed with transition metal itself and thus it can be a very good alternative for noble metal electrocatalysts because of the cost-effective nature of them. These alloys could either be binary alloys or ternary alloys, the authors will discuss examples from both binary and ternary alloys. Ni can form alloy with various other transition metal, but Ni-Mo binary alloy is considered best for the EH_2ER. Zhang et al. worked on the synthesis of $MoNi_4$ supported over the MoO_2 cuboids over the Ni foam. This catalyst has the activity similar to Pt/C with zero onset potential and η_{10} = 15 mV and very low tafel slope of 30 mV dec^{-1} [105]. Ni binary alloys face the problem of corrosion which can be overcome by using the carbon support along with the Ni based alloy. Co also form binary alloys with Fe using N-doped carbon-based support. Also, Co can be alloyed with Mo forming good electrochemical catalyst for EH_2ER such as Co_3Mo having an overpotential of η_{10} = 68 mV and tafel slope of 61 mV dec^{-1} [106]. Ternary alloys in the recent times have gained popularity for EH_2ER electrochemical catalysts as electronic and morphological features of catalyst can be tuned by variation in compositions of various metals and thus it can act as the good promising substitute for the noble-metal-based EH_2ER catalyst. One such example is the use of small amount of Pt (4.6%) to the Fe-Co binary alloy to form the PtCoFe@CN electrocatalyst which demonstrated activity similar to that of commercial 20% Pt/C having an overpotential of η_{10} = 45 mV as depicted in **Figure 10(b)** [107]. Research is still going on to prepare the ternary alloys without using the noble metals in it and that will really be the landmark in this field as it will be a catalyst with cost-effective nature.

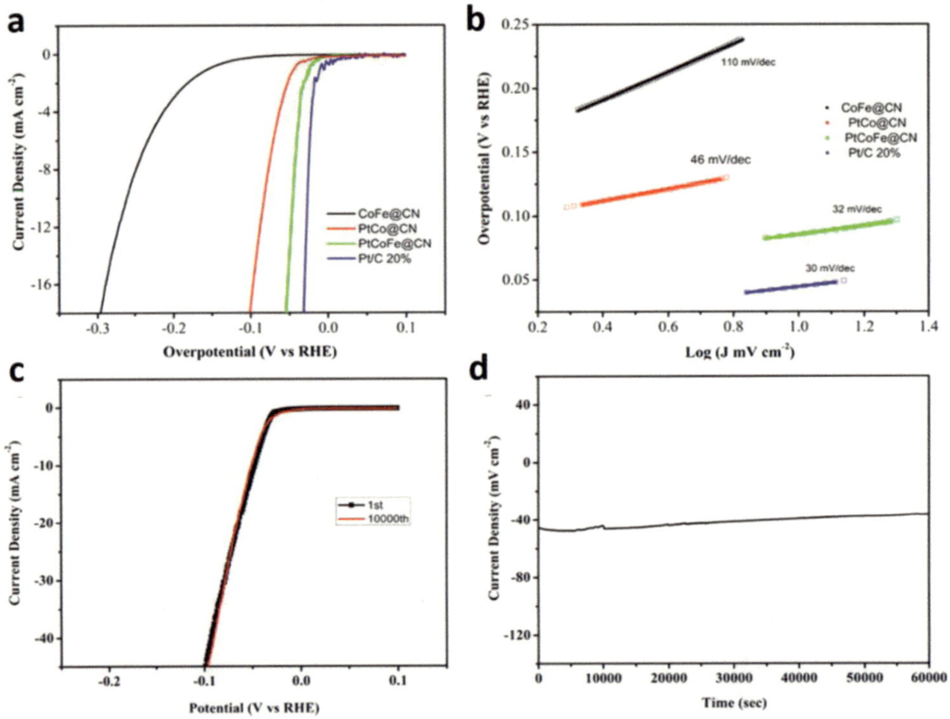

Figure 10.
(a) Polarization curves (b) Tafel plots of samples (c) polarization curves of PtCoFe@CN 1st and 10000th cycles (d) Amperometric i-t curves of PtCoFe@CN. Reprinted with the permission from [107]. Copyright © 2017, American Chemical Society.

5. Conclusions

This chapter discuses a thorough study of recent acheivements by C-NCs-based elelctroctalysts for ECO_2RR and EH_2ER. In this chapter, the authors have tried to summarize role of C-NCs in ECO_2RR and EH_2ER and the scope of these two important electrocatalytic reaction in combating the energy crises for human kind in introduction part. The examples that are discussed in this chapter were taken from recent reports in literature. The first part of this chapter sheds light on colloidal synthesis of nanocystals. The second part of this chapter emphasizes on effect of shape, size and composition in determining the catalytic activity, selectivity and stability of C-NCs for ECO_2RR. Here the effect of ligand functionalization and MOF/NCs hybrid system on ECO_2RR activity is also illustrated. The third part of this chapter addreses the role of C-NCs-based electrocatalysts on EH_2ER, its activity and stability. In this part, indepth study about mechanism of EH_2ER is also discussed. Although a lot has been done in ECO_2RR but ECO_2RR still face a big challenge in selectivity, similarly in EH_2ER a lot of research has been dedicated to find a substitute of state-of-art Pt/C catalyst which is very much expensive. However, researchers have been sucessful in discovering the costeffective electrocatalysts which are as good as Pt/C but they are still facing the stability issues.

Author Contributions

RN, AP (Abhay Prasad), AP*(Ashish Parihar) designed the study, reviewed the literature, wrote and edited the manuscript. MSA and RS provided the inputs while editing the manuscript. The complete review article was edited and finalized by RN, AP, and AP*.

Author details

Roshan Nazir[1,2]*, Abhay Prasad[3], Ashish Parihar[4], Mohammed S. Alqahtani[5] and Rabbani Syed[5]

1 Department of Chemical Engineering, Qatar University, Doha, Qatar

2 Department of Chemistry, Bilkent University, Bilkent, Ankara, Turkey

3 Department of Biological Sciences and Bioengineering, Indian Institute of Technology Kanpur, Kanpur, India

4 Department of Chemistry, Rutgers-The State University of New Jersey, New Brunswick, USA

5 Department of Pharmaceutics, College of Pharmacy, King Saud University, Riyadh, Saudi Arabia

*Address all correspondence to: roshanandrabi@gmail.com

References

[1] Nazir R, Kumar A, Ali S, Saad MAS, Al-Marri MJ. Galvanic exchange as a novel method for carbon nitride supported coag catalyst synthesis for oxygen reduction and carbon dioxide conversion. Catalysts. 2019;9(10):860.

[2] Nazir R, Kumar A, Saad MAS, Ashok A. Synthesis of hydroxide nanoparticles of Co/Cu on carbon nitride surface via galvanic exchange method for electrocatalytic CO2 reduction into formate. Colloids and Surfaces A: Physicochemical and Engineering Aspects. 2020; 598:124835.

[3] Nazir R, Kumar A, Saad MAS, Ali S. Development of CuAg/Cu2O nanoparticles on carbon nitride surface for methanol oxidation and selective conversion of carbon dioxide into formate. Journal of Colloid and Interface Science. 2020; 578:726-737

[4] Nazir R, Basak U, Pande S. Synthesis of one-dimensional RuO2 nanorod for hydrogen and oxygen evolution reaction: An efficient and stable electrocatalyst. Colloids and Surfaces A: Physicochemical and Engineering Aspects. 2019;560:141-8.

[5] Ma S, Kenis PJ. Electrochemical conversion of CO2 to useful chemicals: current status, remaining challenges, and future opportunities. Current Opinion in Chemical Engineering. 2013;2(2):191-9.

[6] Bushuyev OS, De Luna P, Dinh CT, et al. What should we make with CO2 and how can we make it? Joule. 2018;2(5):825-32.

[7] Yaashikaa P, Kumar PS, Varjani SJ, Saravanan A. A review on photochemical, biochemical and electrochemical transformation of CO2 into value-added products. Journal of CO2 Utilization. 2019;33:131-47.

[8] Lu Y, Jiang Z-y, Xu S-w, Wu H. Efficient conversion of CO2 to formic acid by formate dehydrogenase immobilized in a novel alginate–silica hybrid gel. Catalysis Today. 2006;115(1-4):263-8.

[9] Kalyanasundaram K, Graetzel M. Artificial photosynthesis: biomimetic approaches to solar energy conversion and storage. Current opinion in Biotechnology. 2010;21(3):298-310.

[10] Jouny M, Luc W, Jiao F. General techno-economic analysis of CO2 electrolysis systems. Industrial & Engineering Chemistry Research. 2018;57(6):2165-77.

[11] Sun Z, Ma T, Tao H, Fan Q, Han B. Fundamentals and challenges of electrochemical CO2 reduction using two-dimensional materials. Chem. 2017;3(4):560-87.

[12] Cargnello M. Colloidal Nanocrystals as Building Blocks for Well-Defined Heterogeneous Catalysts. Chemistry of Materials. 2019;31(3):576-96.

[13] Wang L, Chen W, Zhang D, et al. Surface strategies for catalytic CO 2 reduction: from two-dimensional materials to nanoclusters to single atoms. Chemical Socicty Reviews. 2019;48(21):5310-49.

[14] Yin Y, Alivisatos AP. Colloidal nanocrystal synthesis and the organic–inorganic interface. Nature. 2005;437(7059):664-70.

[15] Tang Y, Zheng G. Colloidal nanocrystals for electrochemical reduction reactions. Journal of colloid and interface science. 2017;485:308-27.

[16] Ji X, Song X, Li J, Bai Y, Yang W, Peng X. Size control of gold nanocrystals in citrate reduction: the third role of citrate. Journal of

the American Chemical Society. 2007;129(45):13939-48.

[17] Kriegel I, Rodriguez-Fernandez J, Wisnet A, et al. Shedding light on vacancy-doped copper chalcogenides: shape-controlled synthesis, optical properties, and modeling of copper telluride nanocrystals with near-infrared plasmon resonances. ACS nano. 2013;7(5):4367-77.

[18] Puntes VF, Krishnan KM, Alivisatos AP. Colloidal nanocrystal shape and size control: the case of cobalt. Science. 2001;291(5511):2115-7.

[19] Shevchenko EV, Talapin DV, Schnablegger H, et al. Study of nucleation and growth in the organometallic synthesis of magnetic alloy nanocrystals: the role of nucleation rate in size control of CoPt3 nanocrystals. Journal of the American Chemical Society. 2003;125(30):9090-101.

[20] Alivisatos AP, Gu W, Larabell C. Quantum dots as cellular probes. Annu. Rev. Biomed. Eng. 2005;7:55-76.

[21] Heydari N, Ghorashi SMB, Han W, Park H-H. Quantum Dot-Based Light Emitting Diodes (QDLEDs): New Progress. Quantum-dot Based Light-emitting Diodes. 2017:25.

[22] Sugimoto T. Preparation of monodispersed colloidal particles. Advances in Colloid and Interface Science. 1987;28:65-108.

[23] LaMer VK, Dinegar RH. Theory, production and mechanism of formation of monodispersed hydrosols. Journal of the American Chemical Society. 1950;72(11):4847-54.

[24] Vreeland EC, Watt J, Schober GB, et al. Enhanced nanoparticle size control by extending LaMer's mechanism. Chemistry of Materials. 2015;27(17):6059-66.

[25] Polte J. Fundamental growth principles of colloidal metal nanoparticles–a new perspective. CrystEngComm. 2015;17(36):6809-30.

[26] Murray C, Norris DJ, Bawendi MG. Synthesis and characterization of nearly monodisperse CdE (E= sulfur, selenium, tellurium) semiconductor nanocrystallites. Journal of the American Chemical Society. 1993;115(19):8706-15.

[27] Sinatra L, Pan J, Bakr OM. Methods of synthesizing monodisperse colloidal quantum dots. Material Matters. 2017;12:3-7.

[28] Kwon SG, Hyeon T. Formation mechanisms of uniform nanocrystals via hot-injection and heat-up methods. Small. 2011;7(19):2685-702.

[29] Xia Y, Gilroy KD, Peng HC, Xia X. Seed-mediated growth of colloidal metal nanocrystals. Angewandte Chemie International Edition. 2017;56(1):60-95.

[30] Jana NR, Gearheart L, Murphy CJ. Seed-mediated growth approach for shape-controlled synthesis of spheroidal and rod-like gold nanoparticles using a surfactant template. Advanced Materials. 2001;13(18):1389-93.

[31] Xia Y, Xiong Y, Lim B, Skrabalak SE. Shape-controlled synthesis of metal nanocrystals: simple chemistry meets complex physics? Angewandte Chemie International Edition. 2009;48(1):60-103.

[32] Xia Y, Xia X, Peng H-C. Shape-controlled synthesis of colloidal metal nanocrystals: thermodynamic versus kinetic products. Journal of the American Chemical Society. 2015;137(25):7947-66.

[33] Yang TH, Shi Y, Janssen A, Xia Y. Surface Capping Agents and Their Roles in Shape-Controlled Synthesis

of Colloidal Metal Nanocrystals. Angewandte Chemie International Edition. 2020;59(36):15378-401.

[34] Verma S, Kim B, Jhong HRM, Ma S, Kenis PJ. A gross-margin model for defining technoeconomic benchmarks in the electroreduction of CO2. ChemSusChem. 2016;9(15):1972-9.

[35] Hori Y, Wakebe H, Tsukamoto T, Koga O. Electrocatalytic process of CO selectivity in electrochemical reduction of CO2 at metal electrodes in aqueous media. Electrochimica Acta. 1994;39(11-12):1833-9.

[36] Schouten K, Kwon Y, Van der Ham C, Qin Z, Koper M. A new mechanism for the selectivity to C 1 and C 2 species in the electrochemical reduction of carbon dioxide on copper electrodes. Chemical Science. 2011;2(10):1902-9.

[37] Qin B, Wang H, Peng F, Yu H, Cao Y. Effect of the surface roughness of copper substrate on three-dimensional tin electrode for electrochemical reduction of CO2 into HCOOH. Journal of CO2 Utilization. 2017;21:219-23.

[38] Jiang K, Huang Y, Zeng G, Toma FM, Goddard III WA, Bell AT. Effects of Surface Roughness on the Electrochemical Reduction of CO2 over Cu. ACS Energy Letters. 2020;5(4):1206-14.

[39] Kuhl KP, Cave ER, Abram DN, Jaramillo TF. New insights into the electrochemical reduction of carbon dioxide on metallic copper surfaces. Energy & Environmental Science. 2012;5(5):7050-9.

[40] Fan L, Xia C, Yang F, Wang J, Wang H, Lu Y. Strategies in catalysts and electrolyzer design for electrochemical CO2 reduction toward C2+ products. Science Advances. 2020;6(8):eaay3111.

[41] Kleis J, Greeley J, Romero N, et al. Finite size effects in chemical bonding: From small clusters to solids. Catalysis Letters. 2011;141(8):1067-71.

[42] Huang J, Buonsanti R. Colloidal nanocrystals as heterogeneous catalysts for electrochemical CO2 conversion. Chemistry of Materials. 2018;31(1):13-25.

[43] Liu S, Huang S. Size effects and active sites of Cu nanoparticle catalysts for CO2 electroreduction. Applied Surface Science. 2019;475:20-7.

[44] Kim D, Resasco J, Yu Y, Asiri AM, Yang P. Synergistic geometric and electronic effects for electrochemical reduction of carbon dioxide using gold–copper bimetallic nanoparticles. Nature communications. 2014;5(1):1-8.

[45] Sneed BT, Young AP, Tsung C-K. Building up strain in colloidal metal nanoparticle catalysts. Nanoscale. 2015;7(29):12248-65.

[46] Loiudice A, Lobaccaro P, Kamali EA, et al. Tailoring copper nanocrystals towards C2 products in electrochemical CO2 reduction. Angewandte Chemie International Edition. 2016;55(19):5789-92.

[47] Zhu W, Michalsky R, Metin On, et al. Monodisperse Au nanoparticles for selective electrocatalytic reduction of CO2 to CO. Journal of the American Chemical Society. 2013;135(45):16833-6.

[48] Suen N-T, Kong Z-R, Hsu C-S, et al. Morphology manipulation of copper nanocrystals and product selectivity in the electrocatalytic reduction of carbon dioxide. ACS Catalysis. 2019;9(6):5217-22.

[49] Zhou Z-Y, Tian N, Huang Z-Z, Chen D-J, Sun S-G. Nanoparticle catalysts with high energy surfaces and enhanced activity synthesized by electrochemical method. Faraday discussions. 2009;140:81-92.

[50] Lee H-E, Yang KD, Yoon SM, et al. Concave rhombic dodecahedral Au nanocatalyst with multiple high-index facets for CO2 reduction. ACS nano. 2015;9(8):8384-93.

[51] Nørskov JK, Bligaard T, Rossmeisl J, Christensen CH. Towards the computational design of solid catalysts. Nature chemistry. 2009;1(1):37-46.

[52] Huang H, Jia H, Liu Z, et al. Understanding of strain effects in the electrochemical reduction of CO2: using Pd nanostructures as an ideal platform. Angewandte Chemie. 2017;129(13):3648-52.

[53] Kortlever R, Peters I, Koper S, Koper MT. Electrochemical CO2 reduction to formic acid at low overpotential and with high faradaic efficiency on carbon-supported bimetallic Pd–Pt nanoparticles. Acs Catalysis. 2015;5(7):3916-23.

[54] Kwon IS, Debela TT, Kwak IH, et al. Selective electrochemical reduction of carbon dioxide to formic acid using indium–zinc bimetallic nanocrystals. Journal of Materials Chemistry A. 2019;7(40):22879-83.

[55] Guo X, Zhang Y, Deng C, et al. Composition dependent activity of Cu–Pt nanocrystals for electrochemical reduction of CO 2. Chemical Communications. 2015;51(7):1345-8.

[56] Kim M-G, Jeong J, Choi Y, et al. Synthesis of V-doped In2O3 Nanocrystals via Digestive-Ripening Process and Their Electrocatalytic Properties in CO2 Reduction Reaction. ACS Applied Materials & Interfaces. 2020;12(10):11890-7.

[57] Al-Omari AA, Yamani ZH, Nguyen HL. Electrocatalytic CO2 reduction: from homogeneous catalysts to heterogeneous-based reticular chemistry. Molecules. 2018;23(11):2835.

[58] Xiang W, Zhang Y, Lin H, Liu C-j. Nanoparticle/metal–organic framework composites for catalytic applications: current status and perspective. Molecules. 2017;22(12):2103.

[59] Guntern YT, Pankhurst JR, Vávra J, et al. Nanocrystal/Metal–Organic Framework Hybrids as Electrocatalytic Platforms for CO2 Conversion. Angewandte Chemie International Edition. 2019;58(36):12632-9.

[60] Wang Z, Wu L, Sun K, et al. Surface Ligand Promotion of Carbon Dioxide Reduction through Stabilizing Chemisorbed Reactive Intermediates. The journal of physical chemistry letters. 2018;9(11):3057-61.

[61] Zhao Y, Fu G, Zheng N. Shaping the selectivity in heterogeneous hydrogenation by using molecular modification strategies: Experiment and theory. Catalysis Today. 2017;279:36-44.

[62] Liu P, Qin R, Fu G, Zheng N. Surface coordination chemistry of metal nanomaterials. Journal of the American Chemical Society. 2017;139(6):2122-31.

[63] Pankhurst JR, Guntern YT, Mensi M, Buonsanti R. Molecular tunability of surface-functionalized metal nanocrystals for selective electrochemical CO 2 reduction. Chemical science. 2019;10(44):10356-65.

[64] Cao Z, Kim D, Hong D, et al. A molecular surface functionalization approach to tuning nanoparticle electrocatalysts for carbon dioxide reduction. Journal of the American Chemical Society. 2016;138(26):8120-5.

[65] Fujishima A, Honda K. Electrochemical photolysis of water at a semiconductor electrode. nature. 1972;238(5358):37-8.

[66] Khaselev O, Turner JA. A monolithic photovoltaic-photoelectrochemical

device for hydrogen production via water splitting. Science. 1998;280(5362):425-7.

[67] Paracchino A, Laporte V, Sivula K, Grätzel M, Thimsen E. Highly active oxide photocathode for photoelectrochemical water reduction. Nature materials. 2011;10(6):456-61.

[68] Zhu J, Hu L, Zhao P, Lee LYS, Wong K-Y. Recent advances in electrocatalytic hydrogen evolution using nanoparticles. Chemical Reviews. 2019;120(2):851-918.

[69] Das RK, Wang Y, Vasilyeva SV, et al. Extraordinary hydrogen evolution and oxidation reaction activity from carbon nanotubes and graphitic carbons. Acs Nano. 2014;8(8):8447-56.

[70] Kibler LA. Hydrogen electrocatalysis. ChemPhysChem. 2006;7(5):985-91.

[71] Greeley J, Markovic NM. The road from animal electricity to green energy: combining experiment and theory in electrocatalysis. Energy & Environmental Science. 2012;5(11):9246-56.

[72] Conway B, Tilak B. Interfacial processes involving electrocatalytic evolution and oxidation of H2, and the role of chemisorbed H. Electrochimica Acta. 2002;47(22-23):3571-94.

[73] Walter MG, Warren EL, McKone JR, et al. Solar water splitting cells. Chemical reviews. 2010;110(11):6446-73.

[74] Zheng Y, Jiao Y, Zhu Y, et al. Hydrogen evolution by a metal-free electrocatalyst. Nature communications. 2014;5(1):1-8.

[75] Feng Y, Yu X-Y, Paik U. Nickel cobalt phosphides quasi-hollow nanocubes as an efficient electrocatalyst for hydrogen evolution in alkaline solution. Chemical Communications. 2016;52(8):1633-6.

[76] Korzeniewski C, Wieckowski A, Norskov J. Vibrational spectroscopy for the characterization of pem fuel cell membrane materials. Fuel Cell Science: Theory, Fundamentals, and Biocatalysis. 2010:395-413.

[77] Shinagawa T, Garcia-Esparza AT, Takanabe K. Insight on Tafel slopes from a microkinetic analysis of aqueous electrocatalysis for energy conversion. Scientific reports. 2015;5:13801.

[78] Ouyang Y, Ling C, Chen Q, Wang Z, Shi L, Wang J. Activating inert basal planes of MoS2 for hydrogen evolution reaction through the formation of different intrinsic defects. Chemistry of Materials. 2016;28(12):4390-6.

[79] Sabatier P. Le catalyse en chimie organique Berange. Paris; 1920.

[80] Marković N, Ross Jr P. Surface science studies of model fuel cell electrocatalysts. Surface Science Reports. 2002;45(4-6):117-229.

[81] Nørskov JK, Bligaard T, Logadottir A, et al. Trends in the exchange current for hydrogen evolution. Journal of The Electrochemical Society. 2005;152(3):J23.

[82] Parsons R. Volcano curves in electrochemistry. Catalysis in Electrochemistry. 2011:1-15.

[83] Conway B, Jerkiewicz G. Relation of energies and coverages of underpotential and overpotential deposited H at Pt and other metals to the 'volcano curve'for cathodic H2 evolution kinetics. Electrochimica Acta. 2000;45(25-26):4075-83.

[84] Sarkar S, Peter SC. An overview on Pd-based electrocatalysts for the hydrogen evolution reaction. Inorganic Chemistry Frontiers. 2018;5(9):2060-80.

[85] Huang Y-X, Liu X-W, Sun X-F, et al. A new cathodic electrode deposit

with palladium nanoparticles for cost-effective hydrogen production in a microbial electrolysis cell. international journal of hydrogen energy. 2011;36(4):2773-6.

[86] Joshi U, Malkhandi S, Ren Y, Tan TL, Chiam SY, Yeo BS. Ruthenium–Tungsten Composite Catalyst for the Efficient and Contamination-Resistant Electrochemical Evolution of Hydrogen. ACS applied materials & interfaces. 2018;10(7):6354-60.

[87] Li F, Han GF, Noh HJ, Ahmad I, Jeon IY, Baek JB. Mechanochemically assisted synthesis of a Ru catalyst for hydrogen evolution with performance superior to Pt in both acidic and alkaline media. Advanced Materials. 2018;30(44):1803676.

[88] Miles M, Thomason M. Periodic variations of overvoltages for water electrolysis in acid solutions from cyclic voltammetric studies. Journal of the Electrochemical Society. 1976;123(10):1459.

[89] Xu YF, Gao MR, Zheng YR, Jiang J, Yu SH. Nickel/nickel (II) oxide nanoparticles anchored onto cobalt (IV) diselenide nanobelts for the electrochemical production of hydrogen. Angewandte Chemie. 2013;125(33):8708-12.

[90] Qiu HJ, Ito Y, Cong W, et al. Nanoporous graphene with single-atom nickel dopants: an efficient and stable catalyst for electrochemical hydrogen production. Angewandte Chemie International Edition. 2015;54(47):14031-5.

[91] Zou X, Huang X, Goswami A, et al. Cobalt-embedded nitrogen-rich carbon nanotubes efficiently catalyze hydrogen evolution reaction at all pH values. Angewandte Chemie. 2014;126(17):4461-5.

[92] Zhang L, Zhu S, Dong S, et al. Co nanoparticles encapsulated in porous N-doped carbon nanofibers as an efficient electrocatalyst for hydrogen evolution reaction. Journal of The Electrochemical Society. 2018;165(15):J3271.

[93] Eyert V, Horny R, Höck K-H, Horn S. Embedded Peierls instability and the electronic structure of MoO2. Journal of Physics: Condensed Matter. 2000;12(23):4923.

[94] Tang YJ, Gao MR, Liu CH, et al. Porous molybdenum-based hybrid catalysts for highly efficient hydrogen evolution. Angewandte Chemie. 2015;127(44):13120-4.

[95] Jing S, Lu J, Yu G, et al. Carbon-encapsulated WOx hybrids as efficient catalysts for hydrogen evolution. Advanced Materials. 2018;30(28):1705979.

[96] Ham DJ, Lee JS. Transition metal carbides and nitrides as electrode materials for low temperature fuel cells. Energies. 2009;2(4):873-99.

[97] Zhu Y, Chen G, Xu X, Yang G, Liu M, Shao Z. Enhancing electrocatalytic activity for hydrogen evolution by strongly coupled molybdenum nitride@ nitrogen-doped carbon porous nano-octahedrons. ACS Catalysis. 2017;7(5):3540-7.

[98] Zhu Y, Chen G, Zhong Y, Zhou W, Shao Z. Rationally Designed Hierarchically Structured Tungsten Nitride and Nitrogen-Rich Graphene-Like Carbon Nanocomposite as Efficient Hydrogen Evolution Electrocatalyst. Advanced Science. 2018;5(2):1700603.

[99] Lv H, Chen X, Xu D, et al. Ultrathin PdPt bimetallic nanowires with enhanced electrocatalytic performance for hydrogen evolution reaction. Applied Catalysis B: Environmental. 2018;238:525-32.

[100] Yang T, Zhu H, Wan M, Dong L, Zhang M, Du M. Highly efficient and

durable PtCo alloy nanoparticles encapsulated in carbon nanofibers for electrochemical hydrogen generation. Chemical Communications. 2016;52(5):990-3.

[101] Quaino P, Santos E, Wolfschmidt H, Montero M, Stimming U. Theory meets experiment: electrocatalysis of hydrogen oxidation/ evolution at Pd–Au nanostructures. Catalysis today. 2011;177(1):55-63.

[102] Chen J, Xia G, Jiang P, et al. Active and durable hydrogen evolution reaction catalyst derived from Pd-doped metal–organic frameworks. ACS Applied Materials & Interfaces. 2016;8(21):13378-83.

[103] Su J, Yang Y, Xia G, Chen J, Jiang P, Chen Q. Ruthenium-cobalt nanoalloys encapsulated in nitrogen-doped graphene as active electrocatalysts for producing hydrogen in alkaline media. Nature communications. 2017;8(1):1-12.

[104] Zhang C, Liu Y, Chang Y, et al. Component-controlled synthesis of necklace-like hollow Ni x Ru y nanoalloys as electrocatalysts for hydrogen evolution reaction. ACS Applied Materials & Interfaces. 2017; 9(20):17326-36.

[105] Zhang J, Wang T, Liu P, et al. Efficient hydrogen production on MoNi 4 electrocatalysts with fast water dissociation kinetics. Nature communications. 2017;8(1):1-8.

[106] Chen J, Ge Y, Feng Q, et al. Nesting Co3Mo binary alloy nanoparticles onto molybdenum oxide nanosheet arrays for superior hydrogen evolution reaction. ACS applied materials & interfaces. 2019;11(9):9002-10.

[107] Chen J, Yang Y, Su J, Jiang P, Xia G, Chen Q. Enhanced activity for hydrogen evolution reaction over CoFe catalysts by alloying with small amount of Pt. ACS applied materials & interfaces. 2017;9(4):3596-601.